Hybrid Nanomaterials: Design, Synthesis and Applications

Hybrid Nanomaterials: Design, Synthesis and Applications

Edited by **Mindy Adams**

C WILLFORD PRESS

New York

Published by Willford Press,
118-35 Queens Blvd., Suite 400,
Forest Hills, NY 11375, USA
www.willfordpress.com

Hybrid Nanomaterials: Design, Synthesis and Applications
Edited by Mindy Adams

International Standard Book Number: 978-1-68285-073-2 (Hardback)

Printed in the United States of America.

Contents

Preface   VII

Chapter 1   **Cellular Uptake of Tile-Assembled DNA Nanotubes**   **1**
Samet Kocabey, Hanna Meinl, Iain S. MacPherson, Valentina Cassinelli,
Antonio Manetto, Simon Rothenfusser, Tim Lied and Felix S. Lichtenegger

Chapter 2   **Polymorphic Ring-Shaped Molecular Clusters Made of Shape-Variable**
Building Blocks   **15**
Keitel Cervantes-Salguero, Shogo Hamada, Shin-ichiro M. Nomura and
Satoshi Murata

Chapter 3   **Synthesis of Upconversion β-NaYF$_4$:Nd^{3+}/Yb^{3+}/Er^{3+} Particles with Enhanced**
Luminescent Intensity through Control of Morphology and Phase   **25**
Yunfei Shang, Shuwei Hao, Jing Liu, Meiling Tan, Ning Wang, Chunhui
Yang and Guanying Chen

Chapter 4   **Ceramic Nanocomposites from Tailor-Made Preceramic Polymers**   **40**
Gabriela Mera, Markus Gallei, Samuel Bernard and Emanuel Ionescu

Chapter 5   **Alumina Matrix Composites with Non-Oxide Nanoparticle Addition and**
Enhanced Functionalities   **113**
Dušan Galusek and Dagmar Galusková

Chapter 6   **Synthesis, Characterization, and Mechanism of Formation of Janus-Like**
Nanoparticles of Tantalum Silicide-Silicon (TaSi$_2$/Si)   **142**
Andrey V. Nomoev, Sergey P. Bardakhanov, Makoto Schreiber, Dashima Zh.
Bazarova Boris B. Baldanov and Nikolai A. Romanov

Chapter 7   **Recent Advances on Carbon Nanotubes and Graphene Reinforced Ceramics**
Nanocomposites   **152**
Iftikhar Ahmad, Bahareh Yazdani and Yanqiu Zhu

Chapter 8   **Rare Earth Ion-Doped Upconversion Nanocrystals: Synthesis and**
Surface Modification   **177**
Hongjin Chang, Juan Xie, Baozhou Zhao, Botong Liu, Shuilin Xu, Na Ren, Xiaoji Xie,
Ling Huang and Wei Huang

Permissions

List of Contributors

Preface

There has been significant progress in the study of nanomaterials over the last few years and the use of nanomaterials has spread into a multitude of other disciplines as well. This book unfolds the innovative aspects of the study of nanomaterials which will be crucial for the progress of this field in the future. It includes some of the vital pieces of work being conducted across the world, on various topics such as inorganic-organic hybrids and composites, nano-alloys, synthesis and modeling of hybrid nanomaterials and their applications, etc. With its detailed analyses and data, this book will prove immensely beneficial to professionals and students involved in the field of nanomaterials and nanoparticles at various levels.

The researches compiled throughout the book are authentic and of high quality, combining several disciplines and from very diverse regions from around the world. Drawing on the contributions of many researchers from diverse countries, the book's objective is to provide the readers with the latest achievements in the area of research. This book will surely be a source of knowledge to all interested and researching the field.

In the end, I would like to express my deep sense of gratitude to all the authors for meeting the set deadlines in completing and submitting their research chapters. I would also like to thank the publisher for the support offered to us throughout the course of the book. Finally, I extend my sincere thanks to my family for being a constant source of inspiration and encouragement.

Editor

Cellular Uptake of Tile-Assembled DNA Nanotubes

Samet Kocabey [1], **Hanna Meinl** [2], **Iain S. MacPherson** [1], **Valentina Cassinelli** [3], **Antonio Manetto** [3], **Simon Rothenfusser** [2], **Tim Liedl** [1] and **Felix S. Lichtenegger** [2,4,*]

[1] Faculty of Physics and Center for Nanoscience, Ludwig-Maximilians University, Munich 80799, Germany; E-Mails: samet.kocabey@physik.lmu.de (S.K.); iainmacpherson@gmail.com (I.S.M.); tim.liedl@physik.lmu.de (T.L.)

[2] Division of Clinical Pharmacology, Department of Internal Medicine IV, Klinikum der Universität München, Munich 80336, Germany; E-Mails: hanna.meinl@outlook.com (H.M.); simon.rothenfusser@med.uni-muenchen.de (S.R.)

[3] Baseclick GmbH, Tutzing 82327, Germany; E-Mails: v.cassinelli@baseclick.eu (V.C.); a.manetto@baseclick.eu (A.M.)

[4] Department of Internal Medicine III, Klinikum der Universität München, Munich 81377, Germany

* Author to whom correspondence should be addressed;
E-Mail: felix.lichtenegger@med.uni-muenchen.de

Abstract: DNA-based nanostructures have received great attention as molecular vehicles for cellular delivery of biomolecules and cancer drugs. Here, we report on the cellular uptake of tubule-like DNA tile-assembled nanostructures 27 nm in length and 8 nm in diameter that carry siRNA molecules, folic acid and fluorescent dyes. In our observations, the DNA structures are delivered to the endosome and do not reach the cytosol of the *GFP*-expressing HeLa cells that were used in the experiments. Consistent with this observation, no elevated silencing of the *GFP* gene could be detected. Furthermore, the presence of up to six molecules of folic acid on the carrier surface did not alter the uptake behavior and gene silencing. We further observed several challenges that have to be considered when performing *in vitro* and *in vivo* experiments with DNA structures: (i) DNA tile tubes consisting of 42 nt-long oligonucleotides and carrying single- or double-stranded extensions degrade within one hour in cell medium at 37 °C, while the same tubes without extensions are stable for up to eight hours. The degradation is caused mainly by the low concentration of divalent ions in the media. The lifetime in cell medium can be increased drastically by employing DNA tiles that are 84 nt long. (ii) Dyes may get cleaved from the oligonucleotides and then accumulate inside the cell close to the mitochondria, which can lead to misinterpretation of data generated by flow cytometry and fluorescence microscopy. (iii) Single-stranded DNA carrying fluorescent dyes are internalized at similar levels as the DNA tile-assembled tubes used here.

Keywords: DNA nanotechnology; DNA tile; siRNA delivery; stability; folate; cation

---

1. Introduction

Therapeutic agents must overcome multiple barriers to reach their target [1,2]. For example, siRNAs have to reach the target tissue, enter the cells, be released from the endosomal compartment and, finally, silence the target gene via the RISC complex [3]. Up to now, researchers have developed a variety of nanoparticle carrier systems to overcome these barriers, such as polymers [4], liposomes [5] or conjugates [6], with various levels of efficiency and toxicity. Most recently, with improvements in the DNA nanotechnology field, DNA-based nanostructures were developed as carrier systems for a variety of active components, including siRNAs [7], antibodies [8], immunostimulants [9,10] and cancer drugs [11,12]. DNA nanostructures are promising for delivery applications because they can be easily modified with a variety of (bio)chemical moieties for targeting purposes at nanoscale precision; they are monodisperse with well-defined sizes and are non-cytotoxic [10,13–18]. To date, several groups have investigated the targeted delivery of DNA-based nanostructures using different targeting agents, such as cell penetrating peptides or small molecules. Among them, folate is a commonly-used molecule, due to the high expression of its receptors on certain cancer cells. Efficient folate-mediated uptake has been demonstrated using various DNA-based structures, such as DNA nanotubes built from a single palindromic DNA strand [19] or Y-shaped DNA nanostructures prepared by rolling circle amplification [20]. Although the DNA-based nanostructures are promising for targeted delivery applications, as exemplified above, the stability of these structures at 37 °C in blood or tissue is one of the main issues to be considered. In a recent study, the stability of a variety of DNA origami structures with different designs, such as octahedron, six-helix bundle tubes or 24-helix bundle rods, were investigated using *in vitro* conditions, and time-and shape-dependent denaturation and digestion were observed due to the Mg^{2+} depletion in the media and the DNase activity of the serum [21]. As an alternative to the DNA origami method [14,15] and shape-specific designs, such as DNA cubes [22], tetrahedrons [23] or octahedrons [24], single-stranded tile assembly has recently proven to be a versatile and modular design strategy to build a wide variety of two- and three-dimensional shapes [25,26]. In this study, we intended to show efficient folate-mediated uptake and subsequent gene silencing by tile-assembled DNA nanotubes carrying GFP siRNAs *in vitro*. However, we were not able to demonstrate the sought-after effects, but instead observed untimely disassembly of our constructs under certain *in vitro* conditions and, therefore, investigated strategies to maintain the structural integrity in relevant environments. We examined the stability of tile-assembled structures under limited divalent cations and in the presence of nucleases in buffer and in cell media. We then describe a number of artifacts that should be taken into consideration during experiments with DNA-based nanostructures *in vitro*.

2. Results and Discussion

2.1. Design and Self-Assembly of Six-Helix DNA Nanotubes

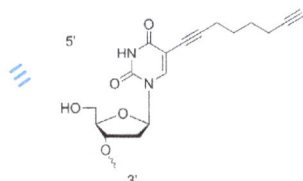

We designed tubule-like DNA nanostructures consisting of 24 oligonucleotides that self-assemble into six parallel helices using the single-stranded DNA tile assembly method introduced by Yin *et al.* (Scheme 1 and Table S1) [25,27]. Six of the oligonucleotides were alkyne-modified during synthesis and conjugated in-house with PEG-folate-azide (Baseclick GmbH, Tutzing, Germany) by a click reaction. Reversed phase high performance chromatography (RP-HPLC) analysis and matrix-assisted laser desorption/ionization (MALDI) mass spectrometry revealed the almost quantitatively conjugation of folate molecules to the alkyne-oligonucleotides. (Figures S1 and S2, Table S2). Another set of six oligonucleotides was extended by an 18 nt-long sequence at the 3' end to allow the attachment via hybridization of six siRNA molecules that potentially silence the expression of GFP upon delivery. To visualize the DNA nanotubes *in vitro*, two different labeling strategies were employed. In the first approach, Atto488-dUTP was enzymatically labeled to the 3' end of a set of 12 tile oligonucleotides using terminal transferase. In the second approach, the same set of oligonucleotides was extended with another 18 nt-long sequence allowing attachment via hybridization of 12 Atto647-modified (via NHS chemistry) oligonucleotides. The nanotubes have a designed length of ~27 nm and an expected diameter of ~6 nm for the dried sample. Note that the tube diameter of a six-helix bundle increases in buffer to 8 nm and that tubes decorated with additional molecules will have a larger effective diameter [28].

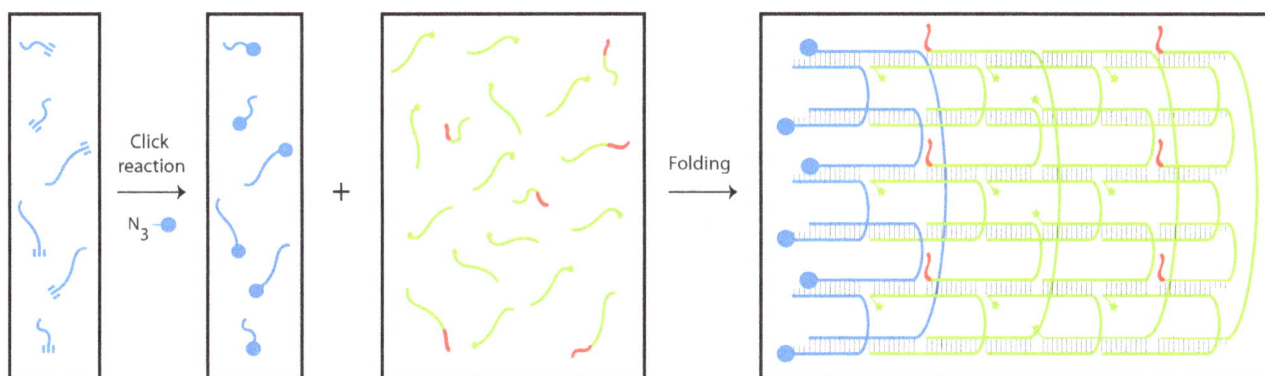

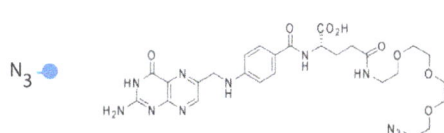

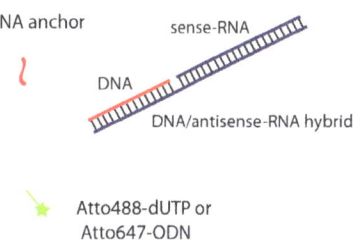

Scheme 1. DNA nanotube assembly. (**Left**) Click reaction of alkyne-modified oligonucleotides with azide-modified PEGylated folate. (**Right**) Self-assembly of 24 oligonucleotides into a six-helix tube after a 17-h annealing process.

The nanotube structures containing the desired subsets of oligonucleotides and modifications were folded in TE-buffer containing 20 mM Mg^{2+} during a thermal annealing process starting at 80 °C and cooling down to room temperature over the course of 17 h. Analysis by gel electrophoresis analysis showed for all designs prominent bands representing the folded structures (Lanes 2 + 3 + 4 in Figure 1). Conjugation of folate and folate + siRNA (Lanes 3 and 4, respectively) to the DNA nanotubes leads to a decrease of their mobility in comparison to nanotubes without folate and siRNA (Lane 2). Transmission electron microscopy (TEM) demonstrates the correct assembly of the nanotubes and the monodispersity of the samples. The measured length of 27 ± 1 nm and the measured diameter of 6 ± 1 nm perfectly match the expected dimensions (Figure 1B–D).

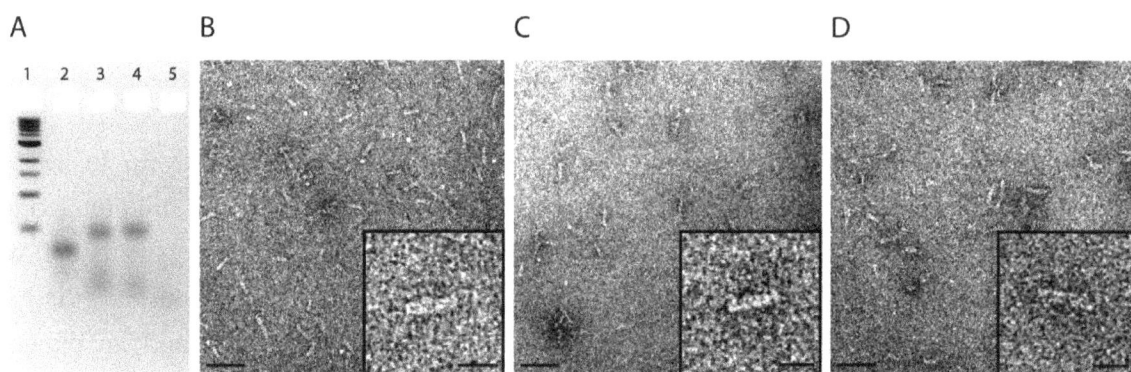

Figure 1. Characterization of nanotubes. (**a**) Gel electrophoresis analysis of assembled nanotubes: (1) 1-kb ladder, (2) nanotube, (3) nanotube + folate, (4) nanotube + folate + siRNA, and (5) individual oligonucleotide. Electron micrographs of (**b**) Nanotubes; (**c**) Nanotubes with folate; and (**d**) Nanotubes with folate and siRNA (scale bars: 50 nm; insets: 20 nm).

2.2. Tubule-Like Tile-Assembled DNA Nanostructures Are Delivered to the Endosome of HeLa Cells Independently of Folic Acid and Are Not Capable of Releasing siRNA into the Cytosol

DNA nanotubes labeled with Atto488 via enzymatic labeling were added to HeLa cell cultures at 10 nM, together with dextran-AF647 as a marker for endosomal uptake. At various time points thereafter, confocal microscopy was performed to evaluate the localization of the construct. After 24 h, we found clear co-localization of the nanotubes with dextran (Figure 2A–C). Observations for up to 72 h did not show any change in localization (Figure S3C,D).

To determine a potential effect of uptake via the folate receptor, which is highly expressed on the surface of HeLa cells, nanotubes with and without folic acid were compared side by side. No influence on the endosomal staining pattern was noticed in the fluorescence microscopy images, neither after 24 h nor after 72 h (Figure S3). For a quantitative analysis of the uptake, we conducted flow cytometry-based measurements of the HeLa cells at different time points after the addition of fluorophore-labeled nanotubes (Figure 2D). A minor signal was already detected after 4 h, which further increased in the course of 24 h. No significant difference was found between the uptake of nanotubes with or without folate.

On a functional level, we tested if the nanostructures released their siRNA cargo successfully to the cytoplasm by analyzing the knockdown capacity of siRNA molecules bound to the DNA nanotube.

Stably *GFP*-transfected HeLa cell lines were used together with siRNA directed against *GFP* (siGFP). The siGFP was either bound to the nanostructure via hybridization or transfected into the cytoplasm by lipofection as a positive control. The GFP signal of the cells was measured by flow cytometry after 96 h (Figure 2E). In the condition with lipofection of GFP-targeting siRNAs, the fluorochrome signal was markedly decreased compared to lipofection of a control siRNA (siCTRL). However, the addition of siGFP to the nanotubes did not result in *GFP*-knockdown, independent of folate labeling, consistent with endosomal trapping of the whole structure, including their siRNA cargo.

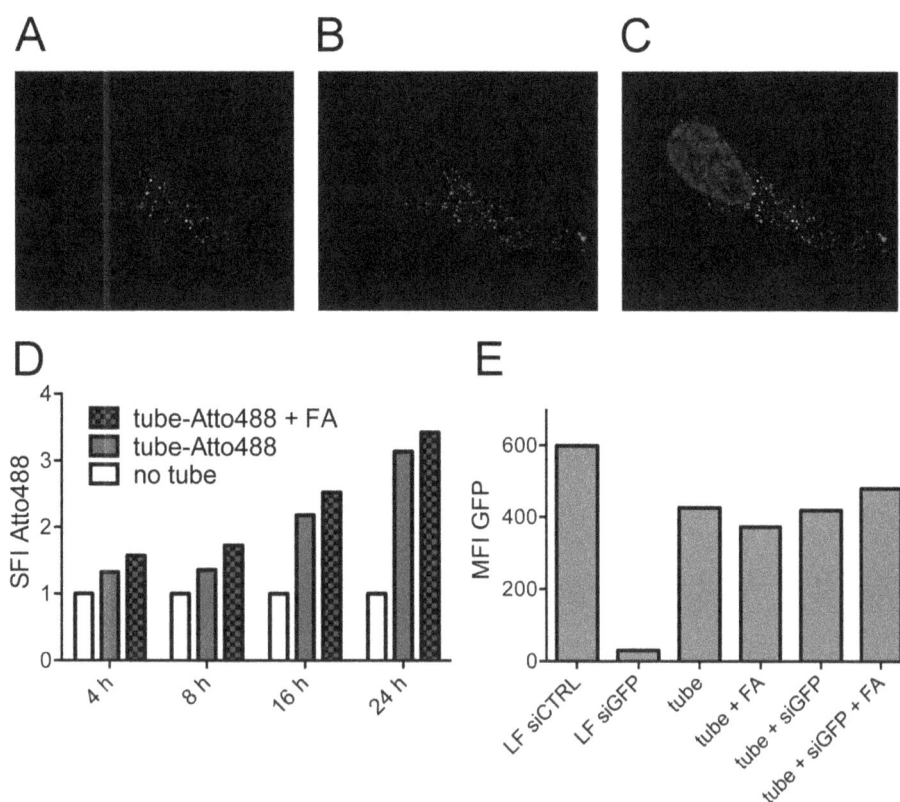

Figure 2. Endosomal uptake of nanotubes in HeLa cells. Endosomal staining of nanotubes with dextran. (**a**) Nanotubes; (**b**) Dextran; (**c**) Merged image from (**a**), (**b**) and a third channel (DAPI, blue). (**d**) Flow cytometry analysis of folate-dependent uptake of Atto488-labeled nanotubes over 24 h. Untreated cells act as the control, and the specific fluorescence intensity (SFI) of the dye is depicted. (**e**) Fluorescence intensity of stably GFP-expressing Hela cells upon the addition of nanotubes carrying *GFP*-targeting siRNAs or upon transfection of a *GFP*-targeting siRNA and a non-targeting siRNA, respectively, as controls using lipofection (LF). The median fluorescence intensity (MFI) of GFP is depicted.

2.3. Stability of DNA Nanotubes Differs in Various Conditions In Vitro

To address the stability of tile-assembled DNA nanostructures *in vitro*, we incubated them in different buffers and cell media. First, we incubated the nanotubes in PBS with different Mg^{2+} concentrations at 37 °C for 2 h. We used PBS as a buffer to simulate the cell media conditions, as both cell media and PBS possess several monovalent and divalent cations at isotonic concentrations. Importantly, for the assembly and stabilization of the DNA nanostructures, usually Mg^{2+} concentrations much higher than those found in PBS and cell media are used. While folding of DNA nanostructures can also be achieved at high Na^+ concentrations [29], the 135 mM NaCl present in PBS are not sufficient to stabilize DNA nanotubes at 37 °C, if the individual DNA tiles are 42 nt long. Gel analysis revealed that the nanotubes without extensions were stable down to 1 mM Mg^{2+}, whereas the nanotubes carrying siRNA started to degrade already below 4 mM Mg^{2+} (Figure 3A,B). This indicates that the addition of extension sequences protruding from the DNA nanotubes destabilizes the structure, which may be explained by distorted stacking of the last base before the extension and with an increase of electrostatic repulsion between the elongated tail and the DNA duplexes in the nanotube [30]. Next, we compared the stability of nanotubes against DNases and incubated the structures in cell medium containing 10% FCS. Gel analysis showed that under these conditions, the plain nanotubes are stable up to 8 h (Figure 3C). However, nanotubes carrying siRNA were degraded in 1 h when the structures were incubated in media containing 10% FCS. These nanotubes were also degraded slightly during 8 h in DMEM medium without FCS, likely due to Mg^{2+} depletion (in all cell media experiments, the concentration of Mg^{2+} was 1.8 mM).

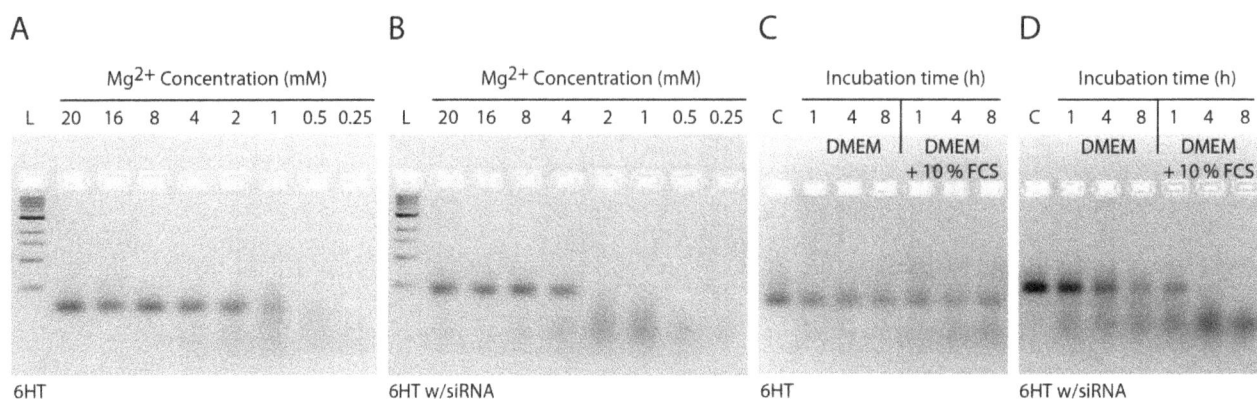

Figure 3. Stability of nanotubes. (**a**) Stability of nanotubes in PBS with different Mg^{2+} concentrations; (**b**) Stability of nanotubes carrying siRNA in PBS with different Mg^{2+} concentrations; (**c**) Stability of nanotubes in DMEM medium in the absence or presence of FCS; (**d**) Stability of nanotubes carrying siRNA in DMEM medium in the absence or presence of FCS (L: 1 kb ladder; C: control. All samples were incubated at 37 °C).

To overcome the problem of premature degradation, DNA tile tubes were assembled from 84 nt-long oligonucleotides. This design allows longer complementary regions (21 bp for the 84mers instead of 10 bp and 11 bp for the 42mers) within the tile assembly, which, in turn, yields much higher thermal stability, but also higher resistance to Mg^{2+} depletion (Figure 4). Our results show that the stability of tile-assembled nanotubes is dependent on sequence design, temperature, salt concentration and structural modifications, such as the addition of single- or double-stranded extensions to the DNA tiles.

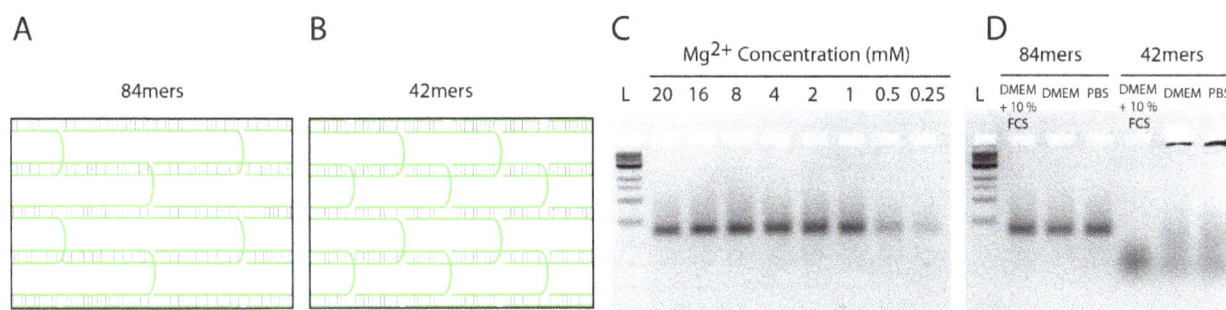

Figure 4. Stability of nanotubes assembled from 84 nt-long oligonucleotides. (**a**) Schematic depiction of a section of the 6HT demonstrating the hybridization of 84mers; (**b**) Schematic depiction of a section of the 6HT demonstrating the hybridization of 42mers; (**c**) Stability of nanotubes (84mers) in PBS with different Mg^{2+} concentrations; (**d**) Stability of nanotubes (84mers and 42mers) in DMEM + 10% FCS, DMEM and PBS. Nanotubes were incubated at 45 °C for 2 h (L: 1 kb ladder; C: control).

2.4. Strong Extra-Endosomal Uptake Can Be Feigned by Dye Cleavage

When nanostructures labeled with Atto647 via hybridization were incubated with HeLa cells, we repeatedly observed a very high fluorescence level in the cells during microscopy- or flow cytometry-based analysis. Furthermore, the fluorochrome did not co-localize with dextran as an endosomal marker (Figure 5A), but instead, mitochondrial localization was detected (Figure 5B). The level of uptake and the mitochondrial staining pattern were associated with the addition of serum to the culture medium (Figure S4). Similarly, when only the oligonucleotide labeled with Atto647 (via NHS chemistry) was added to the HeLa cells, we observed a rapid and strong staining of the cells only in the case when serum was added (Figure 5C). This effect was not observed when the fluorophores were attached via enzymatic binding. We therefore conclude that Atto647 is cleaved off the DNA by some component in the serum and is taken up independently of the nanostructure.

2.5. Single-Stranded DNA Molecules, But Not Deoxynucleotide Triphosphates, Are Internalized at Similar Levels as the Tile-Assembled Nanotube Structures

Specific uptake of the tubule-like tile-assembled DNA nanostructures was analyzed by direct comparison with oligonucleotides and deoxynucleotide triphosphates. All three molecules were labeled with Atto488 and incubated at identical molar concentrations with HeLa cells. Fluorochrome uptake was measured by flow cytometry at various time points (Figure 5D). No intracellular staining was found in the deoxynucleotide triphosphate condition. However, we observed similar uptake of the fluorochrome with the oligonucleotide as with the nanostructure.

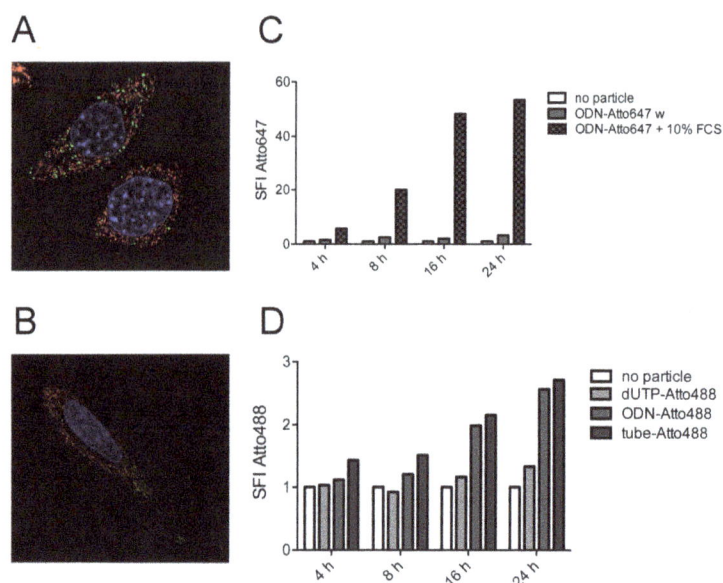

Figure 5. Effect of dye cleavage on cellular uptake. (**a**) Endosomal staining using Alexa Fluor 488-coupled dextran (shown in green) of HeLa cells treated with oligodeoxynucleotide (ODN)-Atto647 (shown in red); (**b**) Mitochondrial colocalization of Atto647 (shown in red) in HeLa cells stained with the mitochondrial dye MitoTracker Green (shown in green). Nuclei are stained with Hoechst 33342; (**c**) Flow cytometry analysis of fluorescence intensity of cells treated with ODN-Atto647 in the absence or presence of FCS; (**d**) Flow cytometry analysis of fluorescence intensity of cells treated with Atto488-dUTP, ODN-Atto488 and nanotube labeled with Atto488. Untreated cells act as the control, and the specific fluorescence intensity (SFI) of the dye is depicted.

3. Experimental Section

3.1. DNA Nanotube Design

DNA nanotubes were designed using Yin's single-strand tile (SST) method [27]. Each tile oligonucleotide is 42 bases long and consists of four domains with ten or eleven bases. Twenty four individual oligonucleotides were used to form 6 helix nanotube. The domains at the ends of the nanotube contain non-pairing poly-A sequences to prevent polymerization. siRNA hybridization to the nanotubes was done by extending 3' ends of six tiles with an 18-nt long overhang sequence (5'-AGGATGTAGGTGGTAGAG-3'). The used siRNA sequences for *GFP* silencing were sense: 5'-GCCACAACGUCUAUAUCAU-3', and antisense: 5'-AUGAUAUAGACGUUGUGGC CTCTACCACCTACATCCT-3'. Six oligonucleotides were modified with PEG-folate azide (Baseclick GmbH, Tutzing, Germany) using click reactions. The underlined sequence shows the complementary overhang. All oligonucleotides were purchased from Eurofins (Ebersberg, Germany) with HPSF or HPLC purification.

3.2. Folate Conjugation and Characterization of Oligonucleotides

Each of the six alkyne-modified oligonucleotides (Baseclick GmbH) were submitted to click reaction, using CuBr as the Cu(I) source. Ten microliters of a freshly prepared CuBr (0.1 M)/THPTA (0.1 M) solution in a 1:2 v/v ratio were added to a 50-µL (0.1 mM, 5 nmol) solution of each alkyne-oligonucleotide. The addition of folate-PEG3-azide (2.5 µL, 10 mM in DMSO) completed the click reaction cocktail. The mixture was mixed for 1.5 h at 45 °C. Finally, the solution was purified via ethanol precipitation. Folate-conjugated oligonucleotides were analyzed by analytical RP-HPLC (e2695 system, Waters, Milford, MA, USA) coupled with a photodiode array detector (PDA 2998, Waters) using a reversed phase XBridge OST C18 column (4.6 mm × 50 mm, Waters). Before injection in RP-HPLC, samples were diluted to 10 µM concentration in HPLC-grade H_2O. Samples (2 µL) were desalted against ddH_2O using a nitrocellulose membrane (MerckMillipore, Frankfurt, Germany), and 0.4 µL were spotted onto a MALDI plate (Bruker Corporation, Millerica, MA, USA) along with 0.4 µL of 3-hydroxypicolinic acid (HPA, Sigma Aldrich, St. Luis, MO, USA). Measurements were carried out with the Autoflex MALDI-TOF (Bruker Corporation).

3.3. Dye Labeling of DNA Nanotubes

Fluorescent dyes were conjugated to the 12 oligonucleotides with two different approaches before the nanotube assembly. In the first approach, 3' ends of twelve tiles were extended by 18 nt-long overhang sequences to hybridize the dye-modified sequence: (Atto647 TTCATTCTCCTATTACTACC). In the second approach, the 3' ends of the same tiles were enzymatically labeled with Atto488-dUTP (Jena Bioscience, Jena, Germany). For this, Atto488-dUTP (80 µM), CoCl$_2$ (5 mM), terminal transferase enzyme (16 U/µL, Roche, Penzberg, Germany) and all DNA tiles (400 pmol) were mixed in a 20 µL, 1× TdT reaction buffer and then incubated at 37 °C for 30 min. Then, 2.5 µL of NaOAc (3 M) were added, and the solution was filled up to 80 µL with ice-cooled ethanol (99%). After 1 h of incubation at −20 °C, samples were centrifuged at 13,000× g for 30 min. Then, samples were washed with 70% ethanol for 10 min again, and the supernatant was discarded. The remaining pellet was dissolved in distilled water.

3.4. DNA Nanotube Assembly and Purification

For DNA nanotube assembly, 400 nM of dye and folate modified tiles, 800 nM of antisense-ODN and 1.6 µM of sense-ODN were mixed in a folding buffer (10 mM Tris-HCl, 1 mM EDTA, pH 8.0, 20 mM MgCl$_2$). For the plain nanotube assembly, 1 µM of unmodified tiles was used. The DNA nanotubes were folded over the course of 16 h (5 min at 80 °C, cooling down to 65 °C at 1 °C/min, cooling down to 25 °C at 2.5 °C/h). Purification of the assembled DNA nanotubes was done using 30K Amicon Ultra 0.5-mL centrifuge filters (30000 MWCO, Millipore, Schwalbach, Germany) to remove excess strands that were not folded into the structures. One hundred microliters of assembled DNA nanotube solution were mixed with 400 µL of folding buffer, filled into the centrifuge filter, and centrifuged 3 times at 13,000× g for 6 min. After every centrifuge step, the flow-through was removed and the filter was refilled up to 500 µL with buffer. After final centrifugation, the remaining solution at the bottom of the filter (~50 μL) was pipetted out, and the concentration of nanotubes was determined by measuring the optical density at 260 nm.

3.5. Gel Electrophoresis and Transmission Electron Microscopy

DNA nanotubes were analyzed by running samples in an agarose gel. For this, 2% agarose was dissolved in 0.5× TAE buffer by heating to boiling. $MgCl_2$ (11 mM) and ethidium bromide (0.5 μg/mL) were added after the cooling, and the solution was poured into a gel cask for solidification. Twenty microliters of each filter-purified DNA nanotube sample were mixed with 4 μL of 6× loading dye before loading into the gel pockets. Six microliters of 1 kb ladder were also loaded adjacent to the samples. The gel was run for 2 h at 70 V in an ice-cold water bath to prevent heat-induced denaturation of DNA nanotubes. To visualize DNA nanotubes, a JEM-1011 transmission electron microscope (JEOL GmbH, Eching, Germany) was used. DNA nanotubes were incubated on plasma-exposed (240 kV for 1 min) carbon-coated grids and then negatively stained with 2% uranyl formate for 10 s.

3.6. Stability of DNA Nanotubes

The stability of DNA nanotubes was tested in PBS buffer, DMEM and DMEM containing 10% FCS, separately. One microliter of DNA nanotubes (50 ng/μL) in 20 μL of buffer/medium was incubated at 37 °C for different time points. Two percent agarose gel with 11 mM $MgCl_2$ was prepared to analyze the samples, as mentioned in Experimental Section 3.5.

3.7. Cell Culture Experiments

HeLa cells were cultured at 37 °C, 5% CO_2 and 95% humidity in Dulbecco's modified Eagle's medium (Life Technologies Thermo Fischer, Waltham, MA, USA) supplemented with 10% heat-inactivated fetal calf serum (FCS, Invitrogen Thermo Fischer, Waltham, MA, USA), 2 mM L-glutamine, 100 U/mL penicillin and 100 μg/mL streptomycin. Stably GFP-transfected HeLa cell lines were generated by retroviral transduction using an eGFP containing pMP71 vector, as described previously [31,32]. To analyze uptake, DNA nanotubes, oligonucleotides and deoxynucleotide triphosphates were added to HeLa cells at a concentration between 10 and 40 nM for the indicated period of time in DMEM with L-glutamine, penicillin, streptomycin and either supplemented or not with 10% FCS. For siRNA-mediated knockdown experiments, GFP-expressing HeLa cells were seeded in 24-well plates and allowed to adhere overnight. On the next day, cells were either incubated with the indicated nanotubes coupled to GFP-targeting siRNAs or were transfected as a control with siRNA oligonucleotides (75 nM) using Lipofectamine RNAiMAX (Invitrogen). After 48 h, GFP-knockdown was measured by flow cytometry. The siRNA sequence used to target GFP (siGFP) was 5' GCCACAACGUCUAUAUCAU 3'. As a control (siCTRL), we used the non-targeting RNA sequence 5' GCGCUAUCCAGCUUACGUA 3' described previously [33]. SiRNAs were purchased from Eurofins and contained dTdT overhangs.

3.8. Flow Cytometry and Confocal Fluorescence Microscopy

Flow cytometry was used to determine the uptake of DNA nanotubes, oligonucleotides and deoxynucleotide triphosphates labeled with the fluorescent dyes, Atto488 or Atto647, into HeLa cells and to assess the knockdown efficiency of *GFP*-targeting siRNAs in stably GFP-expressing HeLa cells. For that, after the indicated time points, single-cell suspensions were prepared and washed several times before analyzing the cells on a FACS Calibur (Becton Dickinson, Franklin Lakes, NJ, USA). FlowJo software was used to analyze the data. GFP expression was depicted by median fluorescence intensity (MFI). For all experiments with fluorescent dye labeling, the data are represented as specific fluorescence intensity (SFI), which was calculated by dividing the MFI of the sample by the MFI of the control. Confocal fluorescence microscopy was used to determine the subcellular localization of nanotubes and RNAs taken up by HeLa cells. For that, HeLa cells were cultured in CELLview cell culture dishes with a glass bottom (Greiner Bio One). After incubation with Atto488- or Atto647-labeled nanotubes for the indicated time points, cells were washed three times with PBS and used for live-imaging on a Leica TCS SP5 confocal microscope (Leica Microsystems GmbH, Wetzlar, Germany). Zero-point-two micrograms per milliliter of Hoechst 33342 and MitoTracker Green (both from Life Technologies) were used according to the manufacturer's protocol to stain nuclei and mitochondria, respectively. In order to visualize endosomes, 20 μg/mL dextran labeled with Alexa Fluor 647 or Alexa Fluor 488 (Life Technologies Thermo Fischer) were added simultaneously with the DNA nanotubes.

4. Conclusions

In this study, we investigated the cellular delivery of tile-assembled DNA nanotubes carrying siRNAs using GFP-expressing HeLa cells via folate targeting. We observed that the nanostructures enter the cells via an endosomal pathway, but the nanostructures and their siRNA cargo are not capable of reaching the cytoplasm for knockdown and gene silencing. Contingently, no significant decrease in GFP expression levels was detectable, and folate modification did not change the uptake kinetics. The stability experiments revealed that unmodified DNA nanotubes are stable at 37 °C up to 8 h in the cell media and that they stay intact in PBS buffer containing 2 mM Mg^{2+} or more. However, the extension of the DNA tile strands with sequences that allow the hybridization of siRNA or dye-modified strands drastically decreases the construct's stability, a fact that may have contributed to the unsuccessful folate targeting experiments. Using DNA tiles that were 84-nt long drastically increased the stability in all cell media and buffers with low Mg^{2+} concentrations. Importantly, we observed that DNA strands alone and cleaved dyes are also uptaken by the cells, which can lead to the misinterpretation of recorded data. Overall, the results presented in this study demonstrate the importance of rigorously testing the stability of DNA nanostructures before applications *in vitro* and *in vivo*.

Supplementary Materials

Supplementary materials can be accessed at: http://www.mdpi.com/2079-4991/5/1/47/s1.

Acknowledgments

This work was supported by the European Commission under the Seventh Framework Programme (FP7), as part of the Marie Curie Initial Training Network, EScoDNA (No. 317110) and the ERC starting grant, ORCA, by the DFG Grant RO2525/5-1 to S.R. and by a Metiphys fellowship of the Medical Faculty of the Ludwig-Maximilian University to F.S.L.. H.M. is supported by the Molecular Medicine Program of the Medical Faculty of the Ludwig-Maximilian University (Förderprogramm für Forschung und Lehre) and DFG Grant GK 1202.

Author Contributions

S.K., S.R., T.L. and F.L. designed the experiments. S.K., H.M., I.S.M., V.C., A.M. and F.L. performed the experiments. S.K., A.M., T.L. and F.L. wrote the manuscript.

Conflicts of Interest

The authors declare no conflict of interest.

References

1. Bareford, L.M.; Swaan, P.W. Endocytic mechanisms for targeted drug delivery. *Adv. Drug Deliv. Rev.* **2007**, *59*, 748–758.
2. Smith, D.; Schüller, V.; Engst, C.; Radler, J.; Liedl, T. Nucleic acid nanostructures for biomedical applications. *Nanomedicine* **2013**, *8*, 105–121.
3. Czech, B.; Hannon, G.J. Small RNA sorting: Matchmaking for argonautes. *Nat. Rev. Genet.* **2011**, *12*, 19–31.
4. Kataoka, K.; Harada, A.; Nagasaki, Y. Block copolymer micelles for drug delivery: Design, characterization and biological significance. *Adv. Drug Deliv. Rev.* **2001**, *47*, 113–131.
5. Miller, A.D. Cationic liposomes for gene therapy. *Angew. Chem. Int. Ed.* **1998**, *37*, 1768–1785.
6. Zhou, J.; Rossi, J.J. Therapeutic potential of aptamer-siRNA conjugates for treatment of HIV-1. *BioDrugs* **2012**, *26*, 393–400.
7. Lee, H.; Lytton-Jean, A.K.; Chen, Y.; Love, K.T.; Park, A.I.; Karagiannis, E.D.; Sehgal, A.; Querbes, W.; Zurenko, C.S.; Jayaraman, M.; *et al.* Molecularly self-assembled nucleic acid nanoparticles for targeted *in vivo* siRNA delivery. *Nat. Nanotechnol.* **2012**, *7*, 389–393.
8. Douglas, S.M.; Bachelet, I.; Church, G.M. A logic-gated nanorobot for targeted transport of molecular payloads. *Science* **2012**, *335*, 831–834.
9. Li, J.; Pei, H.; Zhu, B.; Liang, L.; Wei, M.; He, Y.; Chen, N.; Li, D.; Huang, Q.; Fan, C. Self-assembled multivalent DNA nanostructures for noninvasive intracellular delivery of immunostimulatory cpg oligonucleotides. *ACS Nano* **2011**, *5*, 8783–8789.
10. Schüller, V.J.; Heidegger, S.; Sandholzer, N.; Nickels, P.C.; Suhartha, N.A.; Endres, S.; Bourquin, C.; Liedl, T. Cellular immunostimulation by cpg-sequence-coated DNA origami structures. *ACS Nano* **2011**, *5*, 9696–9702.
11. Zhao, Y.-X.; Shaw, A.; Zeng, X.; Benson, E.; Nyström, A.M.; Högberg, B. DNA origami delivery system for cancer therapy with tunable release properties. *ACS Nano* **2012**, *6*, 8684–8691.

12. Jiang, Q.; Song, C.; Nangreave, J.; Liu, X.; Lin, L.; Qiu, D.; Wang, Z.-G.; Zou, G.; Liang, X.; Yan, H.; *et al.* DNA origami as a carrier for circumvention of drug resistance. *J. Am. Chem. Soc.* **2012**, *134*, 13396–13403.

13. Voigt, N.V.; Torring, T.; Rotaru, A.; Jacobsen, M.F.; Ravnsbaek, J.B.; Subramani, R.; Mamdouh, W.; Kjems, J.; Mokhir, A.; Besenbacher, F.; *et al.* Single-molecule chemical reactions on DNA origami. *Nat. Nanotechnol.* **2010**, *5*, 200–203.

14. Rothemund, P.W. Folding DNA to create nanoscale shapes and patterns. *Nature* **2006**, *440*, 297–302.

15. Douglas, S.M.; Dietz, H.; Liedl, T.; Hogberg, B.; Graf, F.; Shih, W.M. Self-assembly of DNA into nanoscale three-dimensional shapes. *Nature* **2009**, *459*, 414–418.

16. Perrault, S.D.; Shih, W.M. Virus-inspired membrane encapsulation of DNA nanostructures to achieve *in vivo* stability. *ACS Nano* **2014**, *8*, 5132–5140.

17. Mikkilä, J.; Eskelinen, A.-P.; Niemelä, E.H.; Linko, V.; Frilander, M.J.; Törmä, P.; Kostiainen, M.A. Virus-encapsulated DNA origami nanostructures for cellular delivery. *Nano Lett.* **2014**, *14*, 2196–2200.

18. Okholm, A.H.; Nielsen, J.S.; Vinther, M.; Sorensen, R.S.; Schaffert, D.; Kjems, J. Quantification of cellular uptake of DNA nanostructures by qPCR. *Methods* **2014**, *67*, 193–197.

19. Ko, S.; Liu, H.; Chen, Y.; Mao, C. DNA nanotubes as combinatorial vehicles for cellular delivery. *Biomacromolecules* **2008**, *9*, 3039–3043.

20. Hong, C.A.; Jang, B.; Jeong, E.H.; Jeong, H.; Lee, H. Self-assembled DNA nanostructures prepared by rolling circle amplification for the delivery of sirna conjugates. *Chem. Comm.* **2014**, *50*, 13049–13051.

21. Hahn, J.; Wickham, S.F.; Shih, W.M.; Perrault, S.D. Addressing the instability of DNA nanostructures in tissue culture. *ACS Nano* **2014**, *8*, 8765–8775.

22. Seeman, N.C. Construction of three-dimensional stick figures from branched DNA. *DNA Cell Biol.* **1991**, *10*, 475–486.

23. Erben, C.M.; Goodman, R.P.; Turberfield, A.J. Single-molecule protein encapsulation in a rigid DNA cage. *Angew. Chem. Int. Ed.* **2006**, *45*, 7414–7417.

24. Shih, W.M.; Quispe, J.D.; Joyce, G.F. A 1.7-kilobase single-stranded DNA that folds into a nanoscale octahedron. *Nature* **2004**, *427*, 618–621.

25. Wei, B.; Dai, M.; Yin, P. Complex shapes self-assembled from single-stranded DNA tiles. *Nature* **2012**, *485*, 623–626.

26. Ke, Y.; Ong, L.L.; Shih, W.M.; Yin, P. Three-dimensional structures self-assembled from DNA bricks. *Science* **2012**, *338*, 1177–1183.

27. Yin, P.; Hariadi, R.F.; Sahu, S.; Choi, H.M.; Park, S.H.; Labean, T.H.; Reif, J.H. Programming DNA tube circumferences. *Science* **2008**, *321*, 824–826.

28. Schiffels, D.; Liedl, T.; Fygenson, D.K. Nanoscale structure and microscale stiffness of DNA nanotubes. *ACS Nano* **2013**, *7*, 6700–6710.

29. Martin, T.G.; Dietz, H. Magnesium-free self-assembly of multi-layer DNA objects. *Nat. Comm.* **2012**, *3*, doi:10.1038/ncomms2095.

30. Di Michele, L.; Mognetti, B.M.; Yanagishima, T.; Varilly, P.; Ruff, Z.; Frenkel, D.; Eiser, E. Effect of inert tails on the thermodynamics of DNA hybridization. *J. Am. Chem. Soc.* **2014**, *136*, 6538–6541.

31. Kobold, S.; Steffen, J.; Chaloupka, M.; Grassmann, S.; Henkel, J.; Castoldi, R.; Zeng, Y.; Chmielewski, M.; Schmollinger, J.C.; Schnurr, M.; *et al.* Selective bispecific T cell recruiting antibody enhances anti-tumor activity of adoptive T cell transfer. *J. Natl. Cancer Inst.* **2014**, *107*, doi:10.1093/jnci/dju364.

32. Engels, B.; Cam, H.; Schuler, T.; Indraccolo, S.; Gladow, M.; Baum, C.; Blankenstein, T.; Uckert, W. Retroviral vectors for high-level transgene expression in T lymphocytes. *Human Gene Ther.* **2003**, *14*, 1155–1168.

33. Besch, R.; Poeck, H.; Hohenauer, T.; Senft, D.; Hacker, G.; Berking, C.; Hornung, V.; Endres, S.; Ruzicka, T.; Rothenfusser, S.; *et al.* Proapoptotic signaling induced by RIG-I and MDA-5 results in type I interferon-independent apoptosis in human melanoma cells. *J. Clin. Invest.* **2009**, *119*, 2399–2411.

Polymorphic Ring-Shaped Molecular Clusters Made of Shape-Variable Building Blocks

Keitel Cervantes-Salguero [1], Shogo Hamada [2], Shin-ichiro M. Nomura [1] and Satoshi Murata [1,*

[1] Department of Bioengineering and Robotics, Tohoku University, Sendai 980-8579, Japan;
    E-Mails: cervantes@molbot.mech.tohoku.ac.jp (K.C.-S.); nomura@molbot.mech.tohoku.ac.jp (S.M.N.)

[2] Kavli Institute at Cornell for Nanoscale Science, Cornell University, Ithaca, NY 14853, USA;
    E-Mail: sh964@cornell.edu

* Author to whom correspondence should be addressed; E-Mail: murata@molbot.mech.tohoku.ac.jp

Academic Editor: Stephen Ralph

Abstract: Self-assembling molecular building blocks able to dynamically change their shapes, is a concept that would offer a route to reconfigurable systems. Although simulation studies predict novel properties useful for applications in diverse fields, such kinds of building blocks, have not been implemented thus far with molecules. Here, we report shape-variable building blocks fabricated by DNA self-assembly. Blocks are movable enough to undergo shape transitions along geometrical ranges. Blocks connect to each other and assemble into polymorphic ring-shaped clusters via the stacking of DNA blunt-ends. Reconfiguration of the polymorphic clusters is achieved by the surface diffusion on mica substrate in response to a monovalent salt concentration. This work could inspire novel reconfigurable self-assembling systems for applications in molecular robotics.

Keywords: DNA nanostructure; DNA origami; DNA stacking; reconfiguration; substrate

---

1. Introduction

Structures that can change their shapes are of interest at all scales, not only for the potential applications, but also because the design principles may be used across scales [1–8]. Switching structural dimensions [4] and topology [5], and partial detachment/attachment [7] of individual nanostructures, as well as controlling the number of components through geometric complementarity [1,8] and specific logic rules for targeting a configuration [2,3] of interacting macrostructures, are among some approaches for shape-reconfiguration. Of particular interest is the metamorphosis of clusters of mechatronic modules [3], where incorporated degrees-of-freedom allowed shape transitions in each module, generating locomotion and reconfiguration of the modules as a whole. However, there is still a bridge to be crossed between the programmable dynamic behaviour of individual molecular systems and the programmable collective behaviour of top-down fabricated structures. Indeed, the bottom-up fabrication of reconfigurable assemblies of interacting molecular robots remains challenging [9]. Towards this goal, we take a direct approach by designing building blocks with movable parts and argue that the shape properties of these blocks govern their assembly.

The shape properties of building blocks is important in molecular self-assembly, where the shape of a nanostructure or colloidal particle is defined as the surface geometry and the interaction field around the particle surface [10,11]. Although there has been extensive study of static particles [12–16], only a few contributions exist on the fabrication of structures with dynamic shapes [17–20]. Nevertheless, the work on micro- and nano-reconfigurable structures has recently inspired different simulation studies for the assembly and reconfiguration of assemblies of shape-changing nanorods, shape-shifting blocks, patchy nanoparticles with dynamic covalent bonds, and flexible lock-and-key colloids [21–25]. Although these simulations predict notable properties, such as switchable band gaps for optical devices or tunable pores for drug delivery, the field is open for actual molecular structures with variable shapes. Recently, complaint nanostructures made of DNA were demonstrated [26]. In our work, shape-variable building blocks are realized by harnessing the properties of DNA.

DNA does not only plays important roles in molecular biology, but it has also been used to fabricate nanostructures of various shapes and functionalities. Nanostructures ranging from molecular walkers on programmable tracks to static closed topologies, and from pre-stressed tensegrity structures to reconfigurable structures [4,5,17,26–35]. Undoubtedly, DNA has a great potential for engineering novel applications due to its programmability, rigidity, and flexibility [6,36,37]. Here, we show the concept, design, and experimentation of shape-variable building blocks made of DNA that are capable of self-assembly into polymorphic closed clusters. Our experimental setup provides a platform for studying the cluster disassembly and re-assembly (reconfiguration) on substrate in response to a monovalent salt concentration.

2. Experimental Section

The shape-variable building blocks (also called monomers) are restricted to homogenous blocks which self-assemble into clusters (also called x-mers) with finite size or limited number of building blocks (*i.e.*, x). The characteristics of the blocks are inspired by the flexibility of 3-point star DNA tiles [17], which allows the formation of closed nanostructures. Among closed topologies that have been used for DNA nanostructures [17,27–29], a ring is the simplest case because component monomers require only two bonding arms. In our blocks, the angle between bonding arms can adopt any value between a minimum and maximum (Figure 1a). These arms have complementary bonding edges and make bonds to each other in such a way that a 2-mers makes a *cis* configuration (Figure 1b, top). This angle implicitly defines the number of monomers in a cluster, namely the size of the self-assembled ring (Figure 1b, bottom).

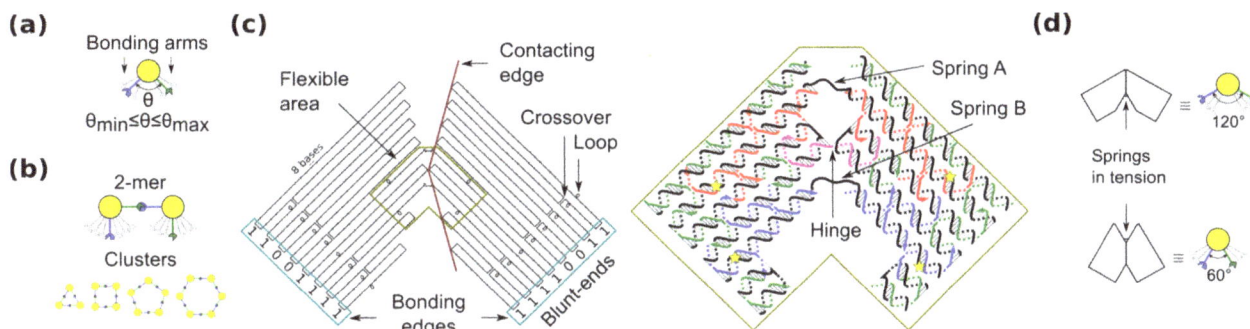

Figure 1. Shape-variable building block and design with DNA origami. (**a**) Abstract representation of the block with two bonding arms; (**b**) 2-mers in *cis* configuration and assembled polymorphic clusters (from 3-mers up to 6-mers); (**c**) (**left**): Origami scaffold (black), flexible area (mustard), contacting edges (red) and blunt-ends with binary codes (sky-blue); (**right**): Detail of the flexible area showing staples in colour (yellow stars indicate scaffold loops for tuning springs); (**d**) Origami profile and its correspondent abstract representation for the 60° and 120° configurations.

The shape-variable building blocks are fabricated with the DNA origami technique [35]. The block is a DNA origami with two symmetrical parts that resembles the two bonding arms explained above. The edge of each arm consists of DNA blunt-ends of which stacking interaction strength can be programmed and it is represented with a binary code [31]. The block incorporates: (1) a flexible area (enclosed area in Figure 1c), which provides degrees-of-freedom to the block; and (2) contacting edges, which set two angular limits of 60° and 120° (Figure 1d). In the flexible area, the block arms are linked by means of a single phosphate of a staple strand (a hinge that allows rotation) and two unpaired scaffold segments. These scaffold segments have two functions. One function is to prevent undesired degrees-of-freedom, such as relative twist of the block arms. Another function of the segments is to act as entropic springs (spring A and B) that tend to move back to states with higher entropy when stretched. The nominal length of each spring can be tuned by spooling the scaffold forming loops in the origami (Figure 1c, yellow stars), similarly as done for tuning the tension in tensegrity structures [30]. For example, by exchanging the blue staples in Figure 1c, it is possible to adjust the nominal length of spring B. DNA sequences of the springs should not form secondary structures and should be flexible. These sequences are chosen along the DNA scaffold in such a way that five thymines are kept in the middle of each spring (Supplementary Information S1). Details of the staples and DNA sequences are shown in Supplementary Information S13.

For simplicity, we define the notation $M(a,b)$ for the monomer with a (spring A) and b (spring B) nucleotides (nt). $a = 11$ nt and $b = 8 - 18$ nt are set in accordance with the flexibility of single stranded DNA (Supplementary Information S2). Monomers are prepared in a solution containing $1 \times$ TAE/Mg^{2+} (40 mM Tris, 20 mM acetic acid, 2 mM EDTA, 12.5 mM Mg acetate), heated and cooled down (Supplementary Information S3).

3. Results and Discussion

We characterize the monomers on mica substrate by using atomic force microscopy (AFM) under $1 \times$ TAE/Mg^{2+}, which is basically practically enough to keep the origami immobilized (Figure S12). AFM images indicate the formation of polymorphic clusters out of M (11,11). Clusters can be in a closed or open state. Figure 2a shows closed clusters, and Figure 2b shows an open 6-mers (no closed 6-mers is found in our AFM observations). The cluster size distribution, the number of monomers contributing to each particular cluster, is obtained by calculating the normalized number of x-mers, in an open or closed state, among several AFM images and multiplying it to the number of monomers in the x-mer. The distribution of 1-mers, 2-mers, and 3-mers changes when varying the length of spring B to 9 nt or 18 nt (Figure 2c). For a given x-mer, we calculate the p-values for those springs (Supplementary Information S12.2). For the analysed springs, 1-mers show a low value at 11 nt ($p < 0.01$) and a high value at 18 nt ($p < 0.01$). 2-mers and 3-mers show a low value at 18 nt ($p < 0.01$ and $p < 0.05$, respectively). An increment of 1-mers is consistent with the idea that a greater spring B gives more flexibility to the monomer and makes the connection with other monomer more difficult, which leads to less 2-mers and as a consequence less 3-mers. A variation in the distribution of 2-mers is also consistent with transition state theory, where the reaction between two particles is described as the relative orientations between them; here, although the surface area of reaction is constant (bonding edges do not change), a flexible monomer generates a different energy landscape compared to a less flexible one. As expected, increasing spring B to 11 nt increases the ratio closed:open 4-mers ($p = 0.04$) (Figure 2d). There are also some kinetic barriers for the formation of x-mers as found by the parallel polarity (Supplementary Information S8.1.4), mismatching and dislocation (Figure S13) of stacking bonds. Closed dimers (Supplementary Information S8.1.2) and larger clusters ($x > 6$; Supplementary Information S8.1.5) are also found. We identify the type of clusters after deposition based on the observation of those kinetic barriers (Supplementary Information S8.1.6).

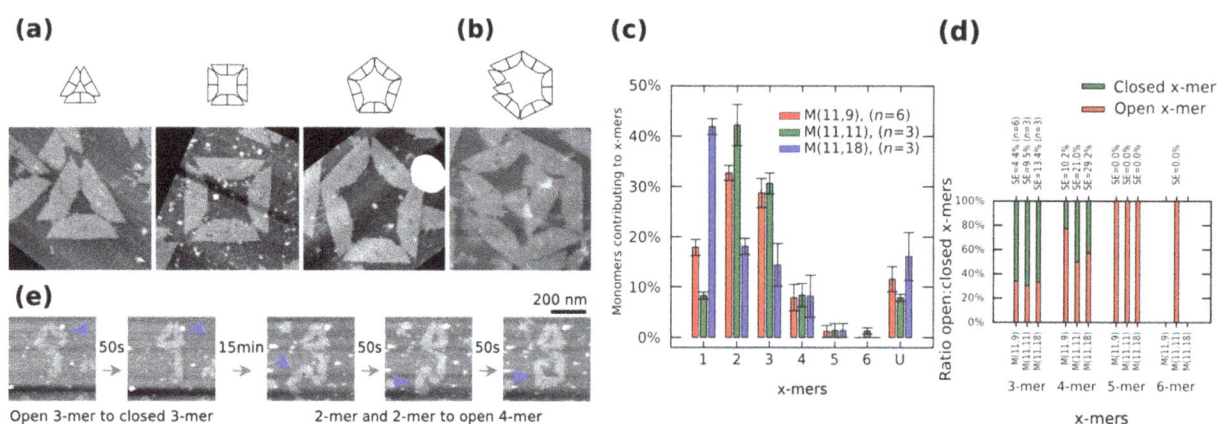

Figure 2. Polymorphic clusters made of shape-variable monomers before "reconfiguration protocol" (**a–d**) and during the "reconfiguration protocol" (**e**). (**a–b**) Clusters of M (11,11). First row: cluster representations. Second row: AFM images. 3-mers, 4-mers and 5-mers (**a**) and open 6-mers (**b**) at 2 nM concentration. (**c–d**) Distribution of monomers contributing to the formation of x-mers in open and closed states (**c**) and ratio open:closed x-mers (**d**) for M (11,9), M (11,11) and M (11,18). U indicates unclear monomers. (**e**) Inset of the frames in movie in Supplementary Information (0.02 fps). A 3-mers closes and two 2-mers self-assemble into an open 4-mer. AFM images in (**a**) are 310 nm × 300 nm in size. Error bars in (**c**) and SE in (**d**) indicate the standard error. n indicates the number of analysed AFM images of 2040 nm × 1680 nm, and the number of counted monomers per each AFM image is shown in Figure S12.

Self-assembled clusters on mica surface can reconfigure in response to monovalent cations. Monovalent cations, such as NaCl are known to reduce the DNA-mica binding interaction [38], causing diffusion of DNA origami on mica substrate [39,40]. A $1 \times$ TAE/Mg^{2+} buffer solution containing 100 mM NaCl is put on top of mica with M (11,11) for four hours (this protocol is called "reconfiguration protocol" hereafter, also in Supplementary Information S6). Then nanostructures are observed by AFM under $1 \times$ TAE/Mg^{2+} buffer solution containing 100 mM NaCl. As a result, monomers and clusters are observed diffusing and re-self-assembling on mica surface (movie in Supplementary Information). Figure 2e shows an inset of the movie. The connectivity between shape-variable monomers and the distribution on mica of the monomers do change after the "reconfiguration protocol": long polymers as well as areas with high populations of monomers appear (Supplementary Information S8.2).

We also restrict our building blocks to fixed shapes. Fixed monomers are prepared in one-pot reaction by bridging the contacting edges with additional staples (Figure 3a.1 and 3b.1). M ($a,b = 0$) (Figure 3a.1) and M ($a = 0,b$) (Figure 3b.1) denote fixed monomers in wide and narrow configuration, respectively.

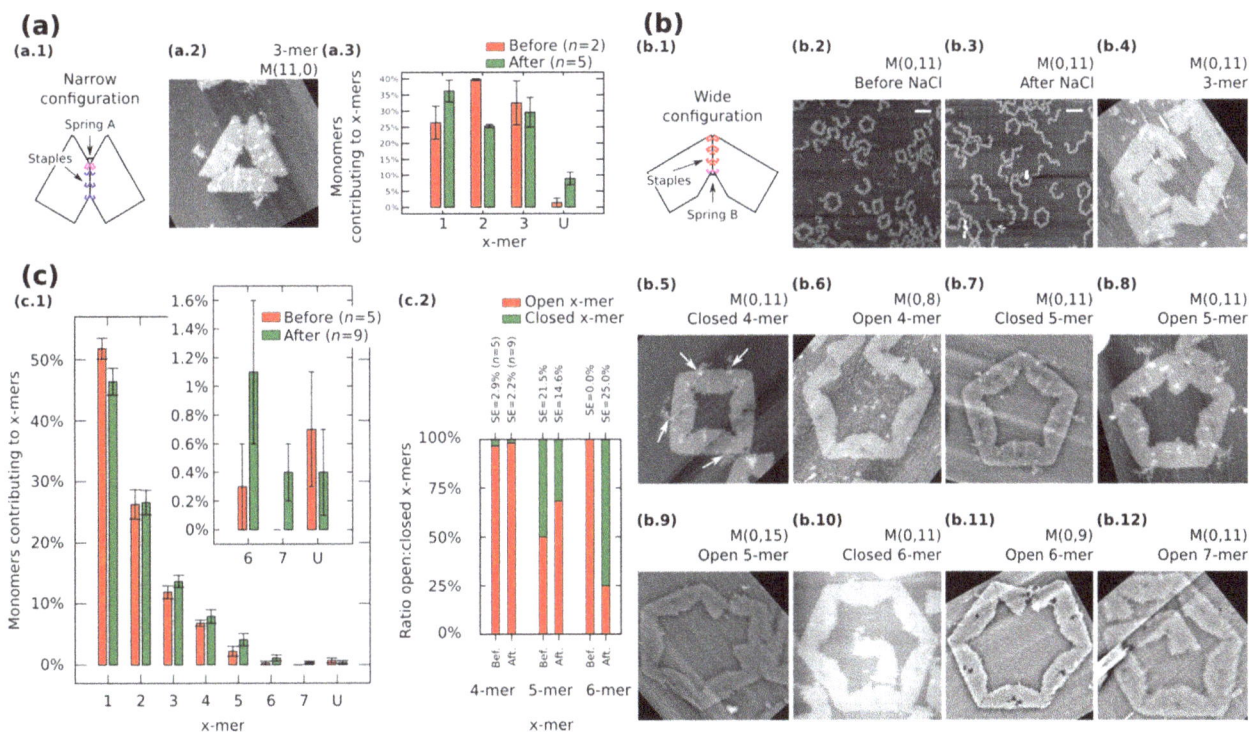

Figure 3. Self-assemble and reconfiguration of fixed monomers. (**a.1**) and (**b.1**): Fixed monomers. (**a.2–3**) 3-mers and cluster distribution before and after reconfiguration of M (11,0). (**b.2–3**) Representative AFM images of M (0,11) before and after reconfiguration. (**b.4–12**) Clusters after reconfiguration for different spring B. White arrows in (**b.5**) show extra M13. (**c**): Distribution of monomers contributing to the formation of x-mers including open and closed states (**c.1**) and ratio open:closed x-mers (**c.2**) for M (0,11). U indicates unclear monomers. AFM images are 310 nm × 300 nm. Width of (**b.9**) is 500 nm. Scale bars are 200 nm. Error bars in (**a.3, c.1**) and SE in (**c.2**) indicate the standard error. n indicates the number of analysed AFM images of 2040 nm × 1680 nm, and the number of counted monomers per each AFM image is shown in Figure S12.

Fixed monomers M (11,0) self-assemble into 3-mers. AFM after sample deposition shows as much as 32.5% of monomers forming 3-mers (Figure 3a.3 and Figure S9). After the "reconfiguration protocol" the shape of histogram changes indicating a decrement of 2-mers ($p < 0.01$).

AFM images of M (0,11) prepared in solution show the formation of polymorphic clusters (Figure 3b.2). 4-mers (28 open and one closed) and 5-mers (four open and four closed) are found in open and closed states among five AFM images of 2040 nm × 1680 nm (number of monomers per image are indicated in Figure S12). However, only one open 6-mers is observed (among 179 monomers in one AFM image), indicating kinetic traps in the formation process of 6-mers. These traps are possibly due to the flexibility of the origami, in other words, the stacking interaction compensates the energy required to bend the DNA helices favoring narrow angles for the monomer.

M (0,11) on mica is prepared by following the "reconfiguration protocol" and observed by AFM under 1 × TAE/Mg^{2+} (Figure 3b.3). Figure 3c.1 shows the distribution of monomers contributing to each particular cluster in open and closed states. In general, small clusters have a high yield and no significant difference exists before and after addition of NaCl except for 1-mers ($p < 0.05$) and the appearance of 7-mers (Figure 3b.12; 0.4% with SE = 0.2%). The ratio of open:closed clusters shows the dominance of closed 6-mers after the addition of NaCl (Figure 3c.2; from 0.0% to 75% with SE = 25%). We can speculate two causes for the formation of closed 6-mers, (1) the diffusion and fluctuation of the origami is restricted to the surface favouring on-plane connections with neighbouring monomers; and (2) the concentration of NaCl in the buffer enhances the electrostatic screening between the arms of the monomer and consequently allowing a wider angle for assembling 6-mers. White arrows in Figure 3b.5 indicate the extra M13 of the origami. This extra M13 suggests connections through parallel stacking polarity (Supplementary Information S8.1.4).

In order to explore the effect of the springs, we tune the length of the spring B and apply the "reconfiguration protocol". Figure 3b.4–12 show representative clusters for different spring B. In the case of M (0,9), monomers in 6-mers show stretched strands and angles less than the designed 120° (Figure 3b.11). We say these clusters are open 6-mers.

In general, the distribution shows the tendency for forming clusters with yields varying according to the spring length (Figure 4a). For a given x-mer, we find critical springs that show significant difference in the yields (Supplementary Information S12.2). For 1-mers, critical springs with 8 nt, 12 nt, 15 nt, 17 nt and 18nt ($p < 0.01$). For 2-mers, critical springs with 8 nt, 13 nt, 15 nt and 18 nt ($p < 0.01$). For 3-mers, critical springs with 8 nt, 11 nt, and 15 nt ($p < 0.01$). For 4-mers, critical springs with 8 nt, 11 nt, 15 nt, and 17 nt ($p < 0.01$). For 5-mers, critical springs with 8 nt and 10 nt ($p < 0.01$). Critical springs with 8 nt and 13 nt for 6-mers ($p < 0.05$), and critical springs with 10 nt, 15 nt and 18 nt for 7-mers ($p < 0.05$). Varying the spring length does not only affect the yield of x-mers, but also the tendency for closed states (Figure 4b). We also find critical springs for the ratio open:closed clusters (Supplementary Information S12.2). 8 nt ($p < 0.01$) and 16 nt ($p < 0.05$) show a high ratio closed:open 4-mers when comparing with near springs. For 12 nt all 6-mers are closed (one 6-mers among 276 monomer, and other 6-mers among 334 monomers from nine AFM images of 2040 nm × 1680 nm), and for springs greater than 11 nt or 12 nt, the number of closed 6-mers seem to decrease ($p < 0.05$). We explain these results considering that short springs pull the bonding arms; and, as a result, the formation of one type of cluster is favoured over the formation of other types, as shown by the critical spring with 11 nt. In addition, closed 4-mers form if the angle of each monomer approximates 90°, which happens if the spring is short, as with the critical spring with 8 nt. The critical spring with 8 nt is comparable with our calculated minimum spring length of 9 nt (Supplementary Information S2). On the other hand, if the spring is longer, the monomer becomes more flexible and it may be difficult for the cluster to close, as shown by the critical springs of 11 nt and 12 nt. These results indicate that the stiffness can be adjusted by tuning the spring length.

The formation of the clusters is time dependent (Supplementary Information S11). Sample $M(11,9)$ is at room temperature for one month and nanostructures are observed by AFM. As a result, the distribution of $M(11,9)$ changes. A similar behavior occurs with sample $M(0,11)$.

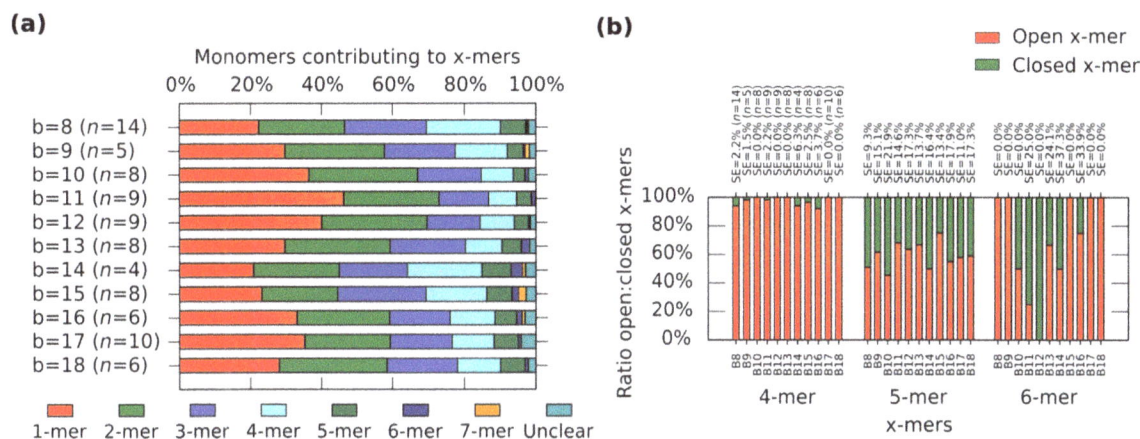

Figure 4. Distributions of the clusters after reconfiguration for $M(0,b)$ ($b = 8-18$). (**a**) Distribution of monomers contributing to the formation of x-mers including open and closed states. Colours indicate each type of cluster. Yellow indicates open 7-mer. Analysed AFM areas are 2240 nm × 1680 nm; (**b**) Ratio open:closed clusters for 4-mers, 5-mers and 6-mers. Error bars in (**a**) and SE in (**b**) indicate the standard error. n indicates the number of analysed AFM images of 2040 nm × 1680 nm, and the number of counted monomers per each AFM image is shown in Figure S12.

4. Conclusions

In summary, we have studied the self-assembly and reconfiguration of clusters made of shape-variable building blocks. We found that the cluster formation depends on the block shape. Moreover, we found that geometric constrains, as well as kinetic barriers, inhibit the formation of large clusters. It may not be difficult to expand the variety of shape-variable blocks for assembling other clusters and crystals, but some issues, such as the degree of flexibility of the block, should be addressed carefully. Finally, the shape-variable building blocks have provided a platform for the implementation of those scenarios proposed in the literatures for reconfiguring molecular assemblies.

Supplementary Materials

Supplementary materials can be accessed at: http://www.mdpi.com/2079-4991/5/1/208/s1.

Acknowledgments

Keitel Cervantes-Salguero is grateful to the Monbukagakusho scholarship given by the Ministry of Education, Culture, Sports, Science and Technology (MEXT) of Japan. This work was supported by two Grants-in-Aid for Scientific Research (No. 22220001 and No. 24104005) by MEXT to S. Murata.

Author Contributions

Keitel Cervantes-Salguero conceived the project idea with input from Shogo Hamada, Shin-ichiro. M. Nomura and Satoshi Murata. All the authors contributed to design the experiments. Keitel Cervantes-Salguero performed the experiments and analysed the data. All the authors contributed to write the manuscript.

Conflicts of Interest

The authors declare no conflict of interest.

References

1.	Mao, C.; Thalladi, V.R.; Wolfe, D.B.; Whitesides, S.; Whitesides, G.M. Mesoscale self-assembly: Capillary interactions when positive and negative menisci have similar amplitudes. *J. Am. Chem. Soc.* **2002**, *124*, 14508–14509.

2.	Klavins, E. Programmable self-assembly. *IEEE Control Syst. Mag.* **2007**, *27*, 4, 43–56.

3.	Murata, S.; Kurokawa, H. Self-reconfigurable robots. *IEEE Robot. Autom. Mag.* **2007**, *14*, 71–78.

4.	Goodman, R.P.; Heilemann, M.; Doose, S.; Erben, C.M.; Kapanidis, A.N.; Turberfield, A.J. Reconfigurable, braced, three-dimensional DNA nanostructures. *Nat. Nanotechnol.* **2008**, *3*, 93–96.

5.	Han, D.; Pal, S.; Liu, Y.; Yan, H. Folding and cutting DNA into reconfigurable topological nanostructures. *Nat. Nanotechnol.* **2010**, *5*, 712–717.

6.	Campolongo, M.J.; Kahn, J.S.; Cheng, W.; Yang, D.; Gupton-Campolongo, T.; Luo, D. Adaptive DNA-based materials for switching, sensing, and logic devices. *J. Mater. Chem.* **2011**, *21*, 6113–6121.

7.	Wei, B.; Ong, L.L.; Chen, J.; Jaffe, A.S.; Yin, P. Complex reconfiguration of DNA nanostructures. *Angew. Chem. Int. Ed.* **2014**, *126*, 7605–7609.

8.	Sacanna, S.; Rossi, L.; Pine, D.J. Magnetic click colloidal assembly. *J. Am. Chem. Soc.* **2012**, *134*, 6112–6115.

9.	Hagiya, M.; Konagaya, A.; Kobayashi, S.; Saito, H.; Murata, S. Molecular robots with sensors and intelligence. *Acc. Chem. Res.* **2014**, *47*, 1681–1690.

10.	Cademartiri, L.; Bishop, K.J.M.; Snyder, P.W.; Ozin, G.A. Using shape for self-assembly. *Philos. Trans. R. Soc. A* **2012**, *370*, 2824–2847.

11.	Sacanna, S.; Pine, D.J.; Yi, G.-R. Engineering shape: The novel geometries of colloidal self-assembly. *Soft Matter* **2013**, *9*, 8096–8106.

12.	Glotzer, S.C.; Salomon, M.J. Anisotropy of building blocks and their assembly into complex structures. *Nat. Mater.* **2007**, *6*, 557–562.

13.	Kim, J.W.; Kim, J.H.; Deaton, R. DNA-linked nanoparticle building blocks for programmable matter. *Angew. Chem. Int. Ed.* **2011**, *50*, 9185–9190.

14. Wang, Y.; Wang, Y.; Breed, D.; Manoharan, V.N.; Feng, L.; Hollingsworth, A.; Weck, M.; Pine, D. Colloids with valence and specific directional bonding. *Nature* **2012**, *491*, 51–56.

15. Zhang, C.; Macfarlane, R.J.; Young, K.L.; Choi, C.H.J.; Hao, L.; Auyeung, E.; Liu, G.; Zhou, X.; Mirkin, C.A. A general approach to DNA-programmable atom equivalents. *Nat. Mater.* **2013**, *12*, 741–746.

16. Walker, D.A.; Leitsch, E.K.; Nap, R.J.; Szleifer, I.; Grzybowski, B.A. Geometric curvature controls the chemical patchiness and self-assembly of nanoparticles. *Nat. Nanotechnol.* **2013**, *8*, 676–680.

17. He, Y.; Ye, T.; Su, M.; Zhang, C.; Ribbe, A.E.; Jiang, W.; Mao, C. Hierarchical self-assembly of DNA into symmetric supramolecular polyhedra. *Nature* **2008**, *452*, 198–202.

18. Sacanna, S.; Irvine, W.T.M.; Chaikin, P.M.; Pine, D.J. Lock and key colloids. *Nature* **2010**, *464*, 575–578.

19. Yoo, J.-W.; Mitragotri, S. Polymer particles that switch shape in response to a stimulus. *PNAS* **2010**, *107*, 11205–11210.

20. Lee, K.J.; Yoon, J.; Rahmani, S.; Hwang, S.; Bhaskar, S.; Mitragotri, S.; Lahann, J. Spontaneous shape reconfigurations in multicompartmental microcylinders. *PNAS* **2012**, *109*, 16057–16062.

21. Nguyen, T.D.; Glotzer, S.C. Reconfigurable assemblies of shape-changing nanorods. *ACS Nano* **2010**, *4*, 2585–2594.

22. Nguyen, T.D.; Jankowski, E., Glotzer, S.C. Self-assembly and reconfigurability of shape-shifting particles. *ACS Nano* **2011**, *9*, 8892–8903.

23. Gang, O.; Zhang, Y.G. Shaping phases by phasing shapes. *ACS Nano* **2011**, *5*, 8459–8465.

24. Guo, R.; Liu, Z.; Xie, X.-M.; Yan, L.-T. Harnessing dynamic covalent bonds in patchy nanoparticles: Creating shape-shifting building blocks for rational and responsive self-assembly. *J. Phys. Chem. Lett.* **2013**, *4*, 1221–1226.

25. Kohlstedt, K.L.; Glotzer, S.C. Self-assembly and tunable mechanics of reconfigurable colloidal crystals. *Phys. Rev. E* **2013**, *87*, 032305.

26. Zhou, L.; Marras, A.E.; Su, H.-J.; Castro, C.E. DNA origami compliant nanostructures with tunable mechanical properties. *ACS Nano* **2014**, *8*, 27–34.

27. Bombelli, F.B.; Gambinossi, F.; Lagi, M.; Berti, D.; Caminati, G.; Brown, T.; Sciortino, F.; Norden, B.; Baglioni, P. DNA closed nanostructures: A structural and Monte Carlo simulation study. *J. Phys. Chem. B* **2008**, *112*, 15283–15294.

28. Yin, P.; Hariadi, R.F.; Sahu, S.; Choi, H.M.T.; Park, S.H.; LaBean, T.H.; Reif, J.H. Programming DNA tube circumferences. *Science* **2008**, *321*, 824–826.

29. Hamada, S.; Murata, S. Substrate-assisted assembly of interconnected single-duplex DNA nanostructures. *Angew. Chem. Int. Ed.* **2009**, *121*, 6952–6955.

30. Liedl, T.; Högberg, B.; Tytell, J.; Ingber, D.E.; Shih, W.M. Self-assembly of three-dimensional prestressed tensegrity structures from DNA. *Nat. Nanotechnol.* **2010**, *5*, 520–524.

31. Woo, S.; Rothemund, P.W.K. Programmable molecular recognition based on the geometry of DNA nanostructures. *Nat. Chem.* **2011**, *3*, 620–627.

32. Ke, Y.; Ong, L.L.; Shih, W.M.; Yin, P. Three-dimensional structures self-assembled from DNA bricks. *Science* **2012**, *338*, 1177–1183.

33. Wickham, S.F.J.; Bath, J.; Katsuda, Y.; Endo, M.; Hidaka, K.; Sugiyama, H.; Turberfield, A.J. A DNA-based molecular motor that can navigate a network of tracks. *Nat. Nanotechnol.* **2012**, *7*, 169–173.

34. Kuzuya, A.; Sakai, Y.; Yamazaki, T.; Xu, Y.; Komiyama, M. Nanomechanical DNA origami "single-molecule beacons" directly imaged by atomic force microscopy. *Nat. Commun.* **2011**, *2*, doi:10.1038/ncomms1452.

35. Rothemund, P.W.K. Folding DNA to create nanoscale shapes and patterns. *Nature* **2006**, *440*, 297–302.

36. Tørring, T.; Voigt, N.V.; Nangreave, J.; Yan, H.; Gothelf, K.V. DNA origami: A quantum leap for self-assembly of complex structures. *Chem. Soc. Rev.* **2011**, *40*, 5636–5646.

37. Pinheiro, A.V.; Han, D.; Shih, W.M.; Yan, H. Challenges and opportunities for structural DNA nanotechnology. *Nat. Nanotechnol.* **2011**, *6*, 763–772.

38. Piétrement, O.; Pastré, D.; Fusil, S.; Jeusset, J.; David, M.O.; Landousy, F.; Hamon, L.; Zozime, A.; Le Cam, E. Adsorption of DNA to mica mediated by divalent counterions: A theoretical and experimental study. *Langmuir* **2003**, *19*, 2536–2539.

39. Aghebat Rafat, A.; Pirzer, T.; Scheible, M.B.; Kostina, A.; Simmel, F.C. Surface-assisted large-scale ordering of DNA origami tiles. *Angew. Che. Int. Ed.* **2014**, *53*, 7665–7668.

40. Woo, S.; Rothemund, P.W.K. Self-assembly of two-dimensional DNA origami lattices using cation-controlled surface diffusion. *Nat. Commun.* **2014**, *5*, doi:10.1038/ncomms5889.

3

Synthesis of Upconversion β-NaYF$_4$:Nd^{3+}/Yb^{3+}/Er^{3+} Particles with Enhanced Luminescent Intensity through Control of Morphology and Phase

Yunfei Shang [1], **Shuwei Hao** [1,*], **Jing Liu** [1], **Meiling Tan** [1], **Ning Wang** [1], **Chunhui Yang** [1,2] and **Guanying Chen** [1,3,*]

[1] School of Chemical Engineering and Technology, Harbin Institute of Technology, Harbin 150001, China; E-Mails: syf19943@sina.com (Y.S.); jing43210@163.com (J.L.); tanmeiling828@163.com (M.T.); wangning9004@sina.com (N.W.); yangchh@hit.edu.cn (C.Y.)

[2] Harbin Huigong Technology Co. Ltd, Harbin 150001, China

[3] Institute for Lasers, Photonics and Biophotonics, University at Buffalo, The State University of New York, Buffalo, NY 14260, USA

*   Authors to whom correspondence should be addressed; E-Mails: haoshuwei031011@126.com (S.H.); chenguanying@hit.edu.cn (G.C.).

Academic Editor: Thomas Nann

Abstract: Hexagonal NaYF$_4$:Nd^{3+}/Yb^{3+}/Er^{3+} microcrystals and nanocrystals with well-defined morphologies and sizes have been synthesized via a hydrothermal route. The rational control of initial reaction conditions can not only result in upconversion (UC) micro and nanocrystals with varying morphologies, but also can produce enhanced and tailored upconversion emissions from the Yb^{3+}/Er^{3+} ion pairs sensitized by the Nd^{3+} ions. The increase of reaction time converts the phase of NaYF$_4$:Nd^{3+}/Yb^{3+}/Er^{3+} particles from the cubic to the hexagonal structure. The added amount of oleic acid plays a critical role in the shape evolution of the final products due to their preferential attachment to some crystal planes. The adjustment of the molar ratio of F$^-$/Ln^{3+} can range the morphologies of the β-NaYF$_4$:Nd^{3+}/Yb^{3+}/Er^{3+} microcrystals from spheres to nanorods. When excited by 808 nm infrared laser, β-NaYF$_4$:Nd^{3+}/Yb^{3+}/Er^{3+} microplates exhibit a much stronger UC emission intensity than particles with other morphologies. This phase- and morphology-dependent UC emission holds promise for applications in photonic devices and biological studies.

Keywords: NaYF$_4$ microcrystals; Nd^{3+} sensitizer; morphology control; upconversion (UC)

1. Introduction

Lanthanide doped upconversion (UC) materials have attracted increasing interest due to their ability to convert low-energy excitations into high-energy emissions at shorter wavelength [1–3]. This unique property makes UC materials suitable for a wide range of applications, such as biological sensing [4,5], *in vivo* imaging [6,7], drug delivery [8,9], and photodynamic therapy [10,11]. Particularly, lanthanide doped fluoride materials revive these applications owing to their higher UC efficiency and excellent physicochemical stabilities [12]. Among them, β-NaYF$_4$ is considered to be one of the most efficient host lattices with low phonon energies (<350 cm^{-1}) to minimize energy losses at the intermediate states of lanthanide ions [7]. Moreover, Yb^{3+} ions were used as an efficient sensitizer to enhance UC emissions by taking advantage of its high absorption cross-section and efficient energy transferring to the activators of Er^{3+}, Ho^{3+}, or Tm^{3+}. However, the absorption ability of Yb^{3+} sensitizers limited the excitation of most current UC materials to be performed at ~980 nm [13,14]. This elicits heating problems, which cause possible damage to cells and tissues due to strong water absorption at the excitation wavelength.

Recently, extensive efforts have been devoted to development of UC materials that can be excited at better excitation wavelengths but cause minimized adverse effects on the biological cells and tissues. Doping of Nd^{3+} ion as a sensitizer into the current UC systems would be appealing towards this purpose, as Nd^{3+} can be excited by the well-established commercial 808 nm excitation source, the wavelength of which is negligibly absorbed by water molecules. Indeed, the absorption coefficient of water at ~808 nm is around 0.02 cm^{-1}, more than 24 times lower than the value of 0.482 cm^{-1} at ~980 nm. The Nd^{3+} sensitizer displays high absorption cross-section (~1.2 × 10^{-19} cm^2) at 808 nm, which is one order of magnitude higher than that of Yb^{3+} sensitizer (~1.2 × 10^{-20} cm^2) [15,16]. Moreover, efficient energy transfers have been established from Nd^{3+} to Yb^{3+} and then from Yb^{3+} to the acceptor (Er^{3+}, Ho^{3+} or Tm^{3+}) [17]. This enables the production of UC materials that can be excited at ~808 nm, and thus draws attention to the need for developing Nd^{3+}-sensitized UC materials for biological applications. Yan *et al.* introduced the Nd^{3+} sensitizer into NaGdF$_4$ nanoparticles and showed that it minimized the overheating effect in *in vivo* bioimaging compared with Yb^{3+}-sensitized nanoparticles [18]. Liu and colleagues presented a new type of core-shell UC nanoparticles based on doping high concentrations of Nd^{3+} ions in the shell structure to enhance the UC emission of activators [13]. Utilizing Nd^{3+}-sensitized UC nanoparticles to remove the constraints in conjunction with conventional UC nanoparticles is an important technology for biological studies; this conclusion has also been independently verified in core-shell UC nanoparticles by Han *et al.* and Wang's group [19,20]. Strategies to prepare Nd^{3+}-sensitized particles and to improve and tailor upconverted luminescence from them are in demand in order to produce impetus for their biomedical applications [21,22]. Thus far, very few studies reported on size- and morphology-controlled synthesis of UC Nd^{3+}-sensitized particles with tailored luminescence when excited at ~808 nm.

The particle morphologies have been found to play a significant role on the luminescent properties of lanthanide-doped UC materials [23–25]. The β-NaLuF$_4$:Yb^{3+}, Er^{3+} nanomaterials with tube shapes display higher luminescent intensity than those of other morphologies [26]. Lin *et al.* [26] reveal that the luminescent intensity was increased to 6 times in microtubes' structure, compared with that of limb-like shape. In our previous studies, the microrods exhibit a better photoluminescence signal than other types that resulted in β-NaYF$_4$ micro- and nanocrystals [27]. Until now, no attempts have been made to prepare a range of Nd^{3+}-sensitized β-NaYF$_4$ particles with tuned morphologies and enhanced UC luminescence. Herein, we report on the controlled synthesis of monodisperse Nd^{3+}-sensitized β-NaYF$_4$ microcrystals and nanocrystals with uniform size and tunable shapes under a hydrothermal condition. We show that variations of the oleic acid ligand and the concentration ratio of F$^-$/Ln^{3+} play a synergistic role in defining the morphologies of resulting particles. The crystal growth mechanisms of β-NaYF$_4$:10% Nd^{3+}, 10% Yb^{3+}, 2% Er^{3+} particles with sphere-like shape, octadecahedral shape, and hexagonal shape with protruding centers, were disclosed by collecting samples at pre-set time intervals. Moreover, spectroscopic investigations revealed that upconverted luminescence intensity is dependent on the resulting particle morphology, size, and crystallinity, manifesting the highest intensity from NaYF$_4$:10% Nd^{3+}, 10% Yb^{3+}, 2% Er^{3+} microplates.

2. Results and Discussion

2.1. Effects of Reaction Time

Because the crystal structure of NaYF$_4$ exhibits cubic (α-) and hexagonal (β-) polymorphic forms, reaction times were changed to probe the phase transformation process. Figures 1 and 2 show the X-ray diffraction (XRD) patterns and field emission scanning electron microscopy (FESEM) images of the resulting NaYF$_4$:10% Nd^{3+}, 10% Yb^{3+}, 2% Er^{3+} samples, respectively. Figure 1 reveals that the phase transformation process ($\alpha \rightarrow \beta$) occurred gradually with an extended reaction time. The sample obtained with a shorter reaction time ($t = 3$ h) shows a mixture of the α-phase (JCPDS No. 06-0342) and the β-phase (JCPDS No. 16-0334). Judging from the full width of the XRD peaks in Figure 1a, the size of α-phase NaYF$_4$ is very small, which agrees well with the SEM result of Figure 2a. As one can see from Figure 2a, the sample includes the aggregates formed by a large quantity of spherical nanoparticles. As the reaction time was extended to 6 h and 12 h, the peak intensity of α-NaYF$_4$ became weak (Figure 1b, 1c). In contrast, the peak intensities corresponding to the β-phase NaYF$_4$ became stronger. In addition, from the images shown in Figure 2b, 2c, it is clearly seen that hexagonal NaYF$_4$ microprisms are obtained, which become more regular and smoother with a longer reaction time. This indicated that the samples transformed gradually from α to β phase as the reaction time was prolonged. When the time was extended to 24 h, only pure β-phase NaYF$_4$ existed. Moreover, the intensities of the diffraction peaks in Figure 1d dramatically increased with respect to those samples of short times, implying that the crystallinity of the sample increases with the increase of reaction time. There is a large quantity of microprisms with uniform, smooth, and flat surfaces shown in Figure 2d, which agrees well with the corresponding XRD result of Figure 1d.

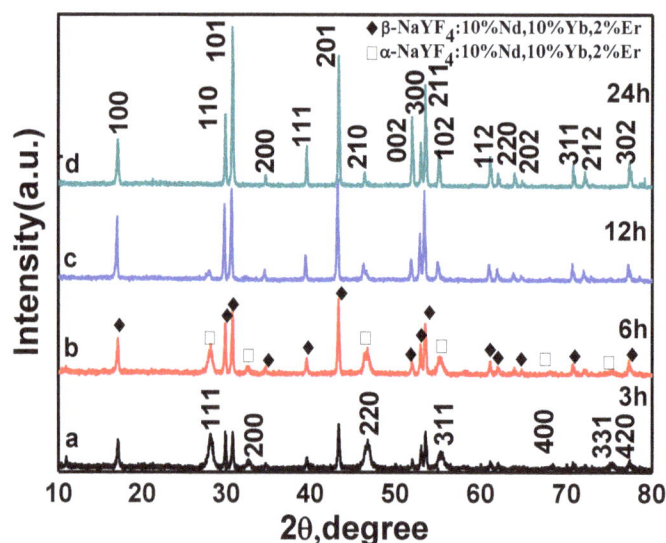

Figure 1. X-ray diffraction (XRD) patterns of NaYF$_4$: 10% Nd^{3+}, 10% Yb^{3+}, 2% Er^{3+} microcrystals synthesized at different hydrothermal times: (**a**) 3 h; (**b**) 6 h; (**c**) 12 h; and (**d**) 24 h (180 °C F$^-$/Ln^{3+} = 5:1 OA/Ln^{3+} = 40:1).

Figure 2. Typical field emission scanning electron microscopy (FESEM) images for the resulting NaYF$_4$:10% Nd^{3+}, 10% Yb^{3+}, 2% Er^{3+} microcrystals prepared with different hydrothermal times: (**a**) 3 h; (**b**) 6 h; (**c**) 12 h; and (**d**) 24 h (Other synthetic parameters were kept identical, 180 °C, F$^-$/Ln^{3+} = 5:1, OA/Ln^{3+} = 40:1).

The above analysis indicates that the hydrothermal reaction time has significant effects on the size, shape, and phase structure of final products. A long hydrothermal time benefits crystallization and the growth of bigger size microcrystals. Moreover, this result suggests that the β-phase NaYF$_4$ crystal particles possess higher thermal energy than the α-phase NaYF$_4$; a long hydrothermal time is needed to accumulate adequate energy to complete the phase transformation process.

2.2. *Effect of the Ratio of OA/Ln^{3+}*

Because OA can rationally control and modify the shape of $NaYF_4$ crystals, we probed the role of variation of the molar ratio of OA/Ln^{3+} to impact the resulted morphology of $NaYF_4$:10% Nd^{3+}, 10% Yb^{3+}, 2% Er^{3+} microcrystals in our synthesis system. Figure 3a shows the SEM images of the sample synthesized with 6:1 of OA/Ln^{3+}. The image indicates that the resulting products are uniform nanolines. The nanoline can be identified as pure Y(OH)$_3$:10% Nd^{3+}, 10% Yb^{3+}, 2% Er^{3+} by XRD (Figure 4a). It might be explained by the fact that the reaction system was alkaline when it contained little OA. The rare earth ions can thus easily combine with hydroxyl to generate Y(OH)$_3$:10% Nd^{3+}, 10% Yb^{3+}, 2% Er^{3+} precipitation. With the ratio of OA/Ln^{3+} exceeding 20:1, pure β-phase $NaYF_4$ was successfully produced, as shown in Figure 4b. Figure 3b shows the SEM images of the sample synthesized with 20:1 of OA/Ln^{3+}. The image reveals that the sample contains uniform nanorods with a length of 1.2 μm and an average lateral diameter of 0.245 μm. As the ratio of OA/Ln^{3+} was increased to 30:1, hexagonal microprisms with a length of 0.25 μm and an average lateral diameter of 0.25 μm were obtained. When the ratio of OA/Ln^{3+} exceeded 40:1, uniform hexagonal microprisms with a length of 0.2 μm and the lateral diameter of 1 μm were produced. Hence the aspect ratio L/D (length/diameter) increases gradually with the molar ratio of OA/Ln^{3+}. This indicates that the OA plays an important role in the morphology evolution from nanorods to hexagonal microplates. The involved underlined growing mechanism will be discussed in Section 2.4.

Figure 3. Typical FESEM images for the as-prepared $NaYF_4$: 10% Nd^{3+},10% Yb^{3+}, 2% Er^{3+} microcrystals at different molar ratios of OA/Ln^{3+}: (**a**) 6:1; (**b**) 20:1; (**c**) 30:1; and (**d**) 40:1 (Other synthetic parameters were kept identical, 180 °C, 24 h, F$^-$/Ln^{3+} = 5:1).

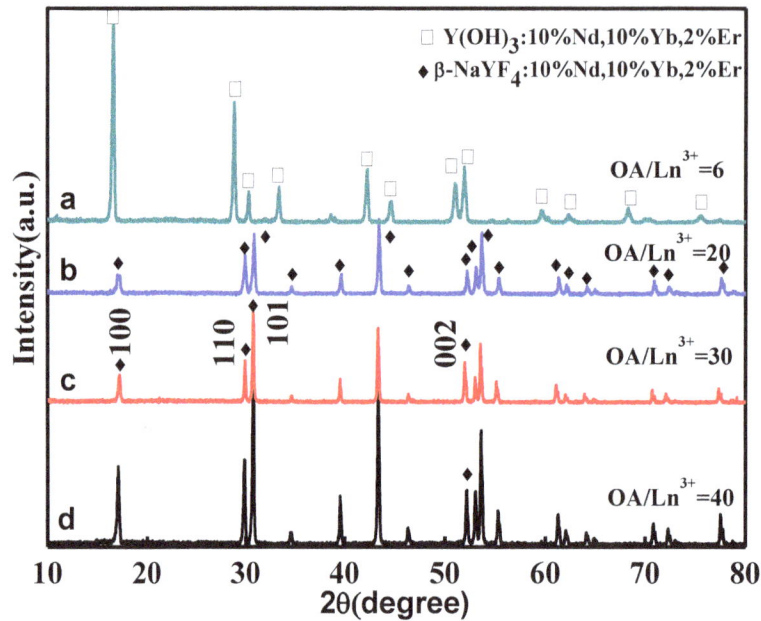

Figure 4. XRD patterns of $NaYF_4$:10% Nd^{3+}, 10% Yb^{3+}, 2% Er^{3+} microcrystals synthesized at different molar ratios of OA/Ln^{3+}: (**a**) 6:1; (**b**) 20:1; (**c**) 3:1; and (**d**) 40:1 (Other synthetic parameters were kept identical, 180 °C, 24 h, F^-/Ln^{3+} = 5:1).

2.3. Effect of the Ratio of F^-/Ln^{3+}

To investigate the effect of the F^-/Ln^{3+} ($F^-/Ln^{3+} \leq 5$) ratio on the morphology, size, and structure of the as-prepared $NaYF_4$:10% Nd^{3+}, 10% Yb^{3+}, 2% Er^{3+} particles, the ratio was taken as 1:1, 2:1, 3:1, and 5:1, respectively. As shown in Figure 5, with a gradual increase of the molar ratio of F^-/Ln^{3+}, a significant change takes place in the morphology and size of $NaYF_4$:10% Nd^{3+}, 10% Yb^{3+}, 2% Er^{3+} microcrystals. When the F^-/Ln^{3+} ratio is fixed at 1:1, the as-prepared microcrystals are spherical with coarse surfaces, implying low crystallization (Figure 5a). When the molar ratio of F^-/Ln^{3+} increased to 2:1 and 3:1, hexagonal microprisms with a protruding center and distortional tubular structure—with the end face of the central convex and concave shape between the center and the edge—were obtained. In addition, with the increase of F^- concentration, the crystallization is also improved. As the molar ratio of F^-/Ln^{3+} exceeded 5:1, uniform nanorods with high crystallization were generated. It can be seen that, with the increase of the molar ratio of F^-/Ln^{3+}, each surface gradually became smooth as along with the higher crystallization, which has been proved by XRD patterns (Figure 6). All the peaks of each sample are characteristic of a pure β-phase of $NaYF_4$; however, the relative intensities are different from each other, implying a different preferential orientation growth.

Figure 5. *Cont.*

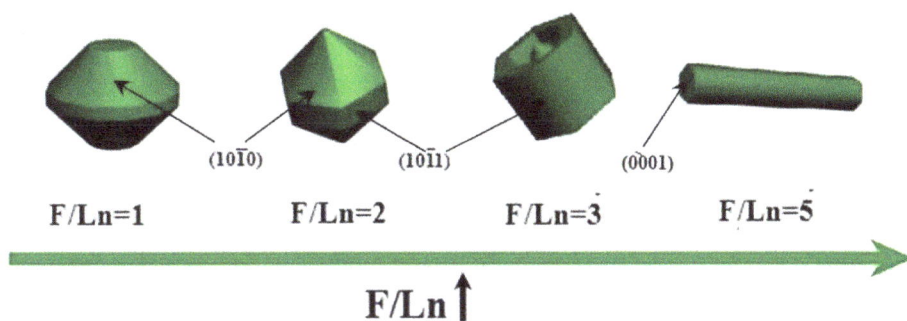

Figure 5. Typical FESEM images for the as-prepared $NaYF_4$: 10% Nd^{3+}, 10% Yb^{3+}, 2% Er^{3+} microcrystals at different molar ratios of F^-/Ln^{3+}: (**a**) 1:1; (**b**) 2:1; (**c**) 3:1; and (**d**) 5:1 (Other synthetic parameters were kept identical, 180 °C, 24 h, OA/Ln^{3+} = 20:1).

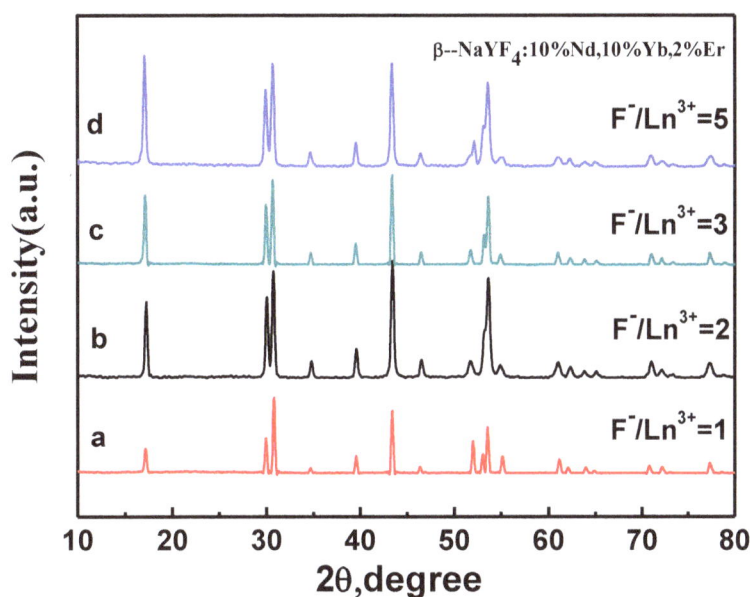

Figure 6. XRD patterns of $NaYF_4$:10% Nd^{3+}, 10% Yb^{3+}, 2% Er^{3+} microcrystals synthesized at different molar ratios of F^-/Ln^{3+}: (**a**) 1:1; (**b**) 2:1; (**c**) 3:1; and (**d**) 5:1 (Other synthetic parameters were kept identical, 180 °C, 24 h, OA/Ln^{3+} = 20:1).

2.4. The Mechanism of NaYF4 Crystal Growth

The hydrothermal growth process of $NaYF_4$ microcrystals includes the nucleation and the growth steps. In the process of nucleation, the OA anions could serve as chelating agent in reaction solution and formed complexes with Y^{3+} ions due to the strong coordination interaction. The chelating ability of the Y^{3+}-OA complex would be weakened under high hydrothermal temperature and pressure, and thus gradually release the Y^{3+} ions to react with Na^+ and F^- ions to generate the $NaYF_4$ nuclei. During the process of growth, OA as stabilizing agents can attach to the surfaces of $NaYF_4$ microcrystal with the alkyl chains left outside, which controls the growth of microcrystals and provides steric forces to prevent their aggregations.

To test our supposition, we have carried out two parts of the synthetic experiment.

Part 1: the molar ratio of F^-/Ln^{3+} was fixed at 5, and then the molar ratio of OA/Ln^{3+} was changed from 20:1 to 40:1. The hexagonal $NaYF_4$ seed generally includes (0001) and $(10\bar{1}0)$ facets, as shown in Scheme 1. It is noted that the facets with smaller surface area will have a higher surface energy and a faster growth rate, thus impacting the shapes of the final products. As for the hexagonal $NaYF_4$, the area of $(10\bar{1}0)$ exceeds that of (0001). This indicates that (0001) facets have higher surface energy, and thus a higher growth velocity, *i.e.*, $v(0001) >> v(10\bar{1}0)$. When the amount of oleic acid (OA) is relatively low $(OA/Ln^{3+} < 20:1)$, OA molecules preferentially adsorb onto the (0001) facets, thus producing uniform microrods (Figure 3b). The short microrods with an average diameter of 0.25 μm and a height of 0.245 μm can be achieved when the OA/Ln^{3+} is increased to 30:1 (Figure 3c). When more OA molecules are introduced, the ratio of $v(10\bar{1}0)/v(0001)$ will be increased, as the number of OA molecules on the (0001) facet will saturate but will get larger on the $(10\bar{1}0)$ facet with larger surface area. When $v(0001) = v(10\bar{1}0)$, the resulting particles will simultaneously grow along the (0001) and $(10\bar{1}0)$ directions to shorter microrods. When this ratio was decreased below 1 as more OA was introduced, particles with microplates formed.

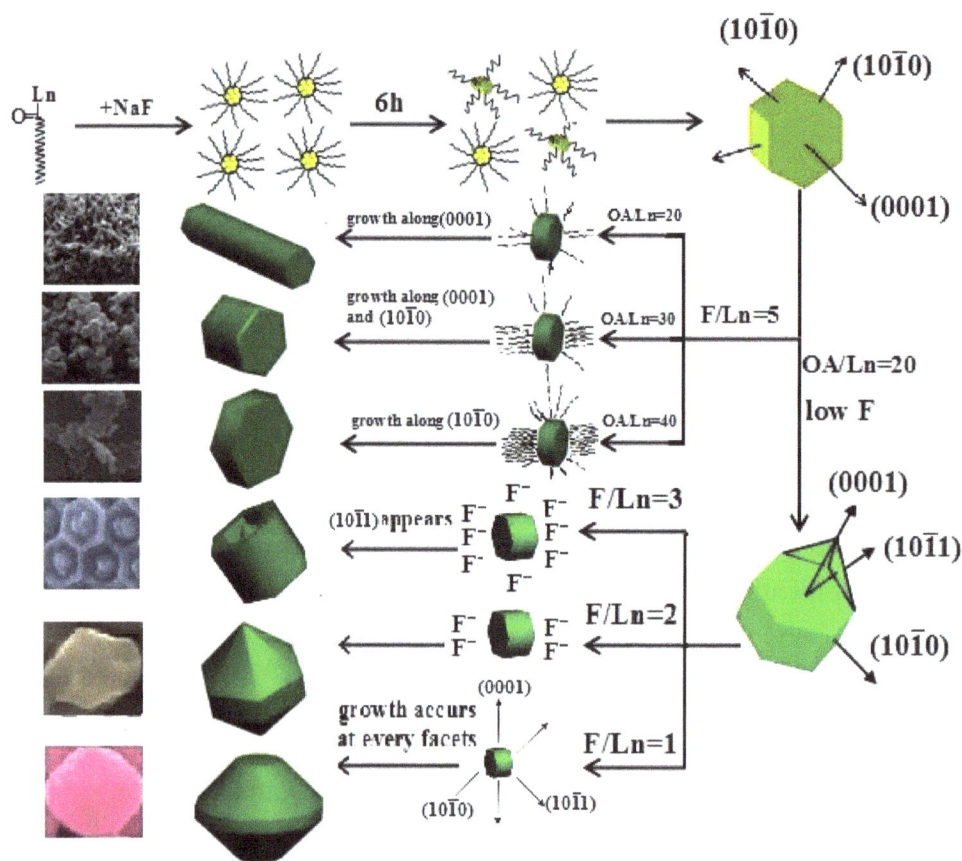

Scheme 1. Schematic illustration of the possible formation mechanism of $NaYF_4$ with various morphologies under different OA/Ln^{3+} and F^-/Ln^{3+} molar ratio conditions.

Part 2: the molar ratio of OA/Ln^{3+} was fixed at 20, while the molar ratio of F^-/Ln^{3+} varied from 1:1 to 5:1. We found that this condition provided an ideal system to induce OA and F^- preferential adsorption onto (0001) facets to control the growth rates of different facets of $NaYF_4$ microcrystals. As can be seen in Figure 5, the hexagonal NaYF$_4$ seed crystals have another type of surface ($10\bar{1}0$) in the case of F$^-$/Ln^{3+} < 5. The nucleation and growth rate was thus controlled through the competition between F$^-$ and OA towards (0001), ($10\bar{1}0$), and ($10\bar{1}1$) facets of NaYF$_4$ microcrystals. The XRD patterns of the products obtained at various ratios of F$^-$/Ln^{3+} (1:1, 2:1, 3:1, and 5:1) are shown in Figure 6a–d, respectively. As can be seen, all the peaks of each sample are characteristic of a pure β-phase of NaYF$_4$, but the relative intensities are different from each other, suggesting different preferential orientation growth. When the amount of OA(OA/Ln^{3+} = 20) is not enough to hinder the growth of (0001) facets, the differences in the molar ratios of F$^-$/Ln^{3+} have a significant effect on the growth of NaYF$_4$ microcrystals. With a low molar ratio of F$^-$/Ln^{3+} (1:1), all of the F$^-$ ions are used in nucleation. Accordingly, the growth of every facet is similar, thus resulting in sphere-like microcrystals (Figure 5a). When the molar ratio of F$^-$/Ln^{3+} is increased to 2:1, the remainder of the F$^-$ ions attacks the high-energy (0001) facet. Simultaneously, owing to the low free energies of the circumferential edges of seeds, the redundant F$^-$ ions preferentially bound to the circumferential edges of seeds, which would result in growth along the circumferential edge direction [27,28]. The concave octadecahedral β-NaYF$_4$ microcrystals can thus be produced. At F$^-$/Ln^{3+} = 3:1, the different crystallographic planes can be recognized in Figure 5c. The presence of conical ends with a raised ridge demonstrates that the ($10\bar{1}1$) facet appears. Moreover, in comparison with octadecahedral β-NaYF$_4$ microcrystals (F$^-$/Ln^{3+} = 2:1), the length enhancement further confirms that the growth rate is in the (0001) facet. However, in the condition F$^-$/Ln^{3+} = 5:1, uniform β-NaYF$_4$ nanorods are produced. This is possibly because of the large amount of F$^-$ ions binding strongly to (0001) surfaces to speed up the growth ofβ-NaYF$_4$ crystallites.

2.5. UC Luminescence Properties

β-NaYF$_4$ is regarded as an efficient host for the UC process due to low phonon energies. Nd^{3+} ions were selected as the sensitizer to investigate the spectral and UC luminescent properties of products with various morphologies. Figure 7 shows the UC PL emission spectra of the resulting β-NaYF$_4$:10% Nd^{3+}, 10% Yb^{3+}, 2% Er^{3+} phosphors prepared at 180 °C and the F$^-$/Ln^{3+} ratio = 5:1 when changing the OA/Ln^{3+} ratio from 6:1 to 40:1 upon irradiation of 808 nm wavelength. UC emission peaks of purple, green, and red centered at 410 nm, 520/540 nm, and 655 nm were observed, which can be assigned to the ^{2}H$_{9/2}$ → ^{4}I$_{15/2}$, ^{2}H$_{11/2}$/^{4}S$_{3/2}$ → ^{4}I$_{15/2}$, and ^{2}H$_{9/2}$ → ^{4}I$_{15/2}$ transitions of Er^{3+} ions, respectively. It is interesting to note that the shapes of the emission spectra are similar in all four samples, with the only difference being in the relative intensities of the bands. Additionally, the microplate crystals show highest emission intensity in the four samples. This difference in UC luminescence intensity may be attributed to the changes in morphology, crystallinity, and particle size [29,30]. Excluding the Y(OH)$_3$ nanoline, the results show that the β-NaYF$_4$: 10% Nd^{3+},10% Yb^{3+}, 2% Er^{3+} microcrystals with high crystallinity and large size emit stronger UC light owing to the low surface-to-volume ratio. The smaller the surface-to-volume ratio of a microcrystal, the fewer active ions are located on the surface of particles. Therefore, microplates with large size and high crystallinity show a small surface defect, which is efficient to restrain the photon quenching. Moreover, in our case, similar results have also been observed by tuning the molar ratio of F$^-$/Ln^{3+} from 1:1 to 5:1 (shown in Figure 8). The strongest emission has been observed for nanorods with relatively large size and high crystallinity.

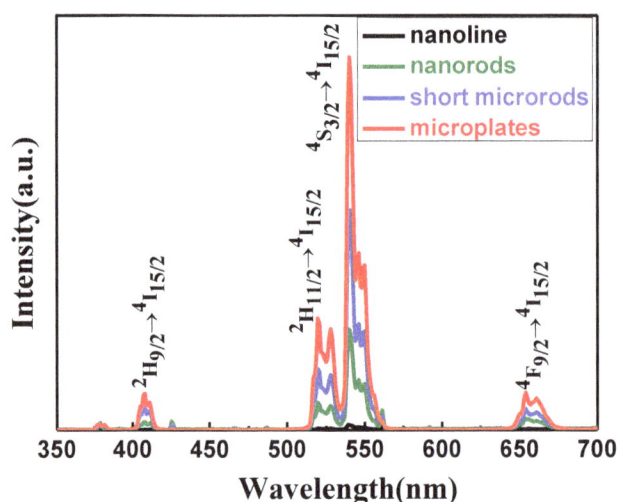

Figure 7. UC Photoluminescence (PL) spectra of NaYF$_4$:10% Nd^{3+},10% Yb^{3+}, 2% Er^{3+} microcrystals with different molar ratios of OA/Ln^{3+} under diode laser excitation at 808 nm. Excitation power density, ~390 W/cm^2.

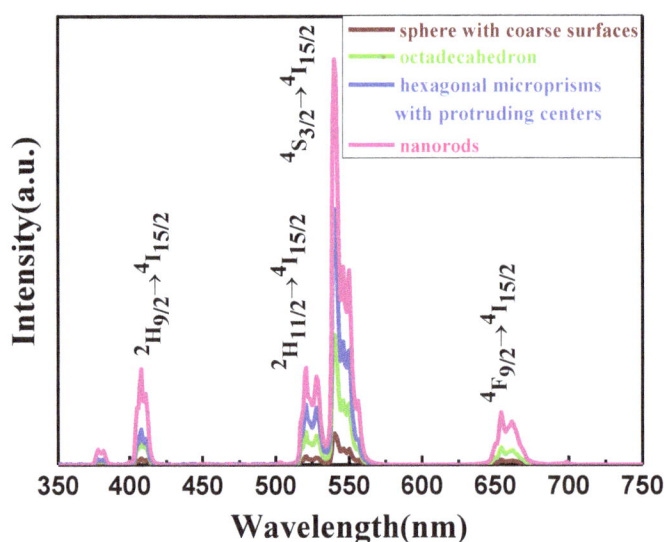

Figure 8. UC Photoluminescence (PL) spectra of NaYF$_4$:10% Nd^{3+},10% Yb^{3+}, 2% Er^{3+} microcrystals with different molar ratios of F$^-$/Ln^{3+} (180 °C, 24 h, OA/Ln^{3+} = 40) under diode laser excitation at 808 nm. Excitation power density, ~390 W/cm^2.

In order to investigate the UC luminescence mechanism, pump power dependences of the luminescence bands centered at 520, 540, and 655 nm were measured and displayed in a double logarithmic plot in Figure 9. For the unsaturated case, the number of photons that are required to populate the upper emitting state can be obtained by the relation $I_f \propto P^n$, where I_f stands for the fluorescent intensity, P is the pump laser power, and n is the number of laser photons required. Fitting the data points yielded slope values of 1.98, 1.93, and 1.96 for the $^2H_{11/2} \rightarrow {}^4I_{15/2}$, $^4S_{3/2} \rightarrow {}^4I_{15/2}$, and $^4F_{9/2} \rightarrow {}^4I_{15/2}$ transitions, respectively. This demonstrates that the generation of UC luminescence through the Nd^{3+}-sensitized system involves a two-photon process. In the Nd^{3+}, Yb^{3+}, and Er^{3+} tri-doped system, the energy transfer from Nd^{3+} to Yb^{3+} and then to Er^{3+} should be taken into account under 808 nm diode laser excitation.

The energy transfer mechanism was displayed in Figure 10. The sensitizer Nd^{3+} ions were firstly excited from the $^4I_{9/2}$ to the $^4F_{5/2}$ state after absorbing the excitation energy at 808 nm, and then relaxed to the $^4F_{3/2}$ level through multiphonon processes. The Nd^{3+} ions transfer their absorbed energy to neighboring Yb^{3+} ions and excite them from the $^2F_{7/2}$ (Yb^{3+}) to the $^2F_{5/2}$ (Yb^{3+}) state. Subsequently, the Yb^{3+} ion at $^2F_{5/2}$ state transfers the received energy to its neighboring Er^{3+} ion, exciting it from $^4I_{15/2}$ (Er^{3+}) to $^4I_{11/2}$ (Er^{3+}) stare. Receiving transferred energy from the Yb^{3+} ion can further excite the Er^{3+} ion to the $^4F_{7/2}$ (Er^{3+}) level. Multiphonon-assisted relaxations from the $^4F_{7/2}$ state can then populate the $^2H_{11/2}$ (Er^{3+}) and $^4S_{3/2}$ (Er^{3+}) levels, which generate the 520 and 540 nm emissions by radiative decay to the ground state. In addition, the red emission around 655 nm can be acquired through transition from $^4F_{9/2}$ (Er^{3+}) to $^4I_{15/2}$ (Er^{3+}). At the same time, a portion of the Er^{3+} ions at $^4F_{9/2}$ state can receive energy transfers from Yb^{3+} ions and be promoted to the $^4G_{11/2}$ state, from which the $^2H_{9/2}$ state can be populated through nonradiative relaxation process. Decay of the excited Er^{3+} ions in $^2H_{9/2}$ state to $^4I_{15/2}$ state results in the purple emission centered at 410 nm.

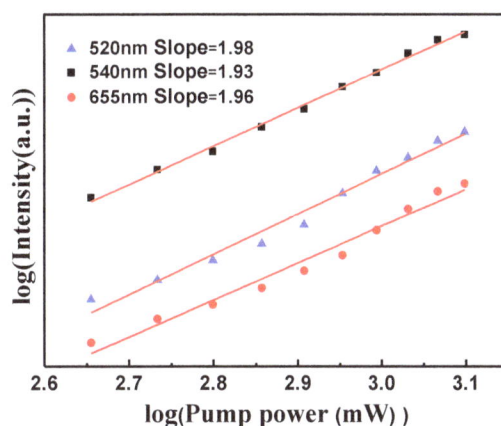

Figure 9. Pump power dependence of the fluorescent bands centered at 520, 540, and 655 nm from NaYF4:10% Nd^{3+}, 10% Yb^{3+}, 2% Er^{3+} on pumping power.

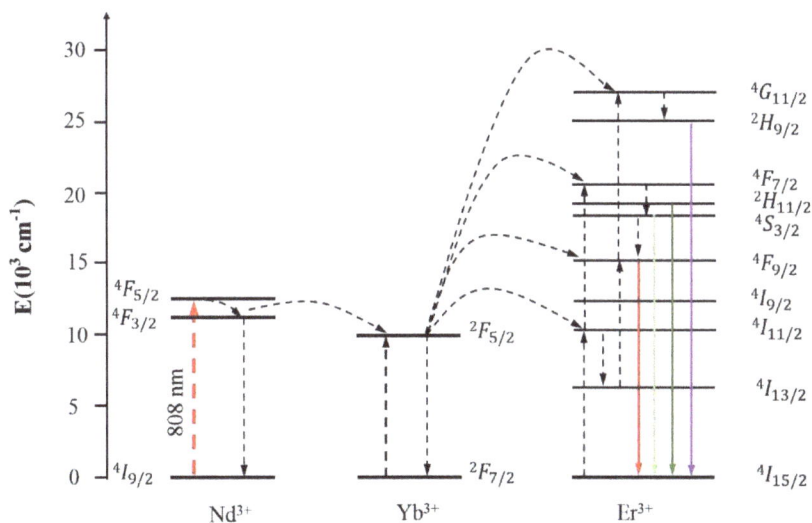

Figure 10. Proposed energy transfer mechanism of Nd^{3+}, Yb^{3+}, Er^{3+} ions following diode laser excitation of 808 nm. The dashed-dotted, dashed, dotted, and full arrows represent the photon excitation, energy transfer, and emission processes.

3. Experimental Section

3.1. Preparation

The NaYF$_4$ microcrystals were hydrothermally prepared using oleic acid as shape modifier, and NaF, Y(NO$_3$)$_3$ as precursors at 180 °C under a basic condition. In a typical synthesis of NaYF$_4$ nanorods, 1.08 g (27 mmol) of NaOH, 9 mL (24 mmol) of oleic acid (OA) (90 wt.%), and 10 mL (120 mmol) of ethanol were well mixed at room temperature to get a white viscous solution. Then, 4 mL (3.6 mmol) of 0.252g NaF (F$^-$/Ln^{3+} = 5:1) solution was added with vigorous stirring until a translucent solution was obtained. Then 6.41 mL (1.2 mmol) of 0.1872 M Y(NO$_3$)$_3$ was poured into the above solution keeping vigorous stirring. After aging for 1 h, the mixture was transferred to a 50-mL Teflon-lined autoclave, and heated at 180 °C for 24 h. The NaYF$_4$ microplates were prepared under similar conditions except that the amount of OA doubled. It is noted that the total molar composition of rare earth metals remains constant. For example, to prepare NaYF$_4$ doped with 10% Nd^{3+}, 10% Yb^{3+}, 2% Er^{3+}, 5 mL (0.936 mmol) 0.1872 M Y(NO$_3$)$_3$, 1 mL (0.12 mmol) 0.12 M Nd(NO$_3$)$_3$, 1 mL (0.12 mmol) 0.12 M Yb(NO$_3$)$_3$, and 1 mL (0.024 mmol) of 0.024 M Er(NO$_3$)$_3$ were mixed together instead of 6.41 mL of 0.1872 M Y(NO$_3$)$_3$. The obtained microcrystals were washed with ethanol and water to remove the oleic acid and other remnants, and then dried in the air at 60 °C for 12 h.

3.2. Characterization

The as-prepared samples were characterized by X-ray powder diffraction (XRD) on a Rigaku D/max-γB diffractometer, which was equipped with a rotating anode and a Cu Kα source (λ = 0.154056 nm). Micrographs of the prepared powders were obtained by using a field emission scanning electron microscope (FESEM, MX2600FE, AIKE Sepp, Oxford, UK). To measure the emitted UC luminescence, the synthesized powders were pressed to form a smooth, flat disk, which was irradiated with a focused 5 W power-controllable 808 nm diode laser (Hi-Tech Optoelectronics Co. Ltd, Beijing, China). The emitted UC luminescence was then collected by a lens-coupled monochromator (Zolix Instruments Co. Ltd, Beijing, China) of 2 nm spectral resolution with an attached photomultiplier tube (Hamamatsu CR131, Hamamatsu Photonics, Hamamatsu, Japan).

4. Conclusions

In conclusion, we have presented our systematic synthesis results on hexagonal NaYF$_4$:10% Nd^{3+}, 10% Yb^{3+}, 2% Er^{3+} micro- and nanocrystals with various morphologies and size. Morphologies with nanorods, short microrods, and microplates can be successfully achieved by varying the amount of OA coordination ligand in the initial precursor solution. This is ascribed to the fact that the OA is able to modulate the growth rate of (0001) and (10$\bar{1}$1) crystallographic facets due to their preferential attachment to them, thus governing the formation of the final morphologies. Moreover, through tuning the molar ratio of F$^-$/Ln^{3+}, sphere-like, octadecahedral, and hexagonal microprisms with protruding centers NaYF$_4$:10% Nd^{3+}, 10% Yb^{3+}, 2% Er^{3+} can be obtained. This shape evolution is a direct cooperative result between the attachment of F$^-$ and OA to the involved crystal facets of (0001), (10$\bar{1}$1), and (10$\bar{1}$0). Furthermore, under 808 nm laser excitation and the same measurement conditions, spectroscopic investigations of the resulting particles revealed a morphology-, size-, and crystallinity-dependent upconverted luminescence; the highest luminescence intensity was from $NaYF_4$:10% Nd^{3+}, 10% Yb^{3+}, 2% Er^{3+} microplates. These Nd^{3+}-sensitized hexagonal $NaYF_4$ nanoparticles with various size and shape have important implications for biophotonic applications with minimized heating effects.

Acknowledgments

This work is supported by the Natural Science Foundation of China (51102066), the Fundamental Research Funds for the Central Universities (Grant No. HIT. NSRIF.2015048 and AUGA5710052614), international scientific and technological cooperation projects (Grant No. 2014DFA50740).

Author Contributions

Y.S and S.H conceived and designed the study. Y.S was responsible for the experimental work. The manuscript was written by Y.S and S.H. S.H, C.Y and G.C reviewed and edited the manuscript. All of the authors read and approved the manuscript.

Conflicts of Interest

The authors declare no conflict of interest.

References

1. Haase, M.; Schäfer, H. Upconverting nanoparticles. *Angew. Chem. Int. Ed.* **2011**, *50*, 5808–5829.
2. Shen, J.; Zhao, L.; Han, G. Lanthanide-doped upconverting luminescent nanoparticle platforms for optical imaging-guided drug delivery and therapy. *Adv. Drug Deliv. Rev.* **2013**, *65*, 744–755.
3. Gu, Z.; Yan, L.; Tian, G.; Li, S.; Chai, Z.; Zhao, Y. Recent advances in design and fabrication of upconversion nanoparticles and their safe theranostic applications. *Adv. Mater.* **2013**, *25*, 3758–3779.
4. Liu, Y.; Tu, D.; Zhu, H.; Ma, E.; Chen, X. Lanthanide-doped luminescent nano-bioprobes: From fundamentals to biodetection. *Nanoscale* **2013**, *5*, 1369–1384.
5. Hao, S.W.; Chen, G.Y.; Yang, C.H. Sensing using rare-earth-doped upconversion nanoparticles. *Theranostics* **2013**, *3*, 331–345.
6. Chen, G.Y.; Shen, J.; Ohulchanskyy, T.Y.; Patel, N.J.; Kutikov, A.; Li, Z.P.; Song, J.; Pandey, R.K.; Agren, H.; Prasad, P.N.; *et al.* (alpha-NaYbF4:Tm^{3+})/CaF2 core/shell nanoparticles with efficient near-infrared to near-infrared upconversion for high-contrast deep tissue bioimaging. *ACS Nano* **2012**, *6*, 8280–8287.
7. Chen, G.Y.; Yang, C.H.; Prasad, P.N. Nanophotonics and nanochemistry: Controlling the excitation dynamics for frequency up- and down-conversion. *Acc. Chem. Res.* **2013**, *46*, 1474–1486.
8. Tian, G.; Gu, Z.J.; Zhou, L.J.; Yin, W.Y.; Liu, X.X.; Yan, L.; Jin, S.; Ren, W.L.; Xing, G.M.; Li, S.J.; *et al.* Mn^{2+} dopant-controlled synthesis of NaYF4:Yb/Er upconversion nanoparticles for *in vivo* imaging and drug delivery. *Adv. Mater.* **2012**, *24*, 1226–1231.

9. Yang, D.; Kang, X.; Ma, P.; Dai, Y.; Hou, Z.; Cheng, Z.; Li, C.; Lin, J. Hollow structured upconversion luminescent NaYF$_4$:Yb^{3+}, Er^{3+} nanospheres for cell imaging and targeted anti-cancer drug delivery. *Biomaterials* **2013**, *34*, 1601–1612.

10. Shan, J.N.; Budijono, S.J.; Hu, G.H.; Yao, N.; Kang, Y.B.; Ju, Y.G.; Prud'homme, R.K. Pegylated composite nanoparticles containing upconverting phosphors and meso-tetraphenyl porphine (TPP) for photodynamic therapy. *Adv. Funct. Mater.* **2011**, *21*, 2488–2495.

11. Zhang, P.; Steelant, W.; Kumar, M.; Scholfield, M. Versatile photosensitizers for photodynamic therapy at infrared excitation. *J. Am. Chem. Soc.* **2007**, *129*, 4526–4527.

12. Zou, W.Q.; Visser, C.; Maduro, J.A.; Pshenichnikov, M.S.; Hummelen, J.C. Broadband dye-sensitized upconversion of near-infrared light. *Nat. Photonics* **2012**, *6*, 560–564.

13. Xie, X.; Gao, N.; Deng, R.; Sun, Q.; Xu, Q.-H.; Liu, X. Mechanistic investigation of photon upconversion in Nd^{3+}-sensitized core–shell nanoparticles. *J. Am. Chem. Soc.* **2013**, *135*, 12608–12611.

14. Xie, X.; Liu, X. Photonics upconversion goes broadband. *Nat. Mater.* **2012**, *11*, 842–843.

15. Weber, M.J. Optical properties of Yb^{3+} and Nd^{3+}-Yb^{3+} energy transfer in YAlO$_3$. *Phys. Rev. B* **1971**, *4*, 3153–3159.

16. Kushida, T.; Marcos, H.M.; Geusic, J.E. Laser transition cross section and fluorescence branching ratio for Nd^{3+} in yttrium aluminum garnet. *Phys. Rev.* **1968**, *167*, 289–291.

17. Liégard, F.; Doualan, J.L.; Moncorgé, R.; Bettinelli, M. Nd^{3+}-Yb^{3+} energy transfer in a codoped metaphosphate glass as a model for Yb^{3+} laser operation around 980 nm. *Appl. Phys. B* **2005**, *80*, 985–991.

18. Wang, Y.-F.; Liu, G.-Y.; Sun, L.-D.; Xiao, J.-W.; Zhou, J.-C.; Yan, C.-H. Nd^{3+}-sensitized upconversion nanophosphors: Efficient *in vivo* bioimaging probes with minimized heating effect. *ACS Nano* **2013**, *7*, 7200–7206.

19. Shen, J.; Chen, G.; Vu, A.-M.; Fan, W.; Bilsel, O.S.; Chang, C.-C.; Han, G. Engineering the upconversion nanoparticle excitation wavelength: Cascade sensitization of tri-doped upconversion colloidal nanoparticles at 800 nm. *Adv. Opt. Mater.* **2013**, *1*, 644–650.

20. Wen, H.; Zhu, H.; Chen, X.; Hung, T.F.; Wang, B.; Zhu, G.; Yu, S.F.; Wang, F. Upconverting near-infrared light through energy management in core-shell-shell nanoparticles. *Angew. Chem. Int. Ed.* **2013**, *52*, 13419–13423.

21. Cui, H.; Hong, C.; Ying, A.; Yang, X.; Ren, S. Ultrathin gold nanowire-functionalized carbon nanotubes for hybrid molecular sensing. *ACS Nano* **2013**, *7*, 7805–7811.

22. Chen, G.Y.; Qiu, H.L.; Prasad, P.N.; Chen, X.Y. Upconversion nanoparticles: Design, nanochemistry, and applications in theranostics. *Chem. Rev.* **2014**, *114*, 5161–5214.

23. Dou, Q.; Zhang, Y. Tuning of the structure and emission spectra of upconversion nanocrystals by alkali ion doping. *Langmuir* **2011**, *27*, 13236–13241.

24. Qu, X.; Yang, H.K.; Pan, G.; Chung, J.W.; Moon, B.K.; Choi, B.C.; Jeong, J.H. Controlled fabrication and shape-dependent luminescence properties of hexagonal NaCeF$_4$, NaCeF$_4$:Tb^{3+} nanorods via polyol-mediated solvothermal route. *Inorg. Chem.* **2011**, *50*, 3387–3393.

25. Hao, S.; Shao, W.; Qiu, H.L.; Shang, Y.F.; Fan, R.W.; Guo, X.Y.; Zhao, L.L.; Chen, G.Y.; Yang, C.H. Tuning the size and upconversion emission of NaYF$_4$:Yb^{3+}/Pr^{3+} nanoparticles through Yb^{3+} doping. *RSC Adv.* **2014**, *4*, 56302–56306.

26. Li, C.; Quan, Z.; Yang, P.; Huang, S.; Lian, H.; Lin, J. Shape-controllable synthesis and upconversion properties of lutetium fluoride (doped with Yb^{3+}/Er^{3+}) microcrystals by hydrothermal process. *J. Phys. Chem. C* **2008**, *112*, 13395–13404.

27. Hao, S.; Chen, G.; Qiu, H.; Xu, C.; Fan, R.; Meng, X.; Yang, C. Controlled growth along circumferential edge and upconverting luminescence of β-NaYF4: 20%Yb^{3+}, 1%Er^{3+} microcrystals. *Mater. Chem. Phys.* **2012**, *137*, 97–102.

28. Lin, M.; Zhao, Y.; Liu, M.; Qiu, M.S.; Dong, Y.Q.; Duan, Z.F.; Li, Y.H.; Murphy, B.P.; Lu, T.J.; Xu, F. Synthesis of upconversion NaYF4:Yb^{3+}, Er^{3+} particles with enhanced luminescent intensity through control of morphology and phase. *J. Mater. Chem. C* **2014**, *2*, 3671–3676.

29. Kim, F.; Connor, S.; Song, H.; Kuykendall, T.; Yang, P.D. Platonic gold nanocrystals. *Angew. Chem. Int. Ed.* **2004**, *43*, 3673–3677.

30. Zhu, L.; Qin, L.; Liu, X.D.; Li, J.Y.; Zhang, Y.F.; Meng, J.; Cao, X.Q. Morphological control and luminescent properties of CeF_3 nanocrystals. *J. Phys. Chem. C* **2007**, *111*, 5898–5903.

4

Ceramic Nanocomposites from Tailor-Made Preceramic Polymers

Gabriela Mera [1,†], **Markus Gallei** [2,†], **Samuel Bernard** [3] **and Emanuel Ionescu** [1,4,*]

[1] Institut für Materialwissenschaft, Technische Universität Darmstadt, Jovanka-Bontschits-Strasse 2, D-64287 Darmstadt, Germany; E-Mail: mera@materials.tu-darmstadt.de

[2] Ernst-Berl-Institut für Technische und Makromolekulare Chemie, Technische Universität Darmstadt, Alarich-Weiss-Strasse 4, D-64287 Darmstadt, Germany; E-Mail: m.gallei@mc.tu-darmstadt.de

[3] Institut Européen des Membranes (UMR 5635-CNRS/ENSCM-UM2) CC 047-Place E. Bataillon, 34095 Montpellier Cedex 05, France; E-Mail: samuel.bernard@univ-montp2.fr

[4] Department Chemie, Institut für Anorganische Chemie, Universität zu Köln, Greinstrasse 6, D-50939 Köln, Germany

[†] These authors contribute equally to this paper.

[*] Author to whom correspondence should be addressed; E-Mail: ionescu@materials.tu-darmstadt.de

Academic Editor: Thomas Nann

Abstract: The present Review addresses current developments related to polymer-derived ceramic nanocomposites (PDC-NCs). Different classes of preceramic polymers are briefly introduced and their conversion into ceramic materials with adjustable phase compositions and microstructures is presented. Emphasis is set on discussing the intimate relationship between the chemistry and structural architecture of the precursor and the structural features and properties of the resulting ceramic nanocomposites. Various structural and functional properties of silicon-containing ceramic nanocomposites as well as different preparative strategies to achieve nano-scaled PDC-NC-based ordered structures are highlighted, based on selected ceramic nanocomposite systems. Furthermore, prospective applications of the PDC-NCs such as high-temperature stable materials for thermal protection systems, membranes for hot gas separation purposes, materials for heterogeneous catalysis, nano-confinement materials for hydrogen storage applications as well as anode materials for secondary ion batteries are introduced and discussed in detail.

Keywords: ceramic nanocomposites; polymer-derived ceramic nanocomposites (PDC-NCs); preceramic polymers; metallopolymers; polymer-to-ceramic conversion

---

1. Introduction

Multifunctional materials are capable of providing two or more primary functions either in a simultaneous manner or sequentially. For instance, one defines the category of multifunctional structural materials, which exhibit additional functions beyond their basic mechanical strength or stiffness (which are typical attributes of structural materials). Thus, they can be designed to possess incorporated electrical, magnetic, optical, sensing, power generative or other functionalities, which work in a synergistic manner [1]. The basic motivation for the development of multifunctional materials relies on their ability to address several mission objectives with only one structure—thus, they are capable of adapting on purpose their performance and response depending on the specific target application. Multi-mission objectives can be addressed simultaneously or consecutively by using multiple structures. However, due to ever-growing number of needed functions, the number of the individual objectives of the respective structures is becoming prohibitive, *i.e.*, the limiting factor concerning the design of suitable functional materials and devices [2]. Also specific aspects concerning the storage, maintenance, interactions, transport of the individual components might become critical.

Consequently, multifunctional materials represent the ultimate solution to address and provide multiple functions with one sole structure. They are usually (nano)composites or (nano)hybrids of several distinct (Gibbsian) phases (*i.e.*, phases with specific, individual chemical composition and physical state), each of them providing a different but essential function. Optimized design of multifunctional materials allows for having no or less "non-function" volume and thus provide significant advantages as compared to the traditional multicomponent "brass-board" systems: They are more weight and volume efficient, exhibit high flexibility with respect to their function(s) and performance, as well as are potentially less prone to maintenance issues [3].

Within the present Review, different classes of ceramic nanocomposite materials prepared from tailored preceramic polymers will be highlighted and discussed. One main emphasis of the critical discussion in the present Review will relate to the intimate relationship between the chemistry and macromolecular architecture of the preceramic polymers and the phase composition, microstructure and properties of the resulting ceramic nanocomposites. Following a Section related to the preparative access to ceramic nanocomposites from suitable preceramic polymers, a detailed description of various structural and functional properties followed by selected examples of prospective applications will be given.

2. Preceramic Polymers

2.1. Silicon-Containing Preceramic Polymers

The current research is moving forward to establishing novel preparative strategies to produce tailor-made silicon-based preceramic polymers [4] as precursors for polymer-derived ceramics

(PDCs) [5,6]. Within this context, studies concerning the intimate relationships between their molecular architecture and the microstructure and properties of the ceramic materials resulting there from are of crucial importance [5,7–9] The thermolysis of Si-based preceramic polymers under specific atmosphere and heat treatment conditions represents a straight-forward and inexpensive additive-free process [5,10] which allows to control and adjust the phase composition and the microstructure and thus the materials properties of ceramic components. Consequently, there is a stringent need in developing designed preceramic polymers, with tailored molecular architecture, physico-chemical properties and suitable ceramization behavior.

The general classes of silicon-based polymers used as precursors for ceramics, *i.e.*, polysilanes, polycarbosilanes, polysiloxanes, polysilazanes and polysilylcarbodiimides are shown in Figure 1 and will be briefly discussed in the following. Metal-containing polysiloxanes, polysilazanes and polycarbosilanes will be highlighted and discussed in Section 2.2 of the present paper.

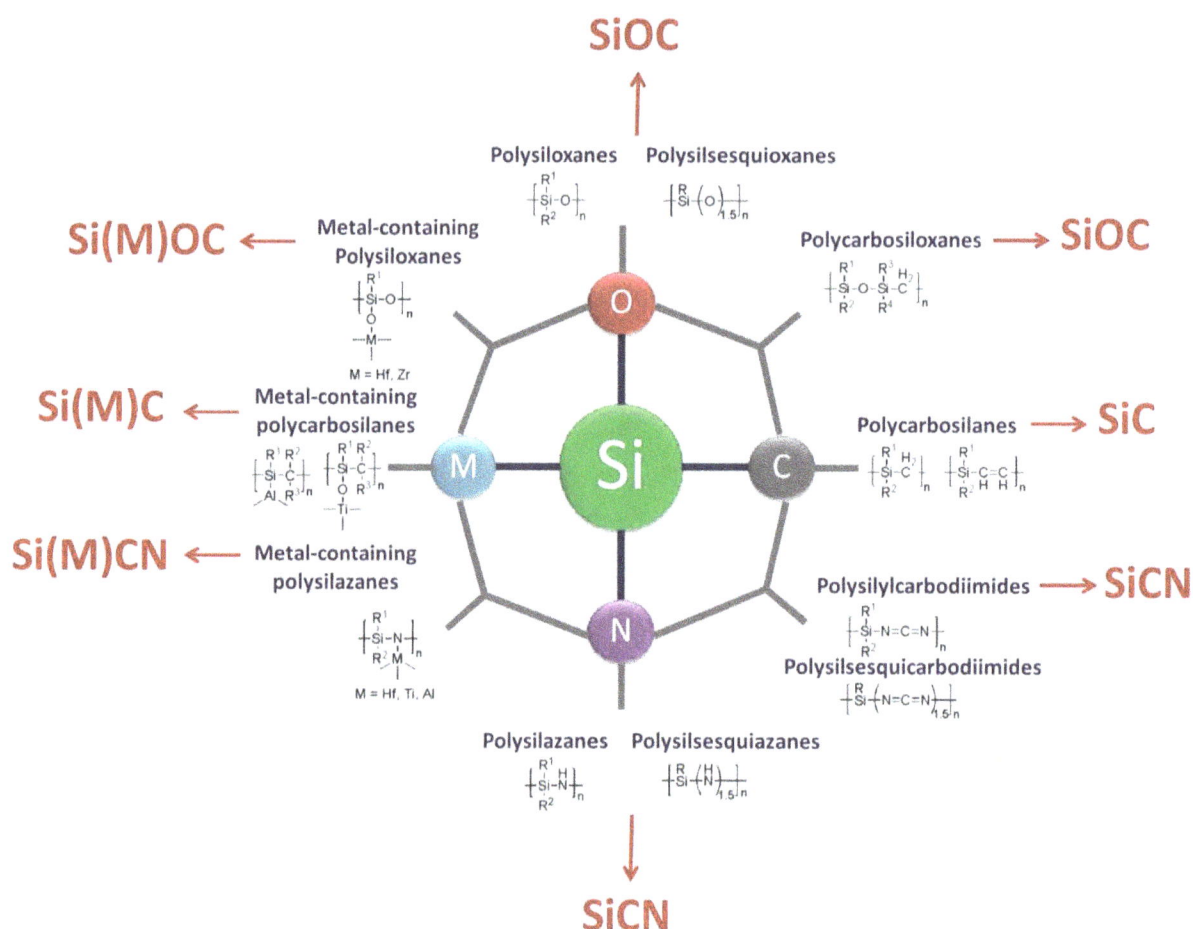

Figure 1. Typical classes of organosilicon polymers used as precursors for ceramics (reprinted with permission from Wiley) [4].

Following requirements should be fulfilled by preceramic polymers in order to be suitable for the production of polymer-derived ceramics: (i) the polymers should possess a sufficiently high molecular weight in order to avoid volatilization of low molecular components; (ii) they should have appropriate rheological properties and solubility for the shaping process and (iii) latent reactivity (presence of reactive, functional groups) for the curing and cross-linking step.

The synthesis of ceramic materials starting from preceramic polymers has several advantages as compared to other preparative methods: (i) pure starting compounds (precursors); (ii) possibility to modify the molecular structure of the precursors by a variety of chemical reactions; (iii) application of shaping technologies well known from plastic forming; (iv) easy machining of the green body; (v) relatively low synthesis temperatures in comparison with classical ceramic powder processing technologies; (vi) preparative access to ceramic systems which cannot be produced by other methods (*i.e.*, ternary and multinary systems such as SiOC, SiCN, *etc.*).

Polysilanes (also named polysilylenes) [11–14] represent a class of polymeric materials composed of an one-dimensional silicon chain and organic groups attached at silicon and exhibit rather unique optoelectronic and photochemical properties related to the extensive delocalization of σ electrons along silicon backbone (σ conjugation). Polysilanes are used as photoresists in microlithography, as well as they found applications as photoconducting polymers, third-order nonlinear optical materials and as valuable precursors for the synthesis of additive-free silicon carbide [11,15].

Polysilanes are typically unstable in air and moisture and suffer from degradation when exposed to UV light. Due to their insolubility, non-meltability and intractability, their processability is rather challenging. Their properties are significantly depending on their molecular weight, as well as on the nature of the organic groups attached at silicon and the conformation (branching) of the polymer chain.

The synthesis of polysilanes is mainly done by using the Wurtz-Fittig reductive coupling reaction of chlorosilanes with sodium or lithium in boiling toluene, benzene or tetrahydrofurane [16,17]. Soluble homo- and copolymers, mainly methyl-containing polysilanes, were reported in the late 70's [18–20]. Despite the fact that this synthesis method for polysilanes is rather old and yields polymers with structures, molecular weight and polydispersities difficult to control, it is still the most common method of choice for the synthesis of polysilanes. Moreover, the Wurtz-Fittig reaction is highly sensitive on the nature and dispersion of the used alkali metals (Na, Li, K, Na/K alloy), on the solvents and additives, on reaction temperature, *etc.* [21–25].

An important salt-free, high-yield synthesis method for high-purity polysilanes is the dehydrocoupling reaction of hydridosilane in the presence of catalysts such as η^2-alkynyl titanocene or -zirconocene [26]. Also zirconocene and hafnocene hydride have been reported as suitable catalyst for the dehydrocoupling polymerization from hydridosilanes [27].

Other suitable methods used for the synthesis of polysilanes are the anionic polymerization of masked disilenes [28] ring-opening polymerization of strained cyclosilanes [29] and recently, reduction reaction of chlorosilanes with Mg in the presence of LiCl and a Lewis acid under mild condition [30].

Among polysilanes, polydimethylsilane represents an important precursor for silicon carbide fibers as reported for the first time by Yajima *et al.* in 1975 [31]. Thermal treatment of polydimethylsilane at *ca.* 400 °C leads to the formation of a soluble and processable polycarbosilane (*i.e.*, polymethylsilylenemethylene), which can be easily spun into fibers or casted. This process, known as Kumada rearrangement or Kumada reaction, consists in a radical-induced methylene migration from one of the methyl substituents attached at Si into the polymer chain (Figure 2) [32]. The fibers obtained from polycarbomethylsilane are subsequently cured (thermal treatment in air or alternatively e-beam curing) and pyrolyzed at ~1100–1300 °C in argon atmosphere to yield SiC fibers known under the commercial name of Nicalon™ or High Nicalon™.

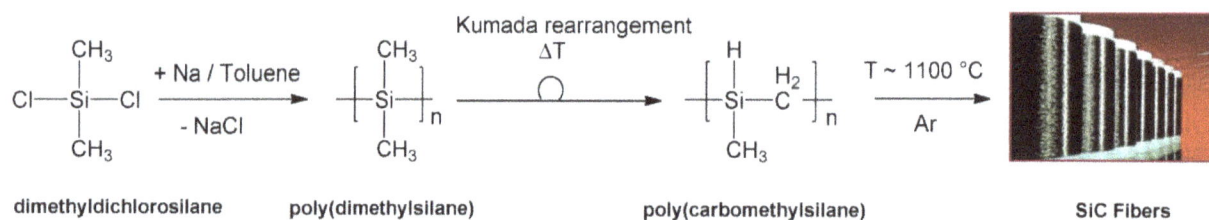

Figure 2. Yajima process for the synthesis of silicon carbide (SiC) ceramic fibers (reprinted with permission from Wiley) [4].

Polycarbosilanes having the general formula–[R^{1}R^{2}Si-C(R^3)(R^4)]$_n$–(R^1, R^2, R^3, R^4 being H or organic groups) can be synthesized by several methods, such as the Kumada rearrangement of polysilanes [32], Grignard polycondensation reactions of chlorosilanes [33], ring-opening polymerization of 1,3-disilacyclobutanes catalyzed by Pt-containing complexes [34,35], dehydrocoupling reaction of trimethylsilane or hydrosilylation of vinylhydridosilanes.

Unsaturated polycarbosilanes represent a special class of hybrid polymers with silicon being bonded to π-conjugated building blocks, such as phenylene, ethenylene, or diethylene [36–40]. The synthesis of these materials has been achieved by using coupling reactions [38,41] thermal cyclopolymerization [42,43] and a variety of ring-opening polymerization reactions including anionic [44–47], thermolytic and catalytic coordination techniques [48], each of them with some limitations. A novel approach for the synthesis of unsaturated polycarbosilanes involves the catalytic acyclic diene metathesis (ADMET) reaction of unsaturated oligosilanes in the presence of a ruthenium carbene complex RuCl$_2$(PCy$_3$)$_2$(=CHPh) (Grubbs catalyst) [49].

Polysiloxanes show excellent chemical and physical properties and have been extensively used as suitable single-source precursors for the synthesis of silicon oxycarbide (SiOC) ceramic materials via pyrolysis in inert or reactive atmosphere [5,7]. They exhibit outstanding thermo-mechanical properties owing to the combination of relatively unique features such as a pronounced elasticity at low temperatures, or high stability at elevated temperatures and in oxidative environments. The low temperature elasticity of polysiloxanes is manifested in some of the lowest glass transition temperatures known to polymers, moreover in low crystalline melting points, fast crystallization processes, specific liquid crystalline behavior and small viscosity-temperature coefficients. These properties rely on the pronounced polymer segmental chain mobility in polysiloxanes, which is correlated to the inherent chain flexibility (large Si–O–Si angles from 140° to 180°) and relatively weak intra- and intermolecular interactions [50].

The high temperature stability of polysiloxanes concerning decomposition is also related to the inherent strength of the siloxane bond as well as to the pronounced flexibility of –(Si-O-Si-O)$_x$– segments. The partial ionic and double bond characters of the Si–O bond lead to its exceptional strength, since both effects increase the binding force between the participating silicon and oxygen atoms. This relies on the unique d$_\pi$–p$_\pi$ bond between Si and O resulting in an Si–O bond dissociation energy of about 108 kcal·mol^{-1}, which is considerably higher than those of single bonds such as C–C (82.6 kcal·mol^{-1}), C–O (85.2 kcal·mol^{-1}) or even C$_{arom}$–C bonds (97.6 kcal·mol^{-1}) [50]. Thus, the Si–O bond withstands exposure to higher temperatures than the bonds normally found in organic polymers, leading to a significantly higher thermal stability of polysiloxanes than that of their organic (C–C) counterparts.

Polysiloxanes are synthesized starting from functionalized silanes, *i.e.*, R_xSiX_{4-x} (with x = Cl, OR, OC(=O)R or NR_2 and R = alkyl, aryl groups). One of the most used functionalized silanes for the industrial synthesis of polysiloxanes is dimethyl dichloro silane, which is obtained via the so-called Direct Process, involving the copper-catalyzed reaction of gaseous chloromethane with silicon in fluidized- or stirred-bed reactors at temperatures of *ca.* 250–300 °C (the Müller-Rochow process). After a subsequent destillative purification of dimethyl dichloro silane from other Me_xSiCl_{4-x} (x = 1, 3, 4) side products, the general route to obtain polysiloxanes consists mainly of two steps: (a) hydrolysis of the dichloro dimethylsilane, which leads to the formation of a mixture of linear and cyclic oligosiloxanes; and (b) polycondensation of hydroxyl-functionalized short-chain polysiloxanes or ring-opening polymerization processes of the cyclic oligomers which lead to high molecular weight polysiloxanes [50].

In order to provide a high ceramic yield, polysiloxanes have to exhibit a high degree of cross-linking. This can be achieved by using suitable functional chain groups in linear polysiloxanes which allow for thermal or irradiation-assisted (e.g., UV light) curing and cross-linking or by using highly cross-linked polysiloxanes. Within this context it has been shown that polysilsesquioxanes, of general formula $RSiO_{1.5}$ (with R being H or an organic group) are suitable for being used as preceramic polymers, since they exhibit a highly branched molecular architecture and consequently lead to high ceramic yields [5,7]. Cross-linked polysiloxanes or silicon resins can furthermore be synthesized through sol–gel processes via hydrolysis and condensation of hybrid silicon alkoxides. This type of precursors has been used since end of the 80s to synthesize silicon oxycarbide glasses [51,52]. They are modified silicon alkoxides of the general formula $R_xSi(OR')_{4-x}$ (R = alkyl, allyl, aryl; R' = methyl, ethyl), which upon gelation convert into silicone resins of the composition $R_xSiO_{(4-x)/2}$. As different hybrid silicon alkoxides can be used for co-hydrolysis and subsequent polycondensation, the sol–gel preparative access to silicone resins allows for controlling and tuning their compositions. Consequently, single-source precursors for stoichiometric silicon oxycarbide, as well as for Si–O–C materials showing excess of carbon or silicon can be prepared [53]. Moreover, this preparative technique allows for introducing additional elements within the preceramic network, e.g., Al, Ti, B, by using suitable metal alkoxides [6]. For instance, functionalized silanes such as $Si(OR)_xR'_{4-x}$ (R, R' = alkyl groups) are reacted with titanium isopropoxide [54], zirconium *n*-propoxide [55] or di-*tert*-amyloxy-vanadate [56] in order to obtain so-called hybrid gels containing homogeneously dispersed transition metals within the gel backbone. The synthesis of hybrid materials comprising of siloxane-type precursors and metal alkoxides has been known since the late 80s. While at the beginning the focus of the investigation was set rather on elucidating the molecular structure and network architecture of the metal-modified sol–gel materials [56–59], later on studies related to their conversion into ceramic nano-composites became more and more numerous and attractive [54,60–63].

Polysilazanes are polymers containing silicon and nitrogen within their backbone and have been used during the last decades to synthesize amorphous silicon nitride and silicon carbonitride ceramics [64].

First attempts were made to synthesize silicon nitride by thermal conversion of $Si[N(CH_2CH_3)_2]_4$ in argon atmosphere. This process was expected to occur similarly to the formation of silica from silicon alkoxides. However, it leads to the formation of silicon carbonitride, SiCN [65]. Carbon-free polysilazanes were synthesized already in 1885 via ammonolysis of tetrachloro silane in liquid phase [66]. This process delivers silicon diimide, which converts upon thermal treatment at *ca.* 1000 °C in inert gas atmosphere, into amorphous silicon nitride and gaseous ammonia [67].

Perhydridopolysilazane (PHPS) represents also a carbon-free precursor for silicon nitride. The synthesis of PHPS is achieved via ammonolysis of dichloro silane (SiH_2Cl_2) in polar solvents. The product consists of low molecular linear and cyclic oligomers, which however cross-link rapidly upon hydrogen loss and lead to highly viscous (up to glassy) polysilazane [68]. Its thermal treatment at 1000 °C in nitrogen atmosphere leads to a mixture of α-Si_3N_4, β-Si_3N_4 and excess silicon. If the thermal treatment of the PHPS is performed in ammonia atmosphere, the formation of elemental silicon is suppressed. Due to the high reactivity of dichloro silane (which is highly flammable and moreover disproportionates into silane, SiH_4 and tetrachloro silane, $SiCl_4$), the process was modified by using $H_2SiCl_2*(NC_5H_5)_2$ [69]. A further improvement of the process involves the co-ammonolysis of dichloro and trichloro silanes, leading to preceramic polymers which are able to be thermally converted into stoichiometric Si_3N_4 [70].

Polycarbosilazanes are usually prepared upon ammonolysis or aminolysis of halogeno-substituted organyl silanes, e.g., R_xSiCl_{4-x}. In a first substitution step a chlorine substituent is replaced by a –NH_2 group; subsequently, condensation reactions occur, leading to the formation of Si–N–Si linkages. Depending on the number of the chlorine substituents as well as on the nature and size of the organic substituents R in R_xSiCl_{4-x}, different types of silazanes can be obtained, such as linear or cyclic, oligomers or highly cross-linked polymers. Starting from R_2SiCl_2, mixtures of cyclic oligomers and low molecular weight linear polymers are obtained, which can be further cross-linked, usually via thermal treatment. In the case of Si–H and N–H containing silazanes, cross-linking can be achieved by means of addition of bases in catalytic amounts (e.g., potassium hydride—KH) [71].

Polysilazanes can be chemically modified by reactions with transition metal alkoxides, as it has been shown for Al or group IV metal alkoxides (M = Ti, Zr, Hf). In different studies, hydrido- [72,73] or methyl-/vinyl-substituted polysilazanes [6,62,74,75] were reacted with group IV metal alkoxides. In the case of hydrido-polysilazane, the reaction with titanium *n*-butoxide takes place at the N–H groups upon formation of Si–Ti linkages. The reaction of HTT1800 with hafnium *n*-butoxide was shown to occur at both N–H and Si–H functional groups as confirmed by Raman and ^{1}H-NMR spectroscopy [74]. Polysilazanes can be modified also with non-oxidic organometallics. Recently, several studies concerning the chemical modification of PHPS as well as vinyl- and hydrido-substituted polysilazanes with Ti [72,73,76,77], Zr and Hf [78,79] amido complexes were reported. The obtained metal-containing single-source precursors were shown to be highly compliant and to provide access to different ultrahigh-temperature stable ceramic nanocomposites (such as MN/Si_3N_4, MN/SiCN, MCN/SiCN, *etc.*, with M = Ti, Zr, Hf), depending on the conditions used for the ceramization.

Polysilylcarbodiimides are valuable precursors for the synthesis of SiCN-based ceramics [5,7,80]. The synthesis of polysilylcarbodiimides was firstly reported by Ebsworth, Wannagat and Birkofer [81–84]. They have been shown to be useful as stabilizing agents for polyurethanes and polyvinylchloride, as insulator coatings, high temperature stable pigments [85] and as irradiation-resistant sealing materials [86]. Moreover, polysilylcarbodiimides have been used for the synthesis of organic cyanamides, carbodiimides and heterocycles [87]. Polysilylcarbodiimides are generally air and moisture sensitive [88]. Upon insertion of bulky aromatic substituents at silicon, their air sensitivity significantly decreases [80]. Carbon-rich poly(phenylsilylcarbodiimide) derivatives, namely –[PhRSi-NCN]$_n$–, (R = H, methyl, vinyl, phenyl) were synthesized by the reaction of phenyl-containing dichlorosilanes with bis(trimethylsilylcarbodiimide) in the presence of pyridine as catalyst [80,89]. These polymers show an increased stability against air and moisture and were shown to be suitable precursors for carbon-rich nanostructured SiCN ceramics [5,7,80,89–93].

Synthesis of polysilylcarbodiimides can be performed upon reacting of di-, tri- and tetrachlorosilanes with silvercyanamide, as reported by Pump and Rochow in 1964 [94], as well as via trans-silylation of bis(trimethylsilylcarbodiimide) with chlorosilanes as shown in 1968 by Klebe and Murray [85]. Alternatively, polysilylcarbodiimides can be obtained by the polycondensation reaction of cyanamide with chlorosilanes [88]. The most appropriate method for the synthesis of polysilylcarbodiimides is the pyridine-catalyzed polycondensation reaction of chlorosilanes (di, tri- and tetra- chlorosilanes) with bis(trimethylsilylcarbodiimide) [5,7,80,88,91,95–97]. The scientific interest on polysilylcarbodiimides increased as Riedel *et al.* [88,95,98–101] reported in the 90s on their thermal transformation to SiCN ceramics.

Starting from dichloro silanes, cyclic or linear polymers can be obtained. Trichloro silanes yield by the reaction with bis(trimethylsilylcarbodiimide) a class of branched polymers, namely polysilsesquicarbodiimides. Depending on the branching of the chain, different microstructures and thermal stabilities were found in their derived ceramics [91]. Interestingly is the reaction of tetrachloro silane with bis(trimethylsilylcarbodiimide) [100]. The decomposition of this highly branched polymer yield up the first ternary crystalline phases in the Si–C–N system, namely SiC_2N_4 and Si_2CN_4 [100].

Due to the pseudochalcogen character of the $(NCN)^{2-}$ anion,[102] polysilylcarbodiimides were shown to exhibit similar properties to those of polysiloxanes [5,7]. Indeed, the non-oxidic sol–gel process of trichlorosilanes with bis(trimethylsilylcarbodiimide) follows the same steps as the sol–gel process of trialkoxysilane with water (Figure 3) [88].

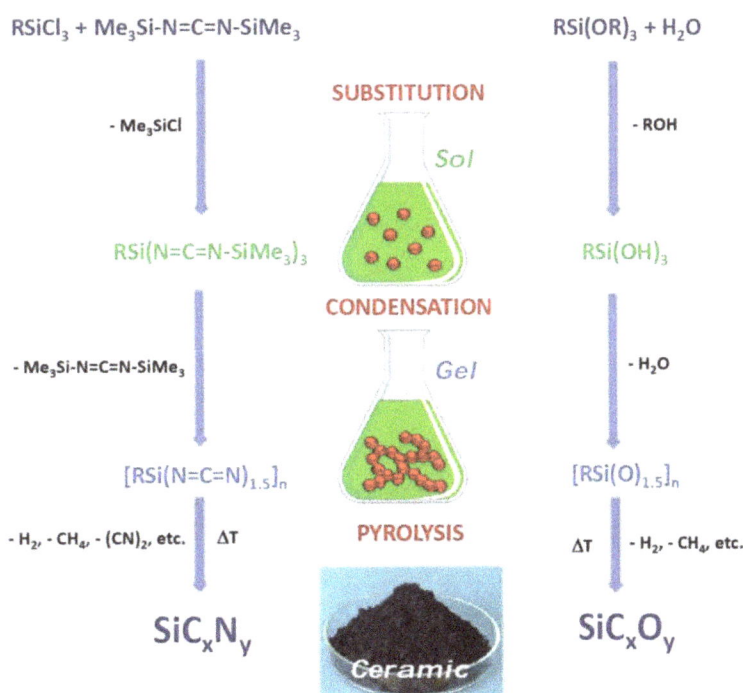

Figure 3. Comparison between the non-oxide sol–gel process for the synthesis of polysilylcarbodiimides and the oxide sol–gel process for the synthesis of polysiloxanes.

2.2. Single-Source Precursors Based on Metallopolymers

Compared to the previous section which comprised feasible tailor-made ceramics featuring high ceramic yields based on the conversion of, e.g., polysiloxanes, polycarbosilanes and polysilazanes (cf. Section 2.1), in this section metallopolymers will be highlighted as suitable ceramic precursors. Recently, metal-containing polymers attracted enormous attention due to their promising combination of redox, mechanical, semi-conductive, photo-physical, optoelectronic, magnetic and catalytic properties [103–109]. Such kinds of polymers can either feature (i) a metal center as integral part of the polymer main chain or (ii) the metal is laterally attached to the polymer chain. As a further structural distinction of metallopolymers—which definitely affect the ceramic yield of the final functional ceramic—these polymers can be distinguished by a linear, dendritic or hyperbranched structure. The preparation of nano- and micro-structured ceramics based on block copolymers and colloidal architectures will be discussed in the ensuing Section 4.3. Selected examples for the conversion of metallopolymers into ceramics will be addressed within this section. They are mainly synthesized to produce carbide nanocomposites.

A soluble poly-yne carbosilane copolymer with sandwiched zirconium moieties was synthesized as a preceramic metallopolymer for the preparation of ZrC/SiC/C ternary composite ceramics. Thermal treatment of the precursor polymer at a temperature of 1400 °C revealed a ceramic yield of over 52% of a highly crystalline material [110]. Another strategy for the preparation of nano-sized ZrC composites with a carbon fiber reinforced carbon matrix was reported by Tao et al. [111]. The authors took advantage of an air-stable and processable zirconium-precursor, polyzirconosaal, for infiltration followed by pyrolysis. Here, the ceramic yield for the final composite yielded 58% of a ZrC-C/C composite after thermal treatment. Very recently, Wu et al. [112] described a method for the preparation of zirconium-based precursor polymers starting from zirconium tetrachloride. While zirconium oxide was initially formed from these polymers at a rather moderate temperature of 1200 °C, crystalline ZrC particles exhibiting face-centered cubic lattice structures (50 to 100 nm) were accessible by subsequent thermal treatment at 1400 °C. ZrC composites have also been prepared by a polymer-analogous route, i.e., by post functionalization of a reactive polycarbosilane [113]. As described in the previous section, hydrosilylation is a powerful method for this purpose. Wang et al. [113] used a zirconocene derivative tethered at a polycarbosilane backbone for the preparation of ZrC/SiC composites after thermal treatment. The final materials were accessible with a remarkable ceramic yield of 78%.

An interesting approach for the formation of mixed lanthanide coordination polymers for the preparation of rare earth oxides was reported by Demars and coworkers [114]. Different shapes of investigated oxides have been found simply by varying the preparation methods of the metal-containing polymeric precursor material. Well-defined cylindrical or spherical micro-morphologies could be adjusted by changing the used solvents (water or tetrahydrofurane, THF) which retained in the final oxides after thermal treatment [114].

The vast majority of reports deal with metallocene-containing polymers, i.e., metal centers sandwiched between cyclic hydrocarbon moieties, as recent synthetic pathways led to stable and well characterized functional materials [115]. Noteworthy, such metallopolymers have been used for the preparation of block copolymers featuring the intrinsic capability of self-assembling into well-ordered nano-scaled structures. Within the field of metallocene-based polymers, especially ferrocene-containing polymers were found to be powerful precursor materials for the preparation of iron-based ceramics with remarkably high ceramic yields. The most prominent example in the field of ferrocene-containing polymers for the conversion into functional ceramic materials was reported by the group of Manners. By ring-opening polymerization (ROP) of ring-strained *ansa*-silaferrocenophanes, polyferrocenylsilanes (PFS) for the preparation of well-defined nanostructured ceramics could be obtained (Figure 4) [116–118].

Figure 4. Ring opening polymerization (ROP) of *ansa*-silaferrocenophanes yielding poly(ferrocenylsilane) (R = alkyl).

As shown on Figure 4, PFS belong to a class of metallopolymers that consist of alternating ferrocene and organosilane units in the polymeric backbone. An important feature of these polymers is their glass transition temperature which reflects their conformational flexibility. For instance, polyferrocenyldimethylsilane is a film-forming thermoplastic and can be melt-processed into various shapes such as nanofibers by electrospinning [119]. Thus, high-molecular weight polyferrocenylsilanes with linear, cyclic, or hyperbranched architecture and their block copolymers have shown potential in the preparation of shaped magnetic ceramics [120–125] and the self-assembly into well-defined hybrid architectures such as micelles [126], spheres [118,127], cylinders [128] and one-dimensional nanostructures [129,130]. Interestingly, PFS-derived ceramics possess tunable magnetic properties between the ferromagnetic and the superparamagnetic state, which can be achieved upon controlling the pyrolysis conditions of PFS.

Clendenning *et al.* [116] reported the usability of PFS films for the preparation of ferromagnetic ceramic films by reactive ion etching using a plasma. For this purpose, the PFS precursor was additionally functionalized with cobalt clusters at the silicon moiety. Ordered arrays of ferromagnetic ceramics could thus be obtained. As another example, PFS-based metallopolymers have been doped with palladium(II) acetylacetonate in order to produce soft processable polymer films [131]. Pyrolysis of this tailor-made preceramic film at 600–900 °C led to ferromagnetic ceramics, while higher applied temperature (1000 °C) led to the formation of FePd alloys [131]. Häußler and coworkers studied hyperbranched poly(ferrocenylene)s as feasible metallopolymers for a pyrolytic transformation into magnetic ceramics [132]. The hyperbranched polymeric framework was advantageous regarding the ceramic yield of magnetic iron nano-clusters featuring high magnetic susceptibilities. Moreover, pyrolysis in an argon atmosphere at 1200 °C led to the formation of iron silicide. The ferrocene-containing polymer poly(2-(methacryloyloxy)ethyl ferrocenecarboxylate) (PFcMA) which is accessible by using controlled polymerization or emulsion polymerization protocols seems to be an excellent candidate for the preparation of iron-based ceramics [133–136]. Mazurowski and coworkers succeeded to convert dense PFcMA brushes attached on organic particles into spherical iron oxides after thermal treatment [137]. Within this work, the chain length of the preceramic polymer as well as the polymer grafting density at the particle surface could be adjusted in a wide range. Recently, Yu *et al.* [138] used the hydrosilylation reaction of vinyl ferrocene with allylhydridopolycarbosilane (AHPCS) to synthesize a processable hyperbranched polyferrocenylcarbosilane. The polymer led to SiC/C/Fe nanocomposites with particular magnetic properties depending on the iron content in the polymer and on the pyrolysis conditions.

By combining the emulsion polymerization protocols of FcMA and the Stöber process, preceramic copolymers based on PFcMA and poly(methyl methacrylate) (PMMA) led to uniform ferrocene-containing particles of adjustable diameters in the range of (100–500 nm) [133]. The core/shell architectures featuring the ferrocene moieties as integral part of the particle shell were proven useful as single-source magnetic ceramic precursors. After thermal treatment, nanorattle-type ceramic architectures featuring a magnetic iron oxide core could be prepared with potential applications in fields of sensing and stimuli-responsive nano-photonics [133].

The pyrolysis of ferrocene-containing phosphonium polyelectrolytes as feasible preceramic polymers yielded iron-rich nanoparticles with carbon-, phosphorus-, and oxygen-rich phases in good ceramic yields (46%) as recently reported by the group of Gilroy [139]. Moreover, this strategy was shown to be feasible to adjust the composition of the final advanced ceramic composite material. Similarly to polycarbosilanes (PCS), PFS can be modified with other elements. Indeed, it is well known that incorporation of boron at atomic scale in polycarbosilane allows improving silicon carbide (SiC) sintering and SiC crystallization rate [140]. When boron is incorporated into PFS to form ferrocenylboranes [141–144], ferrocenylborane polymers [145–149] and ferrocene-containing poly(boro)-carbosilanes [150], multifunctional ceramics with tailorable magnetic properties and high-temperature resistance can be achieved under suitable pyrolysis conditions.

Previous examples described the modification of polycarbosilanes with ferrocene as pioneered by Manners. Final polymer-derived nanocomposites exhibit controlled magnetic properties (See Section 5.2). However, one of the advantages of polycarbosilanes and polycarbosilazanes is the availability of reactive groups within their structure such as Si–H, N–H, Si-vinyl, *etc.*, which allow their reaction with coordination compounds and thus the incorporation of late transition metals within the polymeric architecture. The resulting metal-containing polymers were shown to mainly to produce nanocomposites which exhibit promising catalytic properties (See Section 5.3) [151–156]. During the reaction, covalent bonds between the metal ions and the polycarbosilazanes can be established by the reaction with coordination compounds. During these metal transfer the ligand of the coordination compound is released and after cross-linking may become a part of the preceramic polymer. The synthesis is directed to produce metal-ceramic nanocomposites after pyrolysis during which *in situ* controlled growth of metal occurs in the matrix. As an illustration, a commercially available polycarbosilazanes, *i.e.*, HTT1800, was modified by aminopyridinato metal complexes. The aminopyridinato copper complex [Cu$_2$(Ap^{TMS})$_2$] (Ap^{TMS}H = (4-methylpyridin-2-yl)trimethylsilanylamine) reacted with HTT1800 via transmetalation, *i.e.*, aminopyridine elimination (Figure 5) [152].

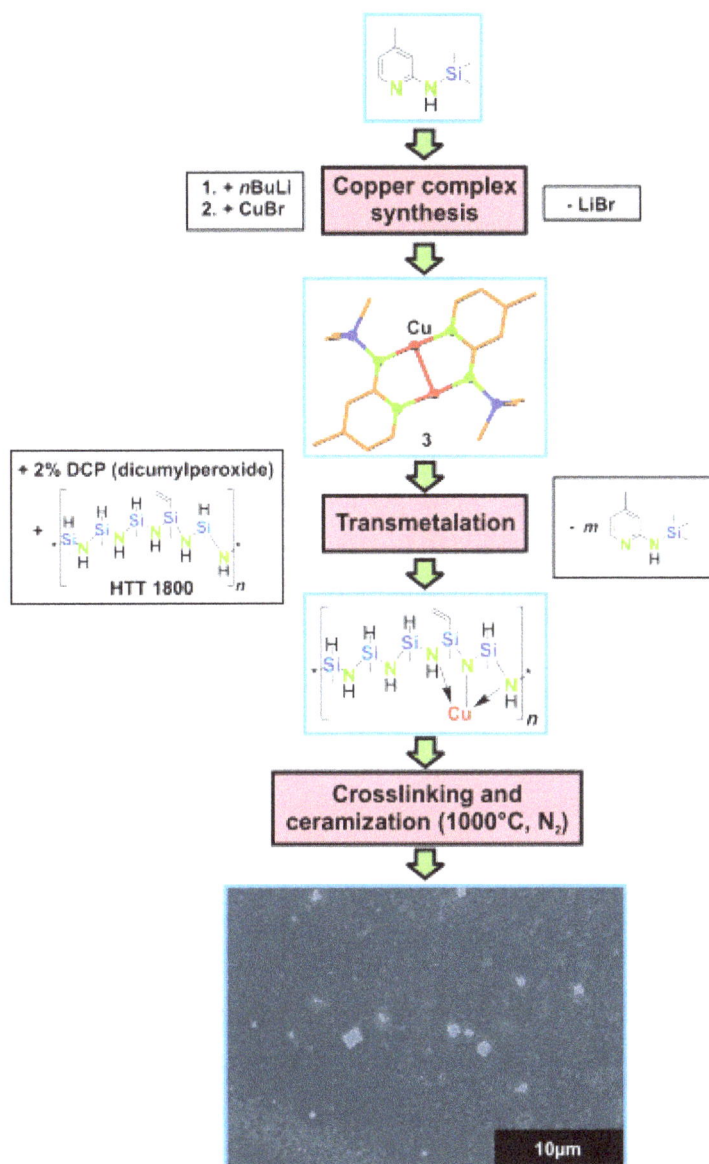

Figure 5. Modification of HTT1800 with an aminopyridinato copper complex leading to Cu-containing SiCN ceramics by pyrolysis to 1000 °C under nitrogen (with permission from Wiley) [152].

This reaction was investigated by ^{1}H and ^{13}C NMR spectroscopy. The driving force of this reaction was considered to be the low coordination number of copper in [Cu$_2$-(Ap^{TMS})$_2$] leading to the establishment of covalent bondings between the copper atoms and the polycarbosilazane. Crosslinking of the copper-modified polycarbosilazane and subsequent pyrolysis led to the copper-containing SiCN ceramics. Using an amido Nickel complex ([Ni(Ap^{TMS})$_2$]$_2$) to react with HTT1800 according to various and controlled Si:Ni ratios in THF was reported to generate a solid and dark polymer. The amido Nickel complex catalyzed the cross-linking of HTT1800 via hydrosilylation at room temperature. After pyrolysis to 600 °C under an inert atmosphere, Ni particles located near the external surface of the SiCN ceramic as well as within the internal voids were obtained [153]. The nanocomposites demonstrated catalytic activity for hydrogenation reactions (See Section 5.3). Recently, by changing the nature of the preceramic polymer and using a commercial allylhydridopolycarbosilane (AHPCS), Ni-containing SiC ceramics could be synthesized through the self-assembly of AHPCS-*block*-polyethylene

(PE) [155]. The block copolymer was synthesized through Ni complex-catalysed dehydrocoupling of Si–H in AHPCS with O–H in the hydroxy-terminated PE. The added nickel complex ([Ni(Ap^{TMS})$_2$]$_2$ also catalyzed the cross-linking of the AHPCS block, for example, through dehydrocoupling reactions of the Si–H bonds. By changing the nature of the metal-containing precursor, *i.e.*, trans-[bis(2-aminoetanol-*N*,O)diacetato-nickel(II)], HTT1800 was chemically modified through (i) reaction between the OH groups in the complex and the Si centers of HTT1800; (ii) hydrosilylation reactions resulting in the formation of carbosilane bonds; and (iii) reduction of Ni^{2+} and *in-situ* formation of Ni nanoparticles in the polymer matrix. After pyrolysis at 700 °C, nanoporous silicon oxycarbonitride ceramics modified by Ni nanoparticles with a BET surface area of 215 m$^2 \cdot$g^{-1} were obtained [156]. Kamperman *et al.* [151] applied micromolding and two-component colloidal self-assembly with cooperative assembly of a five component precursor system (solvent, amphiphilic block copolymer (poly(isoprene-*block*-dimethylaminoethyl methacrylate) (PI-*b*-PDMAEMA)), radical initiator (Dicumyl peroxide), commercial poly(ureamethylvinyl)silazane (PUMVS)) and the coordination compound [(COD)PtMe$_2$] (COD = 1,5-cyclooctadiene) platinum catalyst precursors) to obtain the Pt@SiCN materials after heat treatment to 1000 °C. The authors assumed that platinum segregated in the PDMAEMA domains, as the allyl groups of the PUMVS can efficiently add to Pt in a similar fashion to the double bond coordination of Pt with 1,5-cyclooctadiene in the precursor molecule.

3. Polymer-to-Ceramic Transformation

Controlled thermal decomposition of silicon-based polymers provides nano-structured ceramics with nanostructures strongly influenced by the chemistry and architecture of the precursors, their processing route and the parameters used for their pyrolysis (heating rate, reactive or inert atmosphere and dwelling time). Depending on the temperature, the preceramic polymers suffer different processes during their transformation to ceramics. After polymerization, shaping and cross-linking of the polymer can be easily done at moderate temperatures to obtain complex-shaped green-bodies which can retain their shape upon pyrolysis up to 1400 °C and also during high-temperature annealing up to 2000 °C. All these steps are defined by different chemical processes as discussed in the following (Figure 6).

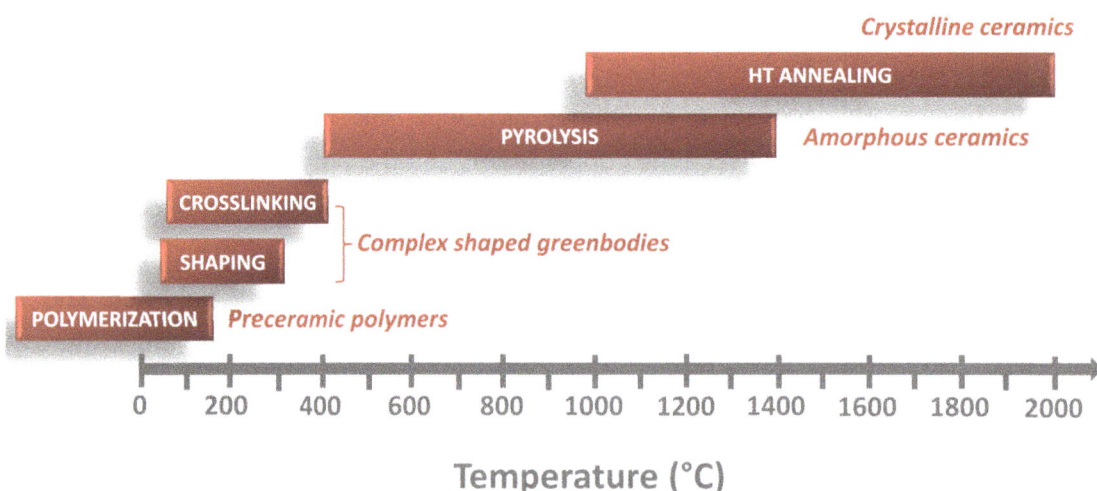

Figure 6. Polymer-to-ceramic transformation of preceramic polymers [6].

The characterization of the individual polymer-to-ceramic conversion steps is usually done by combining several spectroscopic techniques such as multinuclear solid-state NMR, Raman, FTIR and XPS spectroscopy, with diffraction (X-ray diffraction, SAXS, electron diffraction) and microscopic techniques (SEM, TEM), as well as with elemental analysis.

3.1. Cross-Linking

During cross-linking the polymeric precursors are converted into organic/inorganic materials at low temperatures (up to 400–500 °C). This transformation prevents the loss of low molecular weight components of the polymer precursors as well as fragmentation processes during the pyrolysis process, and consequently leads to high ceramic yields. Furthermore, the cross-linking process allows for generating infusible materials (thermosets) which retain their shape during pyrolysis.

Cross-linking of polycarbosilanes can be achieved by thermal curing in air atmosphere or e-beam curing [157–159]. Cross-linking of polycarbosilanes in the presence of oxygen has been found to occur via radical mechanisms: oxidation of Si–H and Si–CH$_3$ bonds occurs with the formation of Si–OH, Si–O–Si and C=O groups, as revealed by IR spectroscopy [160,161], XPS [162] and solid-state ^{29}Si MAS NMR investigations [163]. Oxidative cross-linking of polycarbosilanes leads to SiC materials with oxygen contents of 10%–12%.

Cross-linking of polycarbosilanes in absence of oxygen involves reactions of Si–H bonds with Si–CH$_3$ groups leading to Si–CH$_2$–Si linkages as supported by IR spectroscopy [164] and solid-state ^{29}Si MAS NMR [165] studies. Interestingly, no Si–Si bond formation occurs. Silicon carbide materials synthesized from e-beam cross-linked polycarbosilanes show very low oxygen content (0.2%–0.3%). This consequently leads to a strong improvement of their thermal stability and mechanical properties if compared to oxygen-rich SiC ceramics [10]. Cross-linking of polysiloxanes is achieved via condensation, transition metal catalyzed addition or free radical initiation techniques. In polymers containing methyl or vinyl groups cross-linking can be performed thermally by using peroxides [166].

The hydrosilylation reaction is an effective way to obtain infusible materials which are resistant toward water and elevated temperatures [167,168].

For polysiloxanes having functional groups in the structure, e.g., hydroxy or alkoxy groups, the cross-linking process occurs upon condensation of the silanol groups with *in situ* water release and subsequent hydrolysis reactions of the alkoxy substituents. Using appropriate catalysts, e.g., tetrakis(pentafluorophenyl)borate in the case of a polysiloxanol or [bis(2-ethylhexanoate)tin] in the case of poly(methoxymethylsiloxane) or poly(methylsiloxane), these reactions take place at room temperature [169,170].

Polysilazanes can be cross-linked either thermally and/or using reactive atmosphere such as ammonia and chemical reagents, such as catalysts or peroxides [64]. There are four major processes which can occur during the thermal cross-linking processes of polysilazanes: transamination, dehydrocoupling (between Si–H and N–H resp. Si–H and Si–H groups), vinyl polymerization and hydrosilylation.

Hydrosilylation reactions occur in (poly)silazanes which contain Si–H and vinyl substituents. This is a fast process which occurs at relatively low temperatures (starting from 100–120 °C) and leads to the formation of Si–C–Si and Si–C–C–Si units. This strengthens the polymeric network, since the Si–C and C–C bond are not affected by thermal depolymerization reactions such as transamination or exchange of Si–N bonds. Thus, higher ceramic yields as well as higher carbon contents are possible in the final ceramic materials [171]. Hydrosilylation can be also performed in presence of catalysts, which significantly increase the reaction rate [172].

Dehydrogenation of Si–H/N–H or Si–H/Si–H groups starts at temperatures of ca. 300 °C and leads to the formation of Si-N and Si-Si bonds as well as to hydrogen evolution.

The vinyl polymerization process occurs at higher temperatures and involves no mass loss. Upon UV light irradiation and in the presence of a photo initiator (such as 2,2-dimethoxy-1,2-diphenylethan-1-one), the vinyl polymerization process can occur at temperatures as low as ambient temperature, as it was also shown for other vinyl-substituted polymers [173].

Transamination reactions occur in a temperature range from *ca.* 200 °C to 400 °C and are associated with mass loss (*i.e.*, amines, ammonia, silazane (oligo)fragments), thus leading to a decrease in nitrogen content of the final ceramic materials. Since also redistribution reactions at silicon centers can occur and volatile silicon species, e.g., silanes, can evolve, the ceramic yield and the silicon content of the end ceramics consequently decrease.

3.2. Ceramization

The ceramization process of preceramic polymers involves the thermolysis and release of their organic groups at high temperatures (600–1000 °C) and consequently the organic-to-inorganic transformation of the preceramic materials into amorphous covalent ceramics [5,7,174]. Ceramics which are obtained by using this technique show, however, the disadvantage of high shrinkage and porosity. Greil *et al.* [175] investigated the effect of fillers dispersed within the pre-ceramic polymers on the shrinkage and porosity of the resulting ceramics. Using inert fillers, ceramic materials with less shrinkage can be obtained; the volume change upon polymer decomposition is accommodated however by the appearance of relatively large porosity [176,177]. The use of active fillers can compensate the shrinkage of the polymer matrix by appropriate expansion of the filler phase due to its reactions with the gaseous releases. Appropriate combinations of inert and active fillers might thus lead to dense materials with zero shrinkage, allowing for near-net-shape manufacturing [178,179]. An additional technique to produce dense PDC parts relates to an extensive cross-linking step which increases the ceramic yield of the polymer-to-ceramic transformation. This has been achieved for instance in the case of a polysiloxane, which was cross-linked by means of UV light irradiation. The residual porosity of the SiOC ceramic prepared upon pyrolysis of the UV light cross-linked green body was determined to be below 1% [173]. Also infiltration-pyrolysis cycles of pressureless monoliths have been shown to lead to materials with less residual porosity [180]. Beside pressureless techniques, dense PDC-based parts can be prepared by using pressure-assisted methods, such as uniaxial hot pressing (HP) [181,182], hot isostatic pressing (HIP) [183], or spark plasma sintering (SPS) [184,185], *etc.*

The conversion of preceramic polymers into ceramics involves complex processes which are difficult to investigate, due to the poorly defined structure of the preceramic materials as well as to the amorphous nature of the resulting ceramics. However, the use of solid state NMR, thermogravimetric analysis (TGA) coupled with differential thermal analysis (DTA) and *in situ* evolved gas analysis (EGA, *i.e.*, *in situ* FI-IR spectroscopy and mass spectrometry) as well as other modern structural characterization techniques, it is possible to rationalize the processes which occur during the ceramization of the preceramic polymers.

Pyrolysis of *polycarbosilanes* leads to amorphous silicon carbide-based materials at temperatures between 800 °C and 1000 °C upon release of gaseous species containing Si–H, Si–CH₃, and Si–CH₂–Si groups [165]. ^{29}Si MAS-NMR spectra of the materials annealed in this temperature range showed the presence of one single peak assigned to SiC₄ units [186]. Additional competing decomposition processes lead to the formation of segregated carbon as well as dangling bonds within the amorphous ceramic. Consequently, the ceramics obtained from the pyrolysis of polycarbosilanes at 800 °C can be described as hydrogenated silicon carbide materials containing some segregated carbon [186,187]. As the annealing temperature exceeds 1000 °C, hydrogen is released and the crystallization of the amorphous material into cubic silicon carbide occurs.

The pyrolysis of polysiloxane-based preceramic polymers leads to the formation of silicon oxycarbide (SiOC) [51,188–196]. During the ceramization process, mainly the release of hydrocarbons and hydrogen takes place, in addition numerous distribution reactions between the Si–O, Si–C, and Si–H bonds [197]. They might lead to the evolution of low-molecular-weight silanes (at temperatures in the range from 400 and 600 °C) and consequently to a decrease of the ceramic yield. At higher temperatures (600 to 1000 °C), extensive cleavage processes of C–H, Si–C and Si–O bonds occur and furnish ceramic materials consisting of an amorphous silicon oxycarbide phase and residual segregated carbon [6,198–203]. Studies on the pyrolysis kinetics for the conversion polysilsesquioxane→silicon oxycarbide indicate that processes leading to evolution of hydrocarbons (methane) and hydrogen represent the main mechanism for the removal of carbon during pyrolysis [204]. This mechanism has been found to be of first-order. Furthermore, the reaction rate was found to directly correlate to the amount of the remaining carbon sites. Thus, the nanostructure/architecture of the silicon oxycarbide glassy network which results upon pyrolysis of the preceramic polysilsesquioxanes relies on the geometric configuration of the molecules in the cross-linked preceramic polymer.

Numerous polysiloxane compositions were studied as precursors for SiOC glasses [189–196]. It seems that the composition of the final SiOC glass can be controlled, since the O/Si molar ratio remains almost constant during the pyrolysis process. The tailoring of the polysiloxane composition leads to a minimization of the quantity of excess carbon in the final SiOC glass. This approach was applied by Soraru *et al.* [53] who introduced a proper amount of Si–H groups within the polysiloxane network to minimize the final free C content. Additionally, SiOC glasses with no excess of carbon (so-called "white" SiOC) were prepared via pyrolysis of a polysilsesquioxane in hydrogen atmosphere [205].

Numerous studies were performed in order to assess the conversion of polysilazanes into amorphous silicon carbonitrides (SiCN) [64]. The pyrolysis process of a polysilazane containing methyl and hydrido groups attached to Si, (–[H(Me)Si-NH]ₘ–[Me₂Si-NH]ₙ–) [206,207], was investigated by means of MAS NMR and TGA/EGA. At temperatures exceeding 550 °C, ^{13}C NMR and TG/MS investigations indicate reactions occurring between Si–H and Si–CH₃ groups to form Si–CH₂–Si units with methane evolution. Additionally, reactions involving N–H groups proceed to form SiN₄–units by successive replacement of methyl groups and release of gaseous methane. With higher pyrolysis temperature, the number of Si–N or Si–C bonds successively increase, as observed by means of ^{13}C and ^{29}Si NMR spectroscopy [64]. For vinyl-containing polysilazanes, vinyl polymerization at moderate temperatures (250–350 °C) leads to the formation of carbon chains which subsequently might convert into sp^2 carbon at higher temperatures [149]. Also, here the number of Si-N bonds in the ceramic materials increases with the temperature due to reactions of Si–H and Si–CH$_3$ groups with N–H.

Corroborated TGA/EGA and FTIR spectroscopy studies on the ceramization transformation of a vinyl-substituted polyureasilazane revealed that at lower temperatures vinyl polymerization and hydrosilylation processes are responsible for the cross-linking of the precursor; whereas at higher temperatures transamination reactions occur, accompanied by ammonia gas evolution [208]. Further increase of the temperatures to 600–800 °C leads to a remarkable decrease of the amount of Si–H, Si–CH$_3$ and N–H groups, accompanied by the evolution of hydrogen (dehydro-coupling reactions between Si–H and N–H bonds) and methane (reactions between Si–CH$_3$ and N–H).

Studies on the structure of silicon carbonitride ceramics obtained via pyrolysis of polysilazanes showed that they consist of a single SiCN amorphous phase and some residual excess carbon [209]. Whereas the pyrolysis of polyorganosilylcarbodiimides leads to phase-separated SiCN ceramics, consisting of amorphous silicon nitride and segregated carbon [80,95,96,98,210]. Obviously, the different phase composition and nano/microstructure of polysilazane- and poly(silylcarbodiimide)-derived SiCN relies on the different thermal and pyrolytic behavior of the preceramic polymers. ^{29}Si MAS-NMR and FTIR studies on the thermolysis of polymethylsilylcarbodiimide revealed at moderate temperatures (up to 600 °C) the occurrence of rearrangement and condensation reactions to form SiCNX$_2$, SiCN$_2$X, SiCN$_3$, and SiNX$_3$ sites (with X being NCN or NCHN). At higher temperatures, the decomposition of the carbodiimide units takes place [89,90,152]. In a first step the carbodiimide units rearrange into the isomeric cyanamide structure, followed by the release of C$_2$N$_2$ and N$_2$ and the generation of amorphous Si$_3$N$_4$ [153].

4. Ceramic Nanocomposites with Tailor-Made Phase Compositions and (Micro)Structures

4.1. Tailor-Made Compositions from Single-Source Precursors

The general synthesis strategy for PDC-NCs involves the preparation of suitable single-source precursors which can be converted in a first step into single-phase ceramic materials. Subsequent treatment of the single-phase materials (typically thermal treatment) leads to phase separation and crystallization processes which thus furnish nanocomposite materials with tuned phase compositions and microstructures. In Figure 7, the evolution of the microstructure of a SiHfOC-based material is shown. Pyrolysis at 700 °C delivers an amorphous, single-phasic SiHfOC material, which upon annealing at 1100 °C undergoes a phase separation process leading to an amorphous nanocomposite material consisting of amorphous hafnia nanoparticles homogeneously dispersed within a glassy SiOC matrix. Annealing at 1300 °C induces the crystallization of the hafnia precipitations and thus tetragonal hafnia nanoparticles finely dispersed within SiOC are obtained [4,9]. Within this subsection selected examples related to the ceramization of metal-modified silicon-containing preceramic polymers will be briefly introduced.

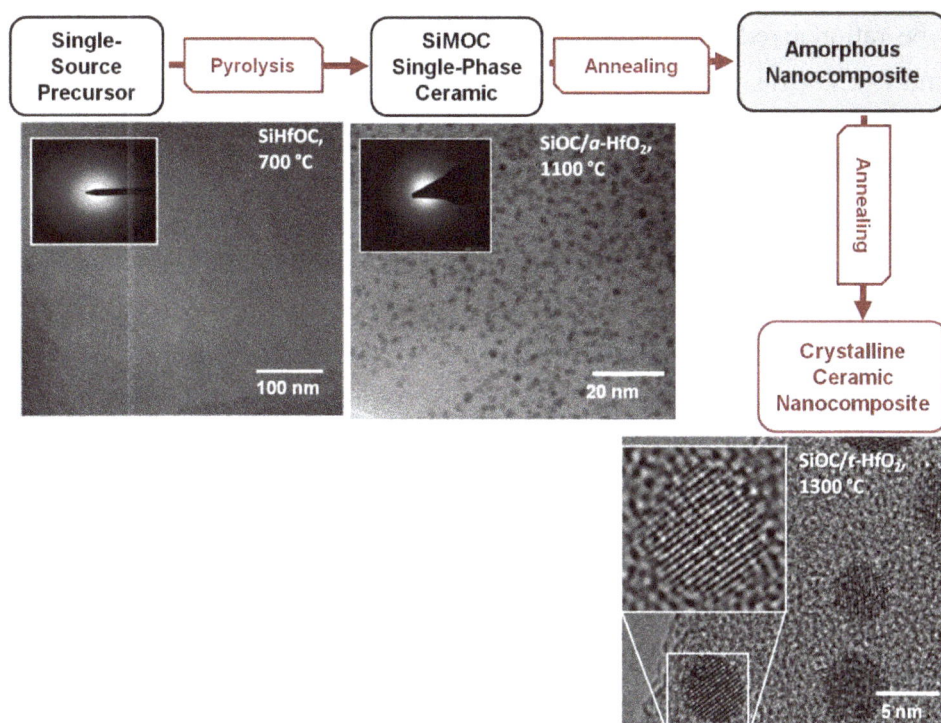

Figure 7. Single-source-precursor synthesis of polymer-derived ceramic nanocomposites (PDC-NCs).

Pyrolysis of suitable alkoxysilanes, $Si(OR)_4$, or polysiloxanes, $-[Si(R)_2-O]_n$, chemically modified with metal alkoxides was shown to give MO_x/SiOC-based PDC-NCs, as reported for M = Al [179,211], Ti [54,58], Zr [203,212], Hf [201,213], *etc.* For M = Zr and Hf a single-phase SiMOC ceramic is obtained upon pyrolysis at rather low temperatures (*ca.* 700 °C), while at higher temperatures amorphous MO_2 nanoparticles precipitate (800–1100 °C). Upon increasing the annealing temperature to 1300 °C, MO_2 nanoparticles crystallize, forming microstructures comprised of tetragonal zirconia/hafnia particles finely dispersed within an amorphous SiOC matrix [201,203,213].

Metal-modified silicon oxycarbonitrides were also synthesized [154]; however, not only MO_x/SiCNO nanocomposites (M = Ti [73], Zr [61], Hf [74,75]), but also M/SiCNO (for Cu [152], Ni [214]) and MSi_x/SiCNO (Fe [215,216], Co [215], Pd [217]) were reported. In the case of MO_x/SiCNO, a similar polymer-to-ceramic transformation as for MO_x/SiOC was proposed (*i.e.*, formation of a single-phase amorphous SiMCNO at low temperatures and subsequent phase separation of MO_x) [75].

Recently, the phase separation and crystallization of metal-modified silicon oxycarbides was systematically assessed [201,203,213,218,219]. Based on a case study related to the temperature-dependent evolution of the environment at the metal (*i.e.*, Hf) in a SiHfOC-based material prepared from a hafnium-alkoxide-modified polysiloxane [220], it was concluded that the Hf sites are coordinated only by oxygen, independently of which temperature was used for the preparation of the samples. This indicates that the amorphous single-phase SiMOC ceramics (M = metal) undergo phase separation and lead to the precipitation of an amorphous metal oxide (MO_x) phase in a first step. Consequently, the (thermodynamic) stability of the MO_x phase with respect to its carbothermal reaction (segregated carbon being typically present within the microstructure of the as-prepared amorphous SiMOC materials) was considered to be of crucial importance with respect to the evolution of the phase composition in SiMOC. Indeed the phase composition of SiMOC materials upon thermal treatment of metal-modified polysiloxanes can be rationalized via assessing the relative change in the free Gibbs energy of the systems M–MO$_x$ and C–CO (as segregated carbon is present within the microstructure of SiMOC, it is considered to determine the oxygen fugacity in the system) [221].

Based on thermodynamic data of the respective oxides, the phase composition of SiMOC ceramics upon annealing at high temperatures can be predicted. The prediction was shown to agree with the experimental results for SiMOC and also for SiMCNO ceramic composites. However, in addition to the stability of the oxides with respect to reduction, some other aspects were shown to be relevant while trying to predict the phase composition of SiMOC/SiMCNO nanocomposites, e.g., thermodynamic stabilization through the conversion of the metal oxide phase into silicates (for MO$_x$ being stable with respect to carbothermal conversion into M) or into silicides or carbides (for MO$_x$ not being stable against carbothermal reduction) [221].

A corroborated MAS NMR and electron microscopy study on the evolution of SiHfCNO indicate also in this case the fact that in a first step the hafnium-alkoxide-modified polysilazane used as a single-source precursor delivers upon pyrolysis an amorphous single-phase SiHfCNO ceramic, which in a subsequent step converts into an amorphous HfO$_x$/SiCN nanocomposite. Annealing at higher temperatures leads to the crystallization of the hafnia phase into tetragonal HfO$_2$ [75].

Also non-oxidic systems (e.g., SiMC, SiMCN, SiMBCN, with M being metal) were shown to be synthetically accessible upon using a similar approach [79,182]. Thus, SiHfC-based nanocomposites were prepared upon thermal conversion of a polycarbosilane modified with a tetrakis amido Hf complex; whereas SiHfCN and SiHfBCN nanocomposites were prepared from Hf-modified polysilazanes and polyborosilazanes, respectively. Interestingly, the high-temperature evolution of those systems seem to be also thermodynamically controlled. Thus, SiHfBCN-based single-phase materials convert into SiC/HfC/HfB$_2$ ceramic nanocomposites via high-temperature exposition in argon atmosphere; whereas annealing in nitrogen leads to HfN(C)/Si$_3$N$_4$/SiBCN nanocomposites [79].

4.2. Hard-Template-Assisted Techniques towards Ordered (Micro)structures

As mentioned in Section 2, the synthesis of ceramic materials from preceramic polymers leads to several benefits as compared to other preparative methods; in particular based on the rheological and solubility capability of preceramic polymers. This offers for instance the opportunity to prepare porous materials. The introduction of pores (controlled or uncontrolled) in ceramic materials leads to hollow frameworks with modified physical and chemical properties in comparison to their dense counterpart. This extends the application potential of these materials in modern society and to particularly investigate their properties in the energy and environment fields.

We can classify porous materials into three categories based on the pore diameter (IUPAC classification): microporous (<2 nm), mesoporous (2–50 nm) and macroporous (>50 nm) [222]. The porosity can be disordered or ordered. It can be also hierarchical combining several types of porosity. Among these materials, the interest in the synthesis, characterization, modification/functionalization and application of ordered mesoporous materials has been developed dramatically over the last 20 years since the separate discovery of ordered mesoporous silica by Japanese scientists and Mobil researchers [223–226]. The important quantity of reports focused on ordered mesoporous materials is related to their particular characteristics which include their high specific surface area and pore volume within a relatively small volume of material, well-defined ordered mesostructure, structural capabilities at the scale of a few nanometers, tunable pore size, and varieties of the framework. This makes these materials attractive for applications in various fields, such as catalysis, adsorption and separation, drug storage and delivery, nanofabrication, sensors, photonics, energy storage and conversion, *etc.* [227]. The majority of mesoporous materials are of oxide type formed by a self-assembly process from combined solutions of sol–gel precursors (e.g., metal alkoxides) and structure-directing amphiphiles, usually block-copolymers or surfactants [222,228]. The most investigated oxide-based materials are siliceous materials. Despite the great success of this soft-templating method for oxides, it was rather challenging to apply this route to the synthesis of oxygen-free compounds satisfying the current technological needs in terms of thermo-structural and thermo-chemical stabilities. This is due mainly to the hydrolysis sensitivity and poor affinity with surfactants of the corresponding precursors. In addition, the generally higher temperatures which are applied to prepare non-oxide ceramics may cause the thermal breakdown of the structural integrity of the material. Only, few ordered mesostructures of non-oxide ceramics are reported using this strategy (see Section 4.3) [229–231]. A more convenient strategy has been proposed by Ryoo and co-workers in 1999 [232]. They reported the preparation of ordered mesoporous carbon using sucrose as the carbon source and the cubic (*Ia3d*) mesoporous silica molecular sieve referred to as MCM-48 as the template via three steps: (i) precursor infiltration inside mesochannels of the silica template; (ii) conversion of the precursor in the nanochannels of the silica template up to 800–1100 °C under vacuum or in an inert atmosphere; (iii) removal of the mesoporous silica template by dissolution typically in aqueous solution containing NaOH and ethanol. The resultant porous carbon material was referred to as CMK-1. This route is known as nanocasting because the entire manufacturing procedure is similar to the traditional casting method but on the nanoscale. This work demonstrated the feasibility of this nanocasting strategy on carbonaceous materials and this "hard template" concept illustrated in Figure 8 from hexagonal SBA-15 to prepare CMK-3 has been mainly applied for preparing mesoporous non-oxide ceramics. In this section, we have selected pioneering works of highly ordered mesoporous PDCs with tailored microstructures using different types of hard templates following the strategy depicted in Figure 8.

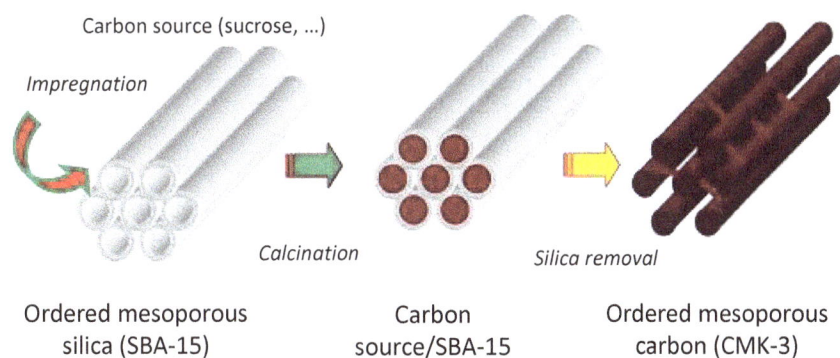

Figure 8. Nanocasting process: Toward the preparation of ordered mesoporous carbon.

It should be mentioned that, like classical ceramics, PDCs can be roughly classified into amorphous and crystalline materials as well as composites and nanocomposites containing one or more phases distinguished from the matrix. As ordered mesoporous materials, the reports are mainly focused on silicon-containing amorphous PDCs.

The first example of ordered mesoporous PDCs concerns the preparation of highly ordered mesoporous silicon carbide (SiC) by nanocasting process using polycarbosilane (PCS) as a SiC precursor and mesoporous silica such as SBA-15 and KIT-6 as hard templates [233]. After successful impregnation, the PCS-to-SiC conversion was achieved at 1200–1400 °C and the resulting SiC–mesoporous silica composites were washed with 10 wt% aqueous hydrofluoric acid several times to remove the silica template. Depending on the nature of the templates, SiC nanowires in two-dimensional (2D) hexagonal arrays (*p6m*) as well as a three-dimensional (3D) bicontinuous cubic mesoporous SiC structures could be elaborated with high BET surface areas (460–720 $m^2 \cdot g^{-1}$) and uniform pore sizes (2–3.6 nm) (Figure 9). Heat-treatment of the sample to 1400 °C under nitrogen resulted in the decrease of the specific surface area (SSA) while crystallization of β-SiC occurred.

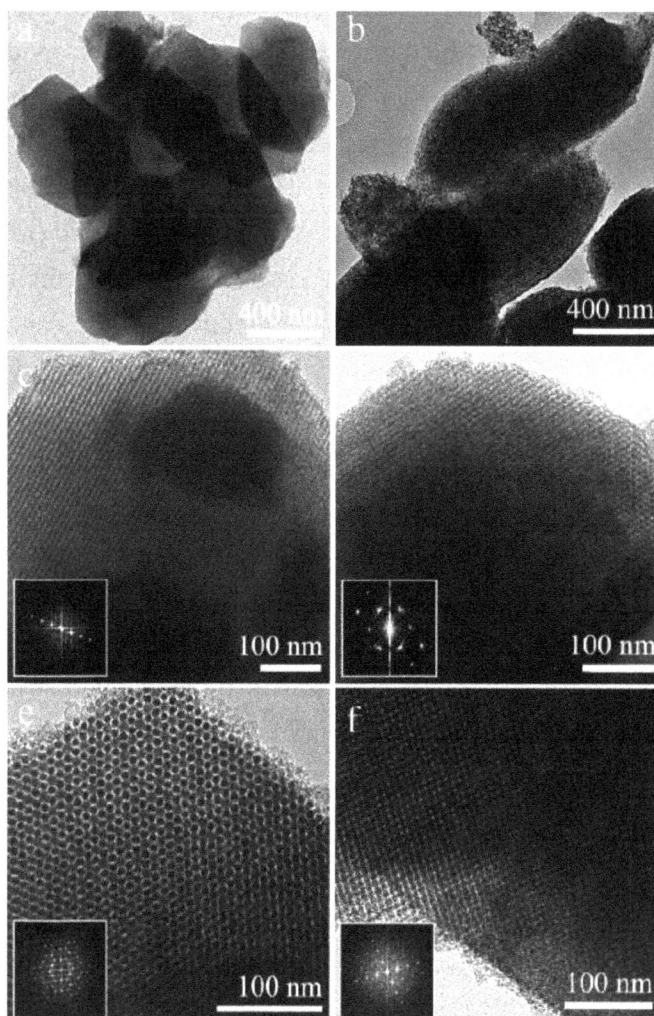

Figure 9. TEM images of ordered mesoporous SiC obtained by pyrolysis at 1200 °C from SBA-15 (silica-based Santa Barbara amorphous material No. 15) at low magnification (**a,b**) and high magnification (**c,d**) taken along the (**c**) (110) and (**d**) (100) directions. TEM images of ordered mesoporous SiC obtained by pyrolysis at 1400 °C from KIT-6 (Korea Advanced Institute of Science and technology No. 6) taken along the (**e**) (111) and (**f**) (531) directions. Insets show the corresponding Fourier diffractograms (reprinted with permission from reference [233]; Copyright Wiley).

The possibility to modify the chemistry of precursors as well as their reactivity allowed tailoring the composition of ordered mesoporous materials. As an illustration, by selecting ammonia instead of inert atmosphere during the polymer-to-ceramic conversion, Zhao et al. [234] reported the preparation of ordered mesoporous amorphous silicon nitride from CMK-8 as a hard template using the reactivity of polycarbosilane (PCS) toward ammonia. Mesoporous silicon nitride samples displayed a 3D bicontinuous cubic mesostructure (Ia3hd) similar to KIT-6, a specific BET surface area of 384 $m^2 \cdot g^{-1}$, a large pore volume of 0.71 $cm^3 \cdot g^{-1}$, as well as a narrow pore size distribution at the mean value of 5.7 nm. A secondary impregnation-pyrolysis cycle could reduce the structural shrinkage and improve the mesostructural regularity. By changing the SiC precursors and selecting allylhydridopolycarbosilane, known as AHPCS, Kim et al. [235] prepared ordered mesoporous SiC from surface-modified SBA-15 by impregnation, pyrolysis to 1000 °C under nitrogen and silica etching using a 10 wt% HF solution in a 50:50 mixture of water and ethanol. Samples exhibited BET surface areas of 260 $m^2 \cdot g^{-1}$ with pore size of 3.4 nm. These examples show that the main strategy to develop the mesoporosity of PDCs and keep the high ordering of the pores is to generate PDCs for which the structure is predominantly amorphous; heat-treatments at higher temperatures involve both the transformation of the amorphous state into crystalline phases and tend to collapse the porous structure.

It is reported that the amorphous state of PDCs can be stabilized by increasing the number of constituents in the Si–C and Si–N systems. Monthioux and Delverdier [236] studied the crystallization behavior of various PDCs. Based on their TEM investigations, they reported that the nucleation of the excess free carbon phase, commonly present in PDCs is always the first crystallization phenomenon to occur, followed by the nucleation of SiC. Depending on the system studied, different onset temperatures for the occurrence of crystallization were monitored. While nucleation within the binary Si–C system started at temperatures as low as 900–950 °C, local crystallization within the ternary Si–C–N and Si–C–O systems was observed at 1100 °C and 1250 °C, respectively. The quaternary Si–C–N–O system remained amorphous up to 1400 °C. However, increasing the number of constituents within polymer-derived ceramics is not the only active parameter in PDCs to raise thermal stability against crystallization. Chemical composition, starting polymer and glass architecture as well as residual porosity have to be considered simultaneously.

Based on these works, recent reports focused on multi-element (more than 2) materials with the objective to develop the thermal stability of ordered mesoporous PDCs. As an illustration, Kim et al. reported for the first time the preparation of highly ordered two-dimensional (2D) hexagonal and three-dimensional (3D) cubic mesoporous Si–C–N ceramics with high surface area (up to 472 $m^2 \cdot g^{-1}$), a narrow pore-size distribution and high thermal stability by nanocasting polycarbosilazane solutions into mesoporous carbon templates of the type CMK-3 and CMK-8 [237]. Interestingly, the BET surface areas of the mesoporous Si–C–N replicas were preserved up to 1000 °C in air. Within the same context, Zhao et al. developed another way which consisted to prepare through pyrolysis at 1400 °C firstly SiC within CMK-3 labeled SiC-C-1400, then in a second step, heat-treatment under air (500 °C) and ammonia (1000 °C) led to carbon elimination while ordered mesoporous Si–O–C (SiOC-1400) and Si–C–N (SiCN-1400) ceramics formed respectively as demonstrated by SAXS experiments (Figure 10) [238]. The ordered mesoporous Si–O–C and Si–C–N ceramics displayed high surface areas (200–400 $m^2 \cdot g^{-1}$), large pore volumes (0.4–0.8 $cm^3 \cdot g^{-1}$), and narrow pore size distributions (4.9–10.3 nm).

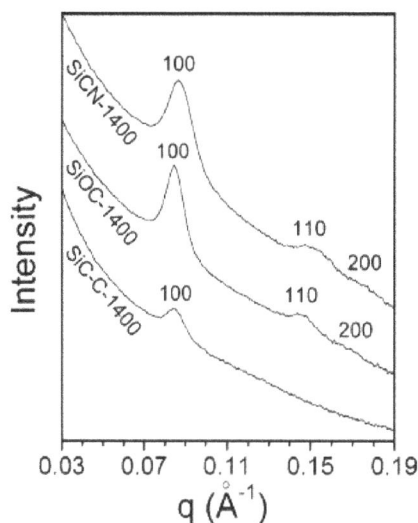

Figure 10. Small angle X-ray scattering (SAXS) patterns for ordered mesoporous SiC–C composites and derived SiOC-1400 and SiCN-1400 samples (reprinted with permission from reference [238]; Copyright 2007 American Chemical Society).

The introduction of boron in the Si–C–N system is known to shift the crystallization onset of the later to high temperature. Bill and co-workers [206,239] investigated the microstructure development of monolithic Si–(B)–C–N ceramics and they concluded that thin turbostratic B(C)N structures, finely dispersed within the amorphous matrix, acted as diffusion barriers preventing SiC and Si_3N_4 nucleation. In 2008, open, continuous, ordered 2D hexagonal mesoporous Si–B–C–N powders have been proposed by a double nanocasting approach using CMK-3 as template and boron-modified polysilazane of the type $[B(C_2H_4SiCH_3NCH_3)_3]_n$ (C_2H_4 = $CHCH_3$, CH_2CH_2) as preceramic polymer [240]. The polymer-to-ceramic conversion was achieved under ammonia up to 200 °C to cross-link the polymer via amine-exchange reactions then under nitrogen up to 1000 °C to generate a SiBCN-carbon composite. CMK-3 was subsequently removed through thermal treatment at 1000 °C in an ammonia atmosphere to generate ordered mesoporous SiBCN structures. Elemental analyses indicated the formation of ordered mesoporous powders with an empirical formula of $Si_{3.0}B_{1.0}C_{4.2}N_{3.5}$, whereas the nitrogen adsorption-desorption isotherm of the specimens showed a clear step at a relative pressure of about 0.5 attributed to capillary condensation in ordered mesoporous structures (Figure 11).

The corresponding specific surface area and the pore volume were calculated to be 600 $m^2 \cdot g^{-1}$ and 0.61 $cm^3 \cdot g^{-1}$, respectively. A narrow pore size distribution (around 3.4 nm) has been found. The material exhibited a relatively good thermal stability in air through heat-treatment to 1400 °C. Changing the Si–B–C–N precursor for a boron-modified polysilazane of the type $[B(C_2H_4SiCH_3NH)_3]_n$ (C_2H_4 = $CHCH_3$, CH_2CH_2) with a higher ceramic yield allowed generating better ordered 2-D hexagonal frameworks [241]. Using a double impregnation cycle combined with a pyrolysis process up to 1000 °C in flowing nitrogen and a carbon removal step at 1000 °C for 3 h in ammonia and nitrogen atmospheres, the ordered mesoporous Si–B–C–N ceramic displayed high surface area (630 $m^2 \cdot g^{-1}$), high pore volume (0.91 $cm^3 \cdot g^{-1}$), and narrow pore size distribution (around 4.6 nm) with a thermal stability which extended up to 1180 °C under nitrogen. From the same polymer and using ordered mesoporous silica, periodic mesoporous silicoboron carbonitride ($Si_{3.0}B_{1.0}C_{3.9}N_{1.8}$) frameworks with *P6mm* hexagonal symmetry could be prepared after a double infiltration/thermal cross-linking cycle followed by a thermal process up to 1000 °C under N_2 and a short chemical etching with dilute HF [242]. The ordered mesoporous $Si_{3.0}B_{1.0}C_{3.9}N_{1.8}$ ceramic displayed a specific surface area of 337 $m^2 \cdot g^{-1}$, a pore volume of 0.55 $cm^3 \cdot g^{-1}$ and a narrow pore-size distribution centered on 4.6 nm by N_2 sorption with an amorphous network remaining stable during continuous heat-treatment to 1480 °C in a nitrogen atmosphere. The study developped in this paper showed that the use of carbonaceous template is preferred for preceramic polymers most probably due to the pore surface chemistry of carbonaceous template that involves more complete pore filling and the expected better chemical compataibility of carbon with ceramic precursors.

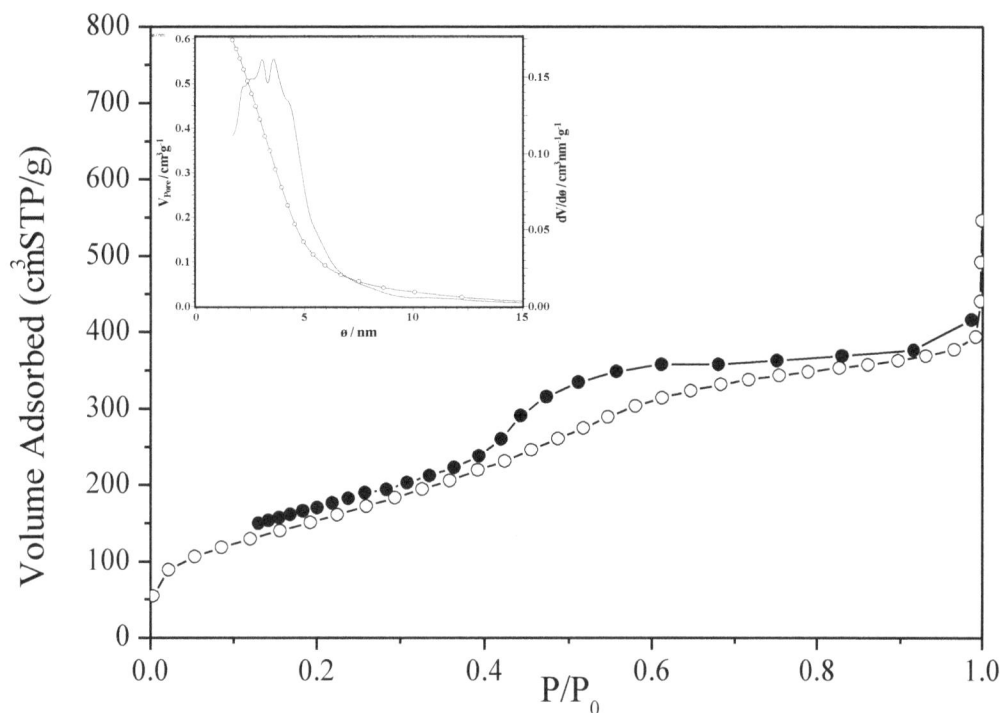

Figure 11. Nitrogen adsorption-desorption ($\circ$ and $\bullet$, respectively) isotherms of the ordered mesoporous SiBCN material. The pore size distribution curves are shown inset.

The Si–Al–C–N system is another interesting system. Indeed, the addition of Al to Si-based ceramics contributes to the improvement of their hydrothermal stability as illustrated through the addition of Al to Si/N/O systems forming Si/Al/O/N ceramics [243–245]. Similarly, An *et al.* reported that the addition of Al to Si/C/N(O) resulted into a non-parabolic oxidation curve (at $T \geq 1000$ °C) which decreased more rapidly with time, down to a negligible level [246]. Authors suggested that the remarkably low oxidation rates of these materials were attributed to the lower permeability of the formed oxide layer to molecular oxygen which resulted from the incorporation of aluminum in the silica network. This passivating Si/O/Al layer is shown to hinder diffusion-controlled oxidation in the bulk. Within this context, the preparation of periodic mesoporous silicon-aluminum-carbon-nitrogen (Si/Al/C/N) frameworks with *P6mm* hexagonal symmetry using mesoporous carbon (CMK-3) as hard template and preceramic polymers containing both $-[R_1R_2Si-N(R_3)]_n-$ and $-[R_4Al-N(R_5)]_n-$ backbones (with $R_1 = R_2 = R_3 = R_4 = H$ and $R_5 = CH_2CH_3$) as ceramic precursors was reported [247]. The preceramic polymers were prepared by blending poly(perhydridosilazane), PHPS, and poly(ethyliminoalane), PEIA, as precursors of silicon nitride/silicon (Si3N4/Si) and carbon-containing aluminum nitride (Al/C/N), respectively. The blended polymers with various and controlled Al:Si ratios were infiltrated into the porous structure of CMK-3 followed by a pyrolysis-template removal cycle performed under nitrogen at 1000 °C (2 h, ceramic conversion) then in an ammonia atmosphere at 1000 °C (5 h, template removal) (Figure 12).

Figure 12. Procedure to prepare ordered mesoporous Si/Al/C/N ceramics by a polymer building block approach.

This procedure resulted in the formation of periodic mesoporous Si/Al/C/N frameworks with surface areas of 182–326 $m^2 \cdot g^{-1}$, a pore size distribution of 4.1–5.9 nm and pore volumes in the range of 0.51–0.65 $cm^3 \cdot g^{-1}$. Amorphous materials did not exhibit weight change up to 1400–1470 °C in flowing nitrogen and their behavior in air up to 1000 °C (with dwelling time of 5 h) depended on the proportion of AlN and Si_3N_4 phases. Mesoporous materials are interesting because their pore size is similar to the dimensions of many molecules, which suggests that these materials could be potentially useful in separation, catalytic or nano-confinement processes (See Section 5.3). However, mesoporous PDCs are in general produced as powders which have some difficulties in practical use and as a consequence a limited application. Practical applications require that the mesoporous material is available in macroscopic form such as monolith.

The current technology for producing porous non-oxide ceramic monoliths involves extrusion or pressing powders together with sacrificial and sintering additives into an engineering shape, removal of all sacrificial additives and finally sintering at high temperature. Sintering additives are usually added to impart high mechanical strength. As an alternative, the elaboration of monolith-type mesoporous PDCs can be achieved by impregnation of silica or carbon foams [248]. As a better alternative, the elaboration of ordered mesoporous powders through the PDCs route may be combined with an approach that adopts the convenience and flexibility of powder-based processes such as spark plasma sintering (SPS). This has been performed on ordered mesoporous Si/B/C/N powders displaying *P6mm* hexagonal symmetry and their processing led to hierarchically porous Si/B/C/N monoliths through SPS without the use of any sintering additives (Figure 13) [249,250].

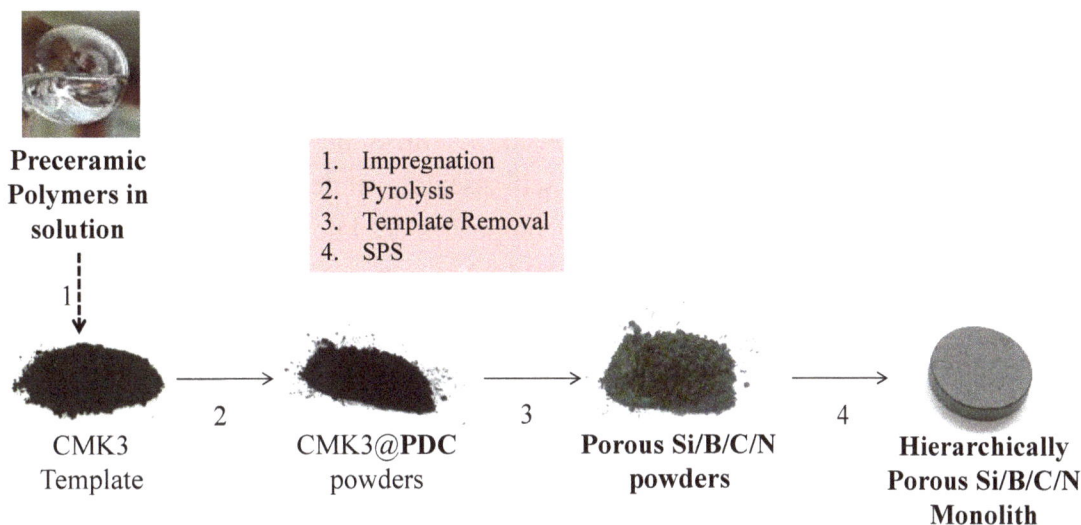

Figure 13. Overall synthetic path employed to generate hierarchically porous Si/B/C/N monoliths coupling the PDCs route with nanocasting and spark plasma sintering (SPS) processes.

The coupled approach allowed obtaining robust monoliths with surface areas of 123–171 $m^2 \cdot g^{-1}$, mesopore diameters of ca. 6.2–6.5 nm and total pore volumes varying from 0.25 to 0.35 $cm^3 \cdot g^{-1}$. The characteristics of the monoliths are related to the use of ordered mesoporous powders as starting materials and to the control and the tailoring of the pore size and the connectivity over a relatively wide range of length scales through the parameters of sintering. SiBCN monoliths displayed porosities from 59% to 69%, a meso-/macroporosity which differed from starting powders. The use of SPS inhibited grain growth during sintering leading to materials which retained the intrinsic properties of the pristine powders.

4.3. Self-Assembly Strategies Based on Preceramic Block Copolymers and Particles

The self-assembly of polymers such as block copolymers or polymeric spherical particles is a feasible tool for the formation of nanostructured (hybrid) materials for a manifold of different potential applications. Compared to the previous section predominantly dealing with hard-templating methodologies, polymer-based templating strategies for the preparation of advanced ceramic are addressed within this section. Two self-assembly techniques of preceramic polymers are of special interest: (i) the self-assembly of block copolymers featuring a ceramic precursor as integral part of the polymer chain for at least one block segment and (ii) colloidal crystallization of polymeric spherical (hybrid) nano-particles. In the recent past, both concepts have attracted significant attention in order to generate ordered ceramic (composite) nano-structures.

Block copolymers which consist of two or more polymer segments covalently connected to each other are capable of undergoing microphase separation. Already in the case of two different block segments, various structures, e.g., spheres, cylinders, lamellae and co-continuous structures as well as porous structures at the nanoscale can be obtained only by variation of the volume fraction of underlying block segments [251–253]. On this account, the exploitation of block copolymer self-assembly in order to template inorganic materials has attracted significant attention. Besides polymeric parameters such as the overall molar mass or the polymer constitution, different other factors, e.g., temperature, solvent vapor or pressure strongly affect the microphase separation. Moreover, block copolymer assembly can be directed by the application of flow fields and electric or magnetic fields [254]. Recent strategies focus on guiding the self-assembly of such polymers on patterned substrates and in confinements [255,256]. Within the block copolymer self-assembly strategies, different approaches for the preparation of ceramic materials inclusively nanocomposites are known. Excellent reviews within that field are given, e.g., by Orilall and Wiesner [257] and other authors [258–262].

Although this review focuses on preceramic polymers, the self-assembly of block copolymers prior to the selective removal of one microphase-separated block segment will be exemplarily discussed focusing on the preparation of block copolymer templated ceramics.

Separated block domains of these nano-structures can be selectively removed. In a subsequent step, the residual nano-structure can be backfilled with inorganic precursors followed by etching or calcination in order to remove the second block copolymer segment. For example, Hsueh et al. [263] reported the preparation of bicontinous anatase (TiO_2) ceramics using this double-templating approach. The authors took advantage of the removal of the microphase-separated poly(L-lactide) segment by hydrolysis followed by backfilling of a titanium alkoxide precursor. Calcination yielded ordered and porous titanium oxide with photocatalytic properties. This route can also by applied for the preparation of porous silica materials featuring a very low refractive index [264]. A bicontinous microemulsion of polyethylene (PE), poly(ethylene-*alt*-propylene) (PEP), and poly(ethylene-*block*-ethylene-*alt*-propylene) (PE-PEP) has been advantageously used for the preparation of porous structures after selective removal of PEP [265]. Backfilling of the voids with poly(ureasilazane) (Ceraset) and thermal treatment yielded 3D continuous SiCN ceramics. The preparation of different mesoporous rare-earth oxide ceramics using evaporation-induced self-assembly of block copolymers has recently been reported [266].

As described above, the multi-step strategy based on the self-assembly of organic block copolymers followed by removal of one block segment is a powerful method. As another feasible route, the block copolymer *co-assembly* and hence a direct incorporation of preceramic polymers or inorganic particles will now be described.

Recently, Rauda and coworkers reported a general method for the production of templated mesoporous materials based on preformed nano-crystal building blocks [267]. Here, soluble diblock copolymers mixed with inorganic particles were capable of undergoing an evaporation-induced self-assembly (also referred to as EISA process) (Figure 14). After thermal treatment, nano-porous metal oxides, e.g., manganese oxides could be obtained [267]. Poly(dimethylsiloxane)-*block*-poly(ethylene oxide) (PDMS-*b*-PEO) has been used as template for the incorporation of methylphenylsiloxane (MPS) in order to fabricate porous MPS/PDMS composites featuring a high framework stability [268].

Block copolymer *co-assembly* of organic block copolymers and a preceramic polymer such as poly(ureamethylvinylsilazane) (PUS)—also referred to as Ceraset—is reported by Wiesner and coworkers [229,230,269]. There, the diblock copolymers poly(isoprene-*block*-dimethylaminoethyl methacrylate) or poly(isoprene-*block*-ethylene oxide) can be used as structure-directing agent for the commercially available silazane-based PUS for the preparation of high temperature SiCN ceramic materials [229,230,269]. The affinity of PEO to PUS has also been investigated by Wan et al. [270]

Upon pyrolysis of the microphase-separated PUS/block copolymer blend (polybutadiene-*block*-PEO), SiCN-based non-oxide ceramics could be obtained.

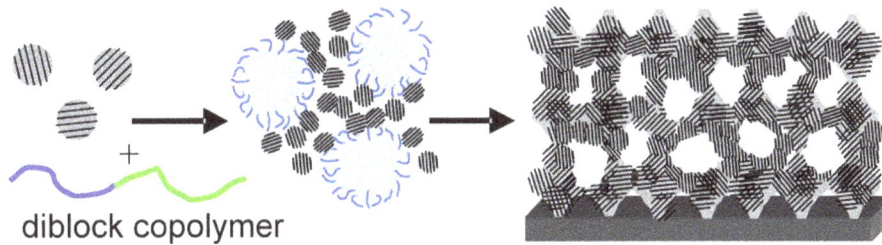

Figure 14. Schematic illustration for the preparation of porous metal oxides by using the evaporation induced self-assembly (EISA) of a dispersion consisting of diblock copolymers and inorganic particles in the first step followed by thermal treatment. By this general methodology mesoporous ceramics can be obtained (reprinted with permission from reference [267]; Copyright 2012 American Chemical Society).

Another efficient strategy for the controlled buildup of ceramic nanostructures focuses on the self-assembly and pyrolysis of preceramic/organic block copolymers wherein the inorganic block segment is directly converted into the ceramic material after thermal treatment. Compared to previously described methodologies, inorganic monomers or metal-containing organic monomers were polymerized in a controlled manner to gain access to well-defined block copolymers. The basic concept is schematically depicted in Figure 15 based on the work of Malenfant *et al.* [230]. The authors took advantage of the self-assembly and pyrolysis of poly(norbornene)-*block*-poly(norbornene decaborane) in which the borane-containing block segment acted as ceramic precursor. Thermal annealing of the block copolymer and pyrolysis at 400–1000 °C in the presence of ammonia was evidenced to maintain the pristine block copolymer morphology for the final boron nitride ceramic. Interestingly, cylindrical morphologies of the preceramic block copolymer could simply be transferred into a lamellar morphology by changing the solvent in the annealing step [230].

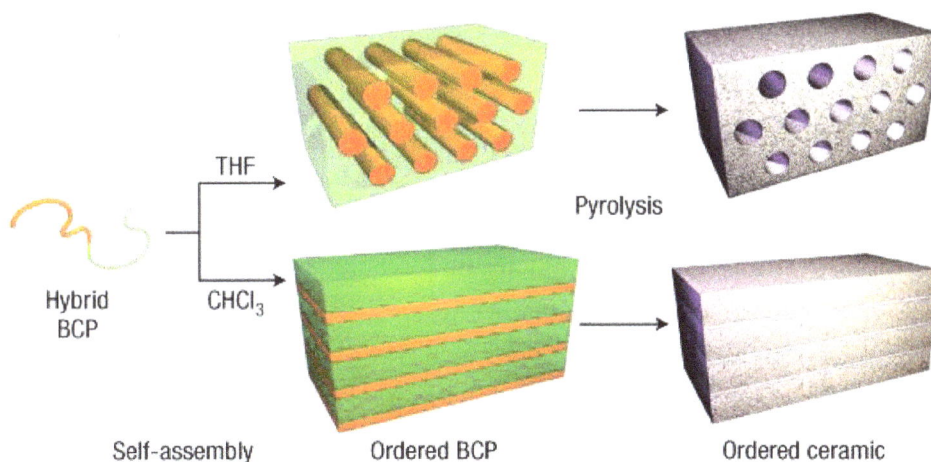

Figure 15. Conversion of a self-assembled hybrid block copolymer featuring an inorganic precursor segment into ordered ceramic structure by pyrolysis (reprinted with permission from Nature Publishing Group) [230].

Within the field of block copolymers featuring a preceramic polymer segment, especially silicon-based precursors have spurred much interest. Nghiem *et al.* [271] reported the synthesis of polycarbosilane-*block*-polystyrene by using living anionic polymerization protocols. Well-ordered SiC-based ceramic materials with a high content of micro-pores could be obtained. Polycarbosilane-based block copolymer micelles have been utilized after platinum-catalyzed cross-linking reaction for the preparation of silicon-containing ceramics [272]. Nguyen *et al.* [273] reported a feasible preceramic diblock copolymer consisting of acrylated poly(vinylsilazane)-*block*-poly(methyl methacrylate) (PVSZ-*b*-PMMA) as precursor for a SiCN-based ceramic patterning. Pyrolysis at 1200 °C yielded mesoporous SiCN ceramics with pore-sizes between 6 and 9 nm dependent on the applied cross-linking strategy.

Fascinating co-continous double gyroid and inverse double gyroid structures of block copolymer architectures consisting of polyisoprene and poly(pentamethylsilylstyrene) have been synthesized and used for the preparation of nano-porous or nano-relief structures as reported by Chan *et al.* [274]. Very recently, self-assembled block copolymers based on silsesquioxane-containing segments have been used for the preparation of ceramic-metal composites as reported by Li and coworkers [275]. For this purpose, platinum nanoparticles were added prior to the solvent-casting step for the phase separating block copolymer. UV-assisted ozonolysis both revealed maintenance of the hybrid block copolymer morphology as well as the formation of ordered inorganic nanocomposites [275]. Silicon oxy carbide nano-ring arrays have been produced after self-assembly and oxygen plasma treatment of polystyrene-*block*-poly(dimethylsiloxane) thin films [276]. The authors expect these interesting structures to be excellent mask materials for patterning applications.

Noteworthy, the usability of block copolymers is not limited for the formation of ceramic films or particles but also for the preparation of SiCN fibers. As an example, Pillai *et al.* [277] reported the reaction of high density polyethylene bearing a reactive hydroxyl moiety with a commercially available polysilazane (HTT1800). Self-assembly from different solvents either led to a cylindrical or lamellar morphology which could be converted into the corresponding SiCN ceramic fibers [277].

As already mentioned in Section 2.2, a manifold of iron-containing ceramics can be obtained by using metallopolymers. These metal-containing block copolymers can be used for the preparation of ceramic structures at the nano-scale as reported, e.g., by the group of Manners. Within that field, especially polyferrocenyldimethylsilane-based (PFS) block copolymers with the intrinsic capability of crystallization have shown a large potential as preceramic polymers replicating the micro-morphologies in the final ceramic material. Rider *et al.* [129] reported the formation of ordered nano-domains by self-assembling polystyrene-*block*-poly(ferrocenylethylmethylsilane) (PS-*b*-PFEMS) in the bulk state or in thin films. Magnetic arrays consisting of C/SiC-containing iron nano-particles were obtained after pyrolysis of the ferrocene-containing block copolymers. Ceramic surface-relief gratings (SRG) based on the ceramization of PFS block copolymers have been reported [278]. Ceramic SRGs are expected to be useful for templating and data storage materials. In earlier studies magnetic nano-lines, nanoscopic iron oxide patterns, separation tunable arrays and cobalt magnetic dot arrays were accessible by using PFS-based block copolymers as preceramic polymers [279–282].

Besides PFS-based block copolymers, also some other ferrocene-containing polymers based on methacrylate derivatives and vinylferrocene were capable of forming fascinating micro morphologies [283–285]. As an example of iron oxide as the final ceramic material, Tang and coworkers investigated the self-assembly and template synthesis of triblock copolymers with a poly(2-(methacryloyloxy)ethyl ferrocenecarboxylate) (PFcMA) block for the preparation of ordered iron oxides [135].

The self-assembly of preceramic polymer and core/shell architectures is another powerful method in order to generate well-defined 3-dimensional (hierarchical) ceramic materials after pyrolysis. It has to be borne in mind that 3D ceramic architectures are typically derived by previously mentioned technique, e.g., sol–gel and infiltration routes. Within this section, some examples for preceramic polymers or polymeric single-source precursors as integral part of colloidal architectures or spherical template route will be addressed. In this particular case, the self-assembly focusses on monodisperse particles featuring the intrinsic capability of colloidal crystallization. Based on such almost monodisperse polymeric colloidal micro- and nanoparticles, fascinating functional materials especially for optical applications can be obtained [286–290]. In general, colloidal crystals can be prepared from their particle dispersions by various techniques of deposition, spin coating or by using the melt-shear technique [291–293].

As mentioned above, PFSs were proven suitable preceramic polymers and also 3-dimensional iron-containing ceramics have been prepared from these precursors. Ozin and Manners *et al.* [294] have shown a gravity sedimentation route of silica spheres followed by thermal polymerization of a [1]silaferrocenophane monomer. Thermal treatment of the preceramic ferrocene polymer led to conversion into predominantly magnetic γ-Fe_2O_3. Well-defined magnetic ceramic opals with high structural periodicities were accessible by this method [294]. SiC inverse opals were accessible by a soft templating route using monodisperse polymeric spheres followed by infiltration with a silylene-acetylene preceramic polymer [295]. These remarkably stable SiC photonic crystals are expected to be feasible for optical applications in harsh environments.

5. Properties of Ceramic Nanocomposites

5.1. Structural Properties

5.1.1. High-Temperature Stability towards Crystallization and Decomposition

Silicon oxycarbides are amorphous ternary ceramics which can be described as consisting of a glassy network of SiO_xC_{4-x} tetrahedral [296,297] which formally can be considered as the result of the incorporation of carbon into silica glass, which has been shown in numerous studies to strongly affect its crystallization resistance at high temperatures [5]. Thus, silicon oxycarbides remain predominantly amorphous up to temperatures of 1300–1350 °C; whereas silica is prone to crystallization at temperatures as low as 1200 °C. The remarkable crystallization resistance of SiOC ceramics has been considered to rely on the incorporation of carbon into the amorphous network. Additionally, excess carbon (which, typically is present within the SiOC microstructure) has been also shown to affect the crystallization behavior of SiOC [297].

At temperatures beyond 1100 °C amorphous silicon oxycarbides undergo a phase separation process to form amorphous nanocomposites consisting of silica, silicon carbide and segregated carbon [53,298,299]. The phase separation in SiOC materials was extensively investigated by means of ^{29}Si MAS NMR as well as high resolution TEM (HR-TEM) and relies on rearrangement processes occurring within SiO_xC_{4-x} tetrahedra. ^{29}Si MAS NMR spectra of SiOC materials as prepared at temperatures of *ca.* 1000–1100 °C exhibit a mixture of SiO_4, SiO_3C, SiO_2C_2, $SiOC_3$ and SiC_4 sites, which reflects the situation present within the glassy SiOC network. However, upon annealing at higher temperatures, significant changes of the network occur. Thus, nearly exclusive SiO_4 and SiC_4 environments are observed, indicating the phase separation of the SiOC to deliver amorphous silica-rich phase beside amorphous silicon carbide nano sized precipitations and segregated carbon [53,300].

Subsequent to the phase separation process, the crystallization of the phase-separated silicon carbide takes place at higher temperatures in the SiOC materials. In silicon oxycarbides annealed at temperatures of *ca.* 1200–1300 °C, nano sized precipitations of β-SiC are dispersed within a silica-rich matrix. Interestingly, the amorphous silica-rich matrix remains amorphous up to high temperatures, (e.g., no cristobalite formation was observed); this relies probably on the low interfacial energy between β-SiC and the amorphous silica-rich phase [301]. In a recent study, the presence of cristobalite was observed upon long-term annealing SiOC-based ceramics at temperatures between 1300 and 1400 °C. Its formation was explained as a result of the presence (or generation) of inner surfaces within the material, which probably promotes the devitrification of silica [302]. Moreover, a case study indicates that probably the molecular architecture of the precursor might also affect the formation of cristobalite: linear polysiloxanes convert into SiOC materials with very high crystallization resistance; whereas cristobalite was detected in silicon oxycarbides derived from cyclic precursors [192].

Annealing of SiOC at temperatures exceeding 1350–1400 °C, leads to its gradual decomposition due to two decomposition processes: at temperatures of 1350–1400 °C the carbothermal reaction of the phase-separated silica with excess carbon occurs, and is accompanied by the formation of β-SiC and gaseous CO. At higher temperatures (*i.e.*, above 1500 °C), silica can react with SiC to gaseous silicon monoxide and CO, thus leading to a severe decomposition of SiOC [300,301].

The incorporation of additional elements into the network of silicon oxycarbides has a substantial effect on their decomposition and crystallization behavior. SiBOC ceramics were shown for instance to be less stable towards decomposition as compared to their boron-free counterparts and thus silicon carbide was found to crystallize SiBOC at lower temperatures [303–306].

The incorporation of transition metals such as Zr or Hf into SiOC was shown in several studies to drastically improve its high-temperature behavior with respect to decomposition and crystallization [6,201,203,307]. This relates to the fact that at temperatures exceeding 1400 °C (*i.e.*, in conditions which lead to the carbothermal degradation of SiOC) the MO_2 phase present within the microstructure of the metal-modified SiOC undergoes a solid state reaction with the phase-separated silica to generate crystalline $MSiO_4$. This was shown not only for SiZrOC and SiHfOC [201,203], but also for SiAlOC (crystallization of mullite at $T > 1300$ °C) [211], SiMnOC (crystallization of $MnSiO_3$ already occurs at $T \approx 1100$ °C) or SiLuOC ($Lu_2Si_2O_7$ crystallization at $T > 1300$ °C) [221]. The preferred reaction of silica with the metal oxide phase is responsible for the improved high-temperature stability of SiMOC (M = Zr, Hf) with respect to decomposition. Due to the fact that silica reacts with MO_2 to form $MSiO_4$, its carbothermal reaction with excess carbon which leads to the crystallization of β-SiC and to CO release (*i.e.*, mass loss) is suppressed [201,203].

Also in *silicon carbonitride*-based ceramics decomposition and crystallization correlate to the chemical composition, architecture and chemical homogeneity of the amorphous SiCN network [5].

The stability of polysilazane-derived SiCN ceramics is determined by the reaction of silicon nitride with excess carbon which leads to silicon carbide and nitrogen gas release at temperatures exceeding 1484 °C [308]. SiCN ceramics which do not have excess carbon in their microstructure show a thermal stability which is greatly improved as it is now limited by the thermal decomposition of silicon nitride (occurring in 1 bar N_2 atmosphere at $T > 1841$ °C [309]).

As already mentioned in Section 4.2, the incorporation of boron into the Si–C–N ceramic network was shown to significantly increase its thermal stability and crystallization resistance [310]. SiBCN ceramics remain amorphous up to *ca.* 1700 °C and show no significant decomposition up to 2000 °C [311]. Their extraordinary thermal stability is believed to rely rather on kinetic than thermodynamic reasons. Structural disorder in Si–B–C–N ceramics, which results in increased activation energies of both crystallization and carbothermal degradation processes, is thought to be responsible for the thermal stability of these materials. Furthermore, the presence of turbostratic $B_xC_yN_z$ phases within the microstructure of SiBCN are considered to kinetically stabilize the crystalline, phase-separated Si_3N_4 (which was still detected in SiBCN materials annealed at temperatures as high as at 2200 °C) and thus provide stable SiBCN composition up to very high temperatures [312,313]. The kinetics of the crystallization of Si_3N_4 in SiBCN ceramics [314] as well as the effect of the boron content on the crystallization behavior of SiBCN [315] was reported. Thermodynamic data of amorphous SiBCN ceramics determined by high-temperature oxide melt solution calorimetry were recently also reported and indicate that SiBCN might be considered as being thermodynamically stable with respect to the crystalline components SiC, Si_3N_4, BN and graphite [316]. As the thermodynamic stability of SiBCN decreases upon increasing the boron content, the effect of boron on the Si_3N_4 crystallization is thought to be exclusively a kinetic effect.

The incorporation of metal such as Al [246,317], Y [72,76], Ti [73], Zr [61,62], or Hf [6,74,318] into SiCN was reported in several studies. Amorphous SiMCNO ceramics are obtained via pyrolysis of metal-alkoxide-modified polysilazanes and undergo phase separation processes at high temperatures to generate amorphous metal oxide nanoparticles dispersed within glassy SiCNO matrix. Annealing at higher temperatures leads to the crystallization of the metal oxide nanoparticles.

5.1.2. High Temperature Oxidation and Corrosion Behavior

Polymer-derived ceramic nanocomposites have been investigated in the last two decades concerning their high temperature oxidation behavior. Typically temperatures exceeding 1000 °C and oxidizing environments such as air or combustion atmosphere are applied. Within this context ternary and multinary materials based on the Si–M–O–C and Si–M–C–N–O systems (M = B, Al, Zr) were tested and were shown to exhibit passive oxidation behavior [319], thus they can be considered as behaving to some extent like silica formers, e.g., silicon, metal silicides, SiC or Si_3N_4.

Polymer-derived SiC ceramics exhibit regular (passive) oxidation behavior in pure oxygen or dry air atmosphere, *i.e.*, a parabolic growth of a pure silica layer in the temperature range from 800 to 1400 °C. The activation energy of the process was found to be *ca.* 100 kJ/mol, indicating the inward diffusion of oxygen through the growing silica layer (accompanied by the outward CO diffusion) as the rate-limiting mechanisms. This behavior is similar to that observed for silicon or pure silicon carbide. The amorphous silica layer crystallizes into cristobalite at temperatures above 1200 °C; however, the formation of cristobalite was shown to not significantly affect the oxidation rates [320,321], despite the diffusion of oxygen through cristobalite is lower than that through glassy silica [322].

SiOC ceramics exhibit parabolic oxidation behavior [323], with oxidation rates strongly depending on the content of the segregated carbon present within their microstructure. Thus, SiOC ceramics with high free carbon content oxidize consequently faster. However, the activity of the free carbon in SiOC is considered to be less than unity; therefore, passive oxidation in the Si–O–C system is possible even for the case of carbon-rich compositions.

Studies on the oxidation behavior of dense SiCNO ceramic materials revealed the formation of a dense and continuous oxide layer with a sharp oxide/SiCN interface and parabolic kinetics from 800 °C to 1400 °C [320]. The parabolic constants and the activation energies were found to be similar with those obtained for silicon carbide and silicon nitride [319].

In the case of SiBCN-based ceramics, exceptionally low oxidation rates were reported [324,325]. However, there are several aspects which were not systematically considered and which probably led to a strong underestimation of the oxidation rates: formation of low-viscosity borosilicate passive layer, volatilization of boron (sub)oxides, *etc.* Moreover, the role of the B incorporation within SiCN on its high-temperature oxidation behavior is not well understood so far. The addition of aluminum (SiAlBCN) [326] or hafnium (SiHfBCN) [327] was shown to negatively affect the oxidation behavior of SiBCN, *i.e.*, higher oxidation rates were reported.

The modification of Si–C–N–(O)-based PDCs with aluminum results in a non-parabolic oxidation behavior at T of *ca.* 1000 °C. At $T = 1400$ °C, a stable parabolic behavior was observed for $t > 20$ h, with oxidation rates about one order of magnitude lower than those reported for Al-free Si–C–N [246,328,329] This behavior was considered to be a consequence of the formation of an Al-containing silica scale, which induces a strong decrease of the oxygen diffusivity if compared to that of the Al-free silica [330].

The Zr addition into Si–C–N–O was shown to reduce the parabolic oxidation rate as compared to those of Si–C–N–O and was related to both the lower content and the lower activity of the segregated carbon in the Zr-containing system [62].

Recently, oxycarbide- and carbonitride-based PDC-NCs were studied with respect to their hydrothermal corrosion behavior [78,181].

Silicon oxycarbide-based ceramics exhibit an active corrosion behavior over the whole investigated temperature range (up to 250 °C), *i.e.*, silica being leached out of the samples. However, the corrosion rates of the SiOC ceramic materials were found to be remarkably lower than those of silicon carbide and were comparable to values reported for silicon nitride. Thus, a corrosion rate of 0.13 mg·cm^{-2}·h^{-1} was determined for the SiOC sample upon corrosion at 250 °C, being *ca.* 5 orders of magnitude smaller than the rates obtained for SiC ceramics corroded under similar conditions. This fact might be related to the presence of a relatively high amount of segregated, "free" carbon within the microstructure of the SiOC sample (*ca.* 37 mol%), which is not affected by hydrothermal corrosion. Interestingly, Zr- and Hf-incorporation within SiOC was shown to lead to a significant improvement of its corrosion resistance. This was attributed to the presence of the ZrO$_2$/HfO$_2$ phase within the PDC-NCs, which exhibits an extremely low solubility in water under the investigated conditions. Thus, the finely disperse oxide nanoparticles acts as a "reinforcing" phase with respect to hydrothermal corrosion and are most probably the reason for the significant improvement in the corrosion resistance of

SiZrOC/SiHfOC as compared to that of SiOC. Interestingly, the SiOC matrix was found to effectively suppress the corrosion-induced phase transformation of the tetragonal ZrO_2/HfO_2 phase into monoclinic ZrO_2/HfO_2, which is a well-known problem in the case of zirconia and hafnia materials exposed to hydrothermal conditions. Thus, the outstanding hydrothermal corrosion behavior of the investigated Zr-/Hf-containing SiOC ceramic nanocomposites relies on a unique synergistic effect between the reinforcing role of the ZrO_2/HfO_2 phase and the protection of the tetragonal oxide precipitates from corrosion-induced phase transformation through the SiOC matrix [181].

Ceramic matrix composites consisting of carbon fibers embedded within a SiCN and SiHfBCN matrix were also investigated concerning their behavior in hydrothermal conditions [78]. The preparation of the CMCs was performed via polymer infiltration and pyrolysis process (PIP) of 2D carbon fabrics with a polysilazane and an Hf- and B-modified polysilazane, respectively. Interestingly, it was observed that the C_f/SiCN CMCs exhibits a weak fiber-matrix interface; whereas the interface between the carbon fibers and the SiHfBCN matrix was strong. Consequently, the mechanical behavior of the C_f/SiCN samples was shown to be affected by the incorporation of Hf and B. The hydrothermal corrosion of the prepared CMCs revealed that C_f/SiHfBCN samples exhibit better resistance as compared to C_f/SiCN due to the improved kinetics upon Hf and B incorporation. Additionally, a tight C_f/matrix interface (which is rather disadvantageous for appropriate mechanical behavior) was found to be beneficial for an improved corrosion behavior in C_f/SiHfBCN [78].

5.1.3. High Temperature Creep Behavior

Polymer-derived ceramics and ceramic nanocomposites were extensively investigated in the last 15 years with respect to their (thermo)mechanical properties. Especially, their high-temperature creep behavior was shown to be outstanding, thus this class of ceramic nanocomposites show near-zero steady state creep even at temperatures exceeding 1000 °C [5].

The creep behavior of silicon oxycarbides was shown to rely on viscous flow, as it is also the case for silica glasses [331–335]. As silicon oxycarbides can be formally considered as silica-based glass having carbon incorporated within the glass network, it was expected that their mechanical properties and refractoriness will be improved if compared to silica [332,333,335]. This effect was expected as a consequence of formally replacing bivalent oxygen atoms within the glassy network with tetravalent carbon atoms, as it was for instance observed in SiAlON glasses (in which bivalent oxygen atoms are partially substituted by trivalent nitrogen atoms) [336]. As a consequence of nitrogen incorporation within the silica network, an increase of the elastic moduli with 30% was observed upon replacing one out of five oxygen ions with nitrogen [337].

Studies on the high-temperature apparent shear viscosity of SiOC materials (from bending or compression creep experiments) indicate that at a specific temperature, the viscosity of SiOC glasses is several orders of magnitude higher than that of amorphous silica. This consequently indicates that the glass transition temperatures for SiOC glasses are significantly higher than that of vitreous silica. Thus, silicon oxycarbides have glass transition temperatures in the range of 1300 to 1350 °C, far beyond the glass transition temperature of vitreous silica (1190 °C). At temperatures exceeding 1350 °C, a creep hardening effect was observed and related to crystallization processes [333].

The unique high-temperature creep behavior of silicon oxycarbides was assumed to rely not only on the refractoriness of the carbon-containing glassy network $Si_xO_yC_z$, but also on their nano/microstructure.

A recent study considers SiOC materials consisting of two continuous (and interpenetrating) phases, *i.e.*, silica and carbon. Thus, silica is considered to be continuous and "embedded" within a continuous carbon "skin". This microstructure consideration was shown to be able to rationalize not only the creep rates of SiOC-based materials, but also the activation energies of creep, which were shown to be strongly affected by the nanodomain size of silica (*i.e.*, the "mesh size" of the carbon network within the microstructure of SiOC) [335]. In the studied temperature range (1000–1300 °C), the mechanism of creep was modeled with the Jeffreys viscoelastic model [338] and supported the proposed microstructural model in SiOC. Thus, two rheological contributions were identified: (i) a high viscous answer, coming from the silica rich network; and (ii) an elastic response from the segregated carbon phase within the samples. Moreover, two distinct effects of the carbon phase on the HT creep behavior of SiOC were identified: the effect of the carbon presence within the SiOC network (the "carbidic" carbon), which was shown to significantly increase the viscosity and in the same time to strongly decrease the activation energy for creep, as compared to vitreous silica; additionally, the segregated carbon phase (the "free" carbon) was shown to affect the viscosity and the activation energy of creep in SiOC and to the creep behavior in phase-separated silicon oxycarbides.

From the HT creep study on SiOC materials it was concluded that single-phase SiOC glasses exhibit relatively large T_g values (1350–1400 °C) and rather low activation energies for creep (*ca.* 280–300 kJ/mol) and consequently they are materials of choice for near-zero-creep applications at HT (see Figure 14). However, they suffer upon long-term exposition to temperatures beyond 1000 °C from phase-separation processes. Phase-separated SiOC glasses were shown to have significantly lower T_g values and larger activation energy for creep than those of their single-phase counterparts. Their creep behavior can be significantly improved by incorporation of segregated carbon. Interestingly, small contents of segregated carbon are sufficient to lead to phase-separated SiOC samples with similar T_g and E_a values as compared to single-phase silicon oxycarbide (Figure 16). Thus, if high-temperature applications are anticipated (*i.e.*, above 1000 °C), SiOC compositions containing segregated carbon are mandatory in order to provide an improved creep resistance [338].

It was shown that the modification of SiOC with additional elements (such as Al, Zr, Hf) induces an increase of the creep rates (*i.e.*, decrease of the shear viscosity). This was explained as a consequence of the strong decrease of the content of the segregated carbon phase [335]. Additionally, in the case of SiZrOC and SiHfOC, the modification of SiOC with Zr/Hf was demonstrated to lead to a significant increase of the activation energy for creep (from 286 kJ/mol for SiOC to 386 and 476 kJ/mol for SiZrOC and SiHfOC, respectively; see Figure 17) [335].

Silicon carbonitride ceramics (SiCN) were also studied with respect to their creep at high temperatures [339–341]. They show similar creep behavior and shear viscosity values as compared to those determined for SiOC materials. Also for SiCN a creep hardening behavior was observed and considered to rely on a nanoscale densification creep mechanism [340]. In comparison to SiOC and SiCN ceramics, SiBCN-based materials are much more refractory with respect to creep [341–343]. For SiBCN it was not possible to determine the glass transition temperature, as for instance the shear viscosity of SiBCN at 1450 °C ($\eta \approx 10^{15}$ Pa·s) was still significantly higher than the typical value observed for T_g (10^{12}–$10^{12.6}$ Pa·s). The T_g of SiBCN is expected to be *ca.* 1600 °C. Some case studies revealed that a thermal annealing of SiBCN prior to creep significantly improve their behavior [343]. However, the creep mechanisms in SiBCN (also generally in PDC-based nanocomposites though) are still poorly understood and have to be investigated in more detail.

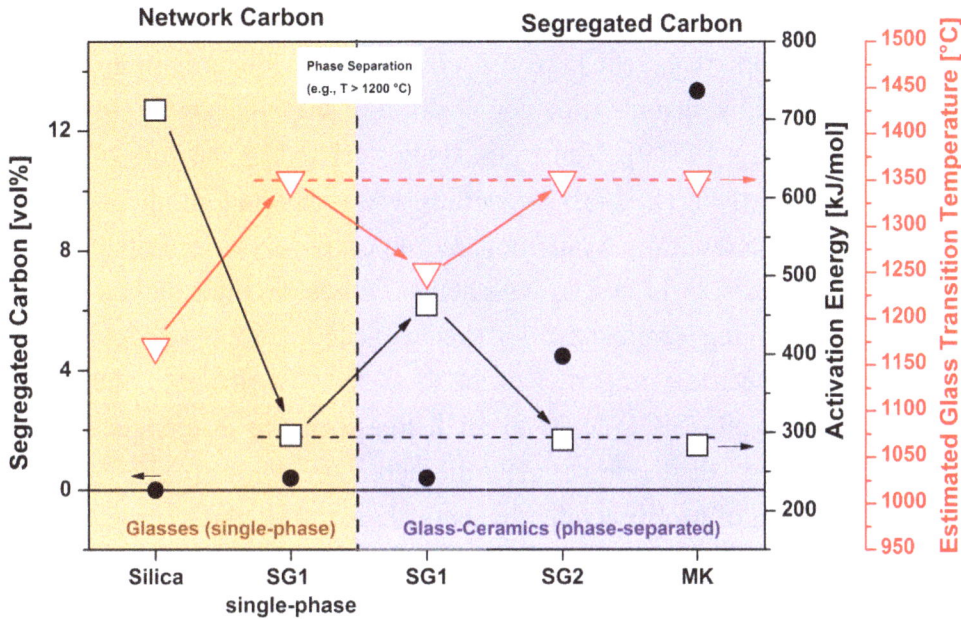

Figure 16. Dependence of the activation energy (red triangles) and T_g (black squares) on the content of segregated carbon (filled circles) in silicon oxycarbides (reprinted with permission from Wiley) [338].

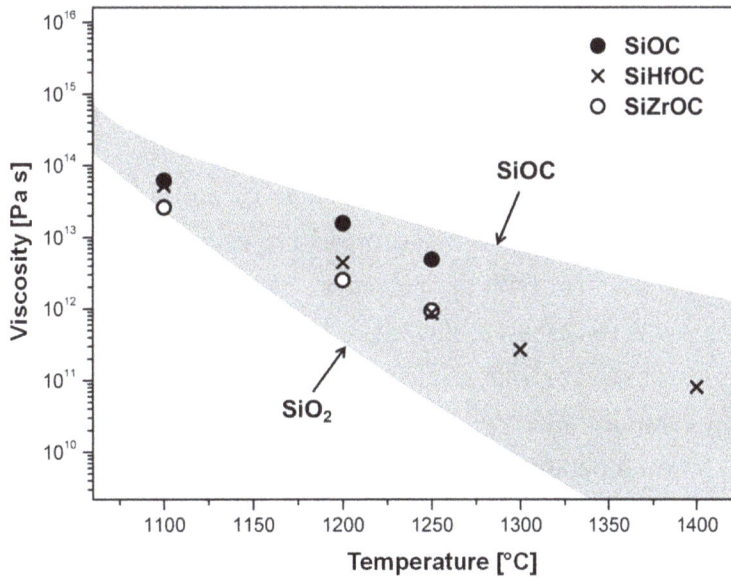

Figure 17. Viscosity of SiOC, SiZrOC, and SiHfOC as a function of temperature. The upper border of the gray region represent the temperature evolution of the viscosity of SiOC; whereas the lower border describes the evolution of the viscosity for vitreous silica (reprinted with permission from Wiley) [335].

5.2. Functional Properties

5.2.1. Electrical Properties

Amorphous PDCs in the systems Si–O–C and Si–C–N exhibit electrical conductivities with values between those of semiconductors (e.g., SiC) and insulators (e.g., silicon nitride). The electrical properties of PDCs can be tuned by changing the composition of the preceramic polymers as well as by altering the annealing conditions [319]. Numerous studies related to the role of the free carbon phase present in PDC-NCs on their electrical properties were reported in the last two decades. According to most studies, the ceramic materials obtained upon pyrolysis at low temperatures (i.e., ca. 1000 °C) are insulators; whereas high pyrolysis temperatures or high carbon content lead to SiOC-/SiCN-based materials having semiconducting behavior [173,185,344,345]. The addition of fillers can drastically affect the electrical properties of the PDC-NCs. For instance, the incorporation of molybdenum disilicide (MoSi$_2$) into SiOC leads to an increase of the dc conductivity with up to 14 orders of magnitudes. For high disilicide contents the behavior of MoSi$_2$/SiOC was found to be metallic, indicating the formation of MoSi$_2$ conductive paths [345]. Also the electrical conductivity of SiCN was shown to increase with increasing pyrolysis/annealing temperature and with the content of segregated carbon [346,347]. AC impedance spectroscopy studies of SiCN/CNT nanocomposites showed that the addition of CNTs remarkably increases their conductivity as compared with that of pure SiCN. The conductivity of SiCN (ca. 10^{-9} S/cm) indicate that the SiCN matrix can be considered as an insulating material; whereas the addition of ca. 1 vol% multi-walled CNTs induced an increase of the conductivity with 5 orders of magnitude higher (10^{-4} S/cm), indicating a low percolation threshold [348].

Boron incorporation into the SiCN system leads to a significant increase of the conductivity. The room temperature conductivity of SiBCN was 4 orders of magnitude higher than that of SiCN. The SiBCN ceramics annealed at high temperatures exhibited p-type conductivity which was related to a compensation mechanism involving carrier generation from both nitrogen and boron [349,350].

Recently it was shown that PDC-NCs exhibit unusually high piezoresistive coefficients (up to values of ~10^3 as for SiCN-based materials [351]). The piezoresistive effect obeys the tunneling-percolation model [352] which is in good agreement with the formation of conductive graphene-like sheets within the PDC matrix. Also silicon oxycarbide-based PDC materials were found to show piezoresistive effect, with strain sensitivities (k factors) of ca. 145 (Figure 18) [353]. However, the piezoresistive effect in SiOC was observed only in ceramics annealed at a rather high temperatures (i.e., 1400 °C), whereas the ceramics pyrolyzed at lower temperatures (1100–1300 °C) are not piezoresistive [353,354].

In two recent papers the high-temperature evolution of the piezoresistive effect in PDC-NCs was reported [355,356]. The determined values of the k factor for SiOC and SiOCN are significantly higher than those of well-ordered carbon and were found to decrease with increasing temperature. Rather high k values (k ~ 10^3) have been reported for C/SiOCN in the temperature range 700 °C < T < 1000 °C [355]. Whereas the gauge factor of C/SiOC was determined in the range of ca. 10^2 for temperatures from 1000 to ca. 1400 °C, its temperature dependence indicates a direct correlation with activated electronic transport ($E_A \geq 0.3$ eV for k) [356]. The values of k at high temperatures for SiOC and SiOCN indicate their potential as robust materials for piezoresistive force and pressure sensing under extreme environments.

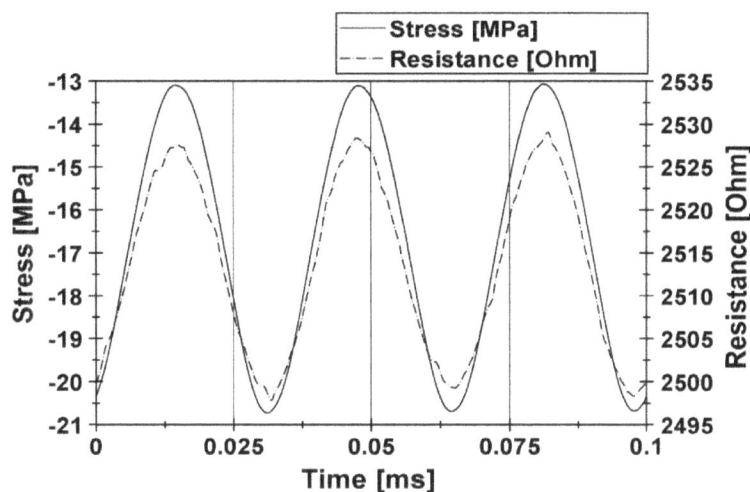

Figure 18. Piezoresistive response in silicon oxycarbide shown as the change of the electrical resistance upon applying a mechanical loading on the sample (reprinted with permission from Wiley) [353].

5.2.2. Magnetic Properties

Several preparative methods for the synthesis of iron-containing PDC-NCs with interesting magnetic properties were reported. For instance, various approaches were reported in the last decades to synthesize iron silicide-based ceramic nanocomposites from tailored single-source precursors [357]. The synthetic principle involves the transformation (usually via heat treatment) of suitable single-source precursors into single-phase amorphous materials which subsequently phase separate and crystallize upon formation of ceramic (nano)composites containing iron silicides dispersed within their microstructure [154]. The single-source-precursor synthesis techniques have several advantages when considering the preparation of ceramic monoliths with tailor-made/tunable properties: By adjusting the molecular structure/architecture of the single-source precursor as well as its Fe and Si content and using appropriate pyrolysis conditions, ceramic composites with tunable magnetic properties can be prepared. For instance, the magnetic properties of SiFeC-based ceramic composites prepared upon ring-opening polymerization followed by cross-linking and pyrolysis of a [1]silaferrocenophanes were shown to be tunable between a superparamagnetic and a ferromagnetic state (see also Sections 2.2 and 4.2.). This was demonstrated to depend on the size of the magnetic nanoprecipitations in the ceramic materials [120]. Additionally, the single-source-precursor technique allows for producing ceramic monoliths, coatings, fibers, *etc.* without additives at relatively moderate temperatures [6].

One convenient access to iron silicide-based ceramic nanocomposites involves the use of polymeric precursors which are modified with suitable iron-containing compounds prior to their conversion into the ceramics (see Table 1). In the case of polycarbosilane-derived ceramic composites, the pyrolysis of iron nitrate- [358] or acetylacetonate-modified precursors [359] leads to iron silicide-based ceramic composites, whereas an iron carbonyl-modified polycarbosilane was shown to convert into Fe/SiC ceramic nanocomposite [360].

Modification of a PSZ with Fe_3O_4 [361] or $FeCl_3$ [362] followed by pyrolysis leads to Fe/SiCN nanocomposites, whereas the reaction of PSZ with $Fe(CO)_5$ [215] or Fe powder [216] produces

$Fe_3Si/SiCN$ composite materials upon pyrolysis. In an additional study, the ceramization of a ferrocene/hexamethyldisilazane-derived copolymer was shown to deliver Fe_3C/Si_3N_4 composites [363].

Table 1. Selected examples of Si–Fe–C–X ceramics (X = O, N) derived from iron-modified preceramic polymers.

Ceramic material	Si-containing precursor	Fe-containing precursor	Pyrolysis conditions	Fe-containing crystalline phase(s)	References
Si–Fe–C–(O)	Polycarbosilane	$Fe(NO_3)_3$	1000–1200 °C (Ar)	Fe_3Si	[358]
		$Fe(acac)_3$	1300 °C (Ar)	Fe_5Si_3 (major), Fe_3Si, FeC	[359]
		$Fe(CO)_5$	1000 °C (N_2)	Fe	[360]
Si–Fe–C–N–(O)	Polysilazane	Fe_3O_4	1000 °C (N_2)	Fe (major), Fe_2SiO_4	[361]
		$FeCl_3$/THF	1000 °C (N_2)	Fe_3Si	[362]
		$Fe(CO)_5$	1100 °C (Ar)	Fe_3Si	[215]
		Ferrocene	1200 °C (Ar)	Fe_3C	[363]
		Fe (powder)	1100 °C (Ar)	Fe_3Si	[216]
Si–Fe–O–C	Polysiloxane	$FeCl_2$	1250 °C (N_2)	Fe_3Si	[364]
			1250–1300 °C (Ar)	Fe_3Si, Fe_5Si_3	
		$Fe(acac)_3$	1100 °C (Ar)	Fe_3Si	[219,221]

Whereas several reports on the synthesis and magnetic properties of Fe/SiC(N) and Fe_3Si/SiC(N) are known, there is not much information about the preparation and properties of ceramic composites consisting of crystalline iron silicide dispersed within a silicon oxycarbide matrix. Recently, the polymer-to-ceramic transformation of a $FeCl_2$-modified polysiloxane was reported. Upon pyrolysis in nitrogen atmosphere the crystallization of Fe_3Si along with Si_2N_2O and Si_3N_4 was found, whereas in argon atmosphere, Fe_3Si, Fe_5Si_3 and β-SiC were formed [364]. Moreover, a study concerning the polymer-to-ceramic transformation of an iron-acetyl-acetonate-modified polysilsesquioxane indicates that probably the single-phase SiFeOC glassy materials obtained at moderate temperatures partitions into FeO and SiOC. As FeO is not stable under the strong carburizing conditions of the pyrolysis process, it converts *in situ* into Fe_3Si. Interestingly, Fe_3C was identified as intermediary phase prior to Fe_3Si formation [219,221]. At temperatures above 1300 °C the amount of Fe_3Si decreased and the crystallization of Fe_5Si_3 and β-SiC was observed. This behavior, which was observed also in the Si–Fe–C–N system, probably relies on the formation of a liquid Fe/Si/C alloy which, depending on the Fe:Si stoichiometry, induces either carbon segregation or SiC crystallization (SLS mechanism) [216]. In the case of SiFeOC and SiFeCN ceramics, the presence of a silicon-rich liquid alloy leads to the crystallization of silicon carbide and Fe_5Si_3 [216,219].

Also cobalt-containing PDC-NCs can be obtained from the complexations of the acetylene moieties of hyperbranched polyynes with cobalt carbonyls $Co_2(CO)_8$ and $CpCo(CO)_2$ [365]. The inorganic-organic hybrid materials are thermally converted into nanostructured cobalt-containing ceramics. The resulting ceramics are highly magnetizable and show near-zero remanence and coercivity. Co-containing SiOC ceramics were prepared via cross-linking and pyrolysis of cobalt phthalate/polymethylphenylsiloxane blends [366]. Cobalt-modified SiCN materials were prepared from a cobalt carbonyl $Co_2(CO)_8$ modified polysilazane upon cross-linking and pyrolysis in inert atmosphere [215].

5.2.3. Mechanical Properties

As previously described, the polymer-derived ceramic route provides a way to fabricate bulk silicon-based ceramics without the use of sintering additives either by plastic forming of preceramic polymers followed by pyrolysis [367–369] or by direct pyrolysis of polymers followed by spark plasma sintering process [370]. The room temperature mechanical properties, *i.e.*, hardness and elastic modulus, are key properties to proof the general applicability of these materials in particular in comparison to standard dense, sintered polycrystalline Si_3N_4 and silicon carbide (SiC). Numerous studies on the micro- and nanoindentation hardness of binary and ternary PDCs have been reported [332,371–378]. Nanoindentation test is one of the most effective and widely used methods to measure the mechanical properties of these materials. This technique uses the same principle as microindentation, but with much smaller probe and loads, so as to produce indentations from less than a hundred nanometers to a few micrometers in size. Galusek *et al.* [375] reported the hardness and elastic modulus of nearly dense cylindrical silicon carbonitride (SiCN) ceramics with a formal chemical composition of $Si_{1.00}C_{0.67}N_{0.80}$ and bulk density of 2.32 g/cm^3 produced by pyrolysis of a warm-pressed poly(hydridomethyl)silazane green body. It was reported that the mean hardness of the SiCN ceramics (13 ± 2 GPa) measured at a load of 250 mN was approximately half of the typical value of polycrystalline Si_3N_4 (24.9 ± 0.6 GPa) according to the presence of microscale defects and structural inhomogeneities (nanopores and clusters of free carbon within the ceramics). However, SiCN ceramics are harder and stiffer than SiO_2 glass. SiCN ceramics display a similar behavior in mechanical properties than SiO_2 glass most probably due to their global amorphous structure. Presumably because of the stronger SiOC and SiON covalent bonding, SiCN ceramics are both harder and stiffer than SiO_2. Later, Cross *et al.* [379,380] studied the mechanical and tribological behavior of polyureamethylvinylsilazane-derived $SiC_xO_yN_z$ ceramics. Specimens in form of coupons were prepared by hot isostatic pressing up to different temperature from 800 to 1400 °C in nitrogen overpressure. All samples were X-ray amorphous and showed an increase of both the elastic modulus from ~100 to 184.0 GPa and the hardness from ~12 to 21.0 GPa with the increase of the pyrolysis temperature. Both values correlated strongly with the density of the specimens and the N/O ratio. More recently, nanoindentation technique was applied to polymer-derived ceramic TiN/Si_3N_4 [77]. Hardness values of 20.4 ± 1.7 GPa and elastic moduli of 171 ± 23 GPa have been measured on cylindrical specimens produced by warm-pressing at 110 °C of a titanium-containing polysilazane with a controlled Si:Ti ratio under different pressure followed by pyrolysis under ammonia up to 1000 °C and further heat-treatment under nitrogen up to 1300 °C [77]. However, since nanocomposites represent heterogeneous specimens in which very small nano-crystals grow, microindentation has been preferred. The Vickers hardness was found to reach 25.1 ± 4.0 GPa whereas a maximum Young's modulus of 183.3 ± 25.9 GPa was measured. The values are closely related to the warm-pressing pressure which is applied during the plastic forming of titanium-containing polysilazanes (Figure 19): Both hardness and Young's modulus gradually increase with the warm-pressing pressure from 74 to 162 MPa, then, mechanical properties decreased most probably due to the fact that the green sample has no open porosity, avoiding gaseous by-products to escape during pyrolysis generating cracks in the materials.

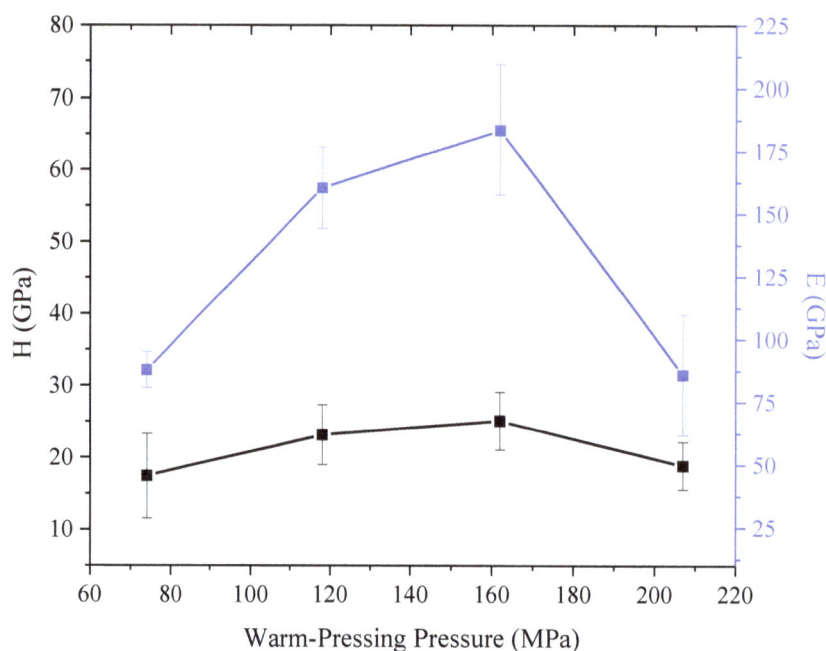

Figure 19. Evolution of the hardness and Young's modulus of TiN/Si$_3$N$_4$ nanocomposites as functions of the applied warm-pressing pressure.

The high hardness of these nanocomposites is clearly related to the *in situ* controlled growth of the TiN nanophase in the amorphous Si$_3$N$_4$ matrix and the generation of very small TiN nano-crystals homogeneously distributed in the matrix.

SiBCN monoliths prepared by Spark Plasma Sintering of amorphous polymer-derived SiBCN powders under nitrogen exhibited Vickers hardness values which are closely related to the sintering temperature as well as the proportion of carbon in the powders [185]. Starting from Si$_{3.0}$B$_{1.3}$C$_{4.1}$N$_{1.9}$ powders, the mechanical properties of the derived monoliths increased from the sample sintered at 1500 °C to the sample sintered 1850 °C and remained almost constant in specimens sintered at 1850 and 1900 °C. A maximum Vickers hardness of 5.4 ± 1.2 GPa was measured whereas a maximum Young's modulus of 102 ± 5 GPa was for samples sintered at 1900 °C. The modulus (and the hardness) is strongly correlated to the density of the specimens. The interesting point here is that the modulus increased linearly with the density of the sintered materials. Vickers hardness showed an increasing trend with the increase in processing temperature. This is mainly due to the lower residual porosity in the structure at higher SPS temperature resulting in the increase of the hardness. Furthermore, mechanical properties showed a significant variation when the raw powders exhibit lower carbon content. Hardness increased by to 15 ± 2 GPa and the Young's modulus reached 150 ± 5 GPa for specimens obtained by SPS of Si$_{3.0}$B$_{1.3}$C$_{0.6}$N$_{7.1}$ powders at 1800 °C. This clearly shows that free carbon plays a prominent role with regard to the mechanical properties of polymer-derived (nano)composites. It could be concluded that ceramic closed to the "Si–B–N" composition (mainly composed of Si$_3$N$_4$ and BN) exhibit significantly higher hardness than those close to the "Si–B–C–N" system (composed of Si$_3$N$_4$ and SiC nanophases embedded in a turbostratic BN(C) matrix).

5.3. Applications of Ceramic Nanocomposites

5.3.1. Ceramic Membranes

Microporous precursor-derived ceramic membranes for gas separation purposes were prepared in the recent past and they were shown to exhibit relatively high gas permeances and a good stability at high-temperatures which make them superior to polymer-based membranes concerning for instance dehydrogenation of hydrocarbons or steam reforming reaction for hydrogen synthesis [381]. Moreover, the single-source precursor route provides access to tailor-made micro-/meso-porous structure development [382].

Microporous membranes based on amorphous were prepared on permeable alumina porous supports having graded and layered porous structure [383,384]. The gas permeance for small molecules (such as He or H_2) measured for amorphous silica-based membranes deposited on mesoporous anodic alumina capillary were significantly higher than those for bulkier gas molecules, indicating their potential in gas separation applications [385]. Amorphous silica-based membranes synthesized upon air pyrolysis of a polysilazane deposited on porous silicon nitride substrate revealed a hydrogen permeance of 1.3×10^{-8} mol·m^{-2}·s^{-1}·Pa^{-1} at 300 °C and a H_2/N_2 selectivity of 141, which is comparable with the permselectivity of other amorphous silica- or silicon oxycarbide-based membranes [386].

Also amorphous Si–C, Si–N, as well as Si–C–N and Si–B–C–N based ceramic membranes were prepared from appropriate single-source precursors. The possibility of using amorphous SiC ceramic membrane as a molecular sieve was firstly reported for a material prepared from polysilastyrene [387]. Moreover, SiC-based ceramic membranes were synthesized also upon thermal [384], e-beam [388] or chemical [389] curing of polycarbosilanes followed by pyrolysis in inert atmosphere. SiOC-based membranes with enhanced hydrogen permselective were prepared via curing of polycarbosilanes in air and subsequent pyrolysis in argon [390,391].

Amorphous silicon nitride-based ceramic membranes were prepared by ammonolysis of a polysilazane at 650 °C. The as-synthesized membrane showed a hydrogen permeance of 1.3×10^{-8} mol·m^{-2}·s^{-1}·Pa^{-1} at 200 °C and a H_2/N_2 selectivity of 165, whereas after hydrothermal treatment at 300 °C the permeance was higher than 1.0×10^{-7} mol·m^{-2}·s^{-1}·Pa^{-1} at 300 °C with H_2/N_2 selectivity beyond 100 [382].

Other preceramic polymers such as polysilazane, polyborosilazanes or polysilylcarbodiimides were used for the preparation of high-temperature stable amorphous ceramic membranes in the systems Si–C–N and Si–B–C–N. Mesoporous SiCN-based ceramic membranes (pore sizes in the range from 2 to 5 nm) were prepared via pyrolysis of polysilylcarbodiimide-based precursors which were synthesized from a non-oxidic sol–gel process based on reactions of bis(trimethylsilyl)carbodiimide with chlorosilanes [392]. Amorphous SiBCN-based ceramic membranes exhibiting a trimodal pore size distribution (0.6, 2.7 and 6 nm) were prepared on macroporous alumina supports via dip-coating and pyrolysis of a polyborosilazane [393]. Also, a multilayered a-SiBCN/γ-Al$_2$O$_3$/α-Al$_2$O$_3$ membrane with gradient porosity was prepared and investigated with respect gas separation behavior. A permeance of 1.05×10^{-8} mol·m^{-2}·s^{-1}·Pa^{-1} and a H_2/CO permselectivity of 10.5 were determined, showing its potential for applications such as hydrogen purification (Figure 20) [394].

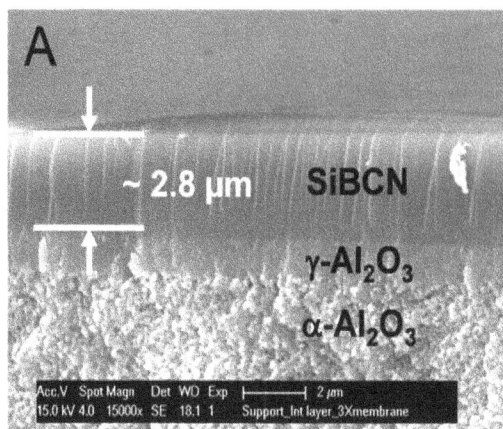

Figure 20. Cross section of a multilayered a-SiBCN/γ-Al₂O₃/α-Al₂O₃ membrane (reprinted with permission from reference [394]; Copyright 2010 Wiley).

In order to enhance the hydrothermal stability of the PDC-NCs gas separation membranes, single-source precursors based on metal-modified preceramic polymers were used, as for the case of Ni/SiCNO, WC(O)/SiOC or WN/SiOCN nanocomposites. The *in situ* precipitation of the metallic or ceramic nano-particles within the SiOC/SiOCN matrix was found to be responsible for the improved stability of the ceramic membranes in aggressive environments [395,396].

A recent systematic study reports on specific synthetic parameters which can be used in order to tune the chemical composition, (micro/meso)porosity and the specific surface area of PDC-NCs-based membranes. Thus, different preceramic polymers (e.g., polysiloxanes, polysilazanes) were shown to be able to develop stable microporosity and large specific surface area upon ceramization at moderate temperatures (600–700 °C) in ammonia atmosphere [397].

Due to their tunability as well as because of their robustness at high temperatures and in hostile environments, PDC-NC-based membranes might be the devices of choice for high-temperature gas separation applications.

5.3.2. Porous Materials for Hydrogen Generation and Storage

Proton exchange membrane fuel cell-based (PEMFC) systems are attractive alternatives to current energy conversion technologies due to their potential to directly convert chemical energy such as hydrogen into electrical energy. They consist of three subsystems—fuel cell stack, hydrogen generator, and hybrid power management system. They display high efficiency, fast responses to loads and they have potentially zero emissions (except water). One of the most key challenges is the controlled release of hydrogen to meet the overall energy requirements for civil vehicle applications such as for drones. For this purpose, boron- and nitrogen-based chemical hydrides are attractive to be potential hydrogen carriers for PEM fuel cells owing to some very advantageous features: they can store and produce hydrogen "on-demand", they are stable during long periods of storage without usage, they are non-flammable and non-toxic, they are hydrogen-rich compounds (large gravimetric energy densities) and side-products can be recyclable depending on the hydrides. Here we distinguish the liquid-phase hydrogen carriers such as the alkaline solution of sodium borohydride (NaBH₄) to generate hydrogen by hydrolysis more suitable for portable applications and hydrogen storage materials such as ammonia borane (AB) which can release pure hydrogen at moderate temperatures by thermolysis for automotive applications.

Polymer-derived ceramics can be used in both types of applications as catalyst supports for hydrogen release from $NaBH_4$ during its hydrolysis and as host material to confine AB with the objective to lower the temperature threshold for the H_2 release and the emission of gaseous by-products such as ammonia and borazine.

Concerning the first aspect, PDCs find their interest according to the severe conditions of the reaction: hydrolysis is a strong exothermal heterogeneous reaction and the sodium borate which is formed is a strong base, $cf.$ $NaBH_4$ (aq) + 4 H_2O → 4 H_2 + $NaB(OH)_4$ (aq) + Heat.

As soon as it is produced, the solution becomes basic. As a consequence, the application of MOFs, zeolites and metal oxides in this catalytic hydrolysis is somehow limited because their hydrothermal stability may turn to be poor in the severe conditions imposed by the reaction, leading in general to the collapse of the porous structure. To address the issue of highly stable supports to produce H_2 from the alkaline solution of $NaBH_4$, the use of ordered mesoporous polymer-derived SiAlCN powders as a support of Pt (which is required for an efficient hydrolysis reaction) allowed generating important volume of hydrogen after 2 h of reaction at 80 °C [247]. The performance was strongly correlated to the specific surface areas and pore volumes of the support: the highest specific surface area and pore volume are, the highest volume of hydrogen is generated. However, ordered mesoporous powders have necessary some difficulties in practical use, $i.e.$, in the scale of the demonstrator reactor. Within this context, further investigations concerning the preparation and application of monolithic PDCs and nanocomposites with tailored porosity are required.

Concerning the second aspect, nanoconfinement, appears to be one of the most efficient strategies to release pure hydrogen at low temperature for AB [398]. Gutowska et $al.$ [399] showed that AB confined in the mesoporosity of silica SBA-15 has improved dehydrogenation behaviour in comparison to the pristine hydride, with an onset at 70 °C and the liberation of borazine-free H_2. Since this pioneering work, various nano-scaffolds have been studied and reported: among others, carbonaceous hosts such as mesoporous CMK-3 [400] or nanotubes [401], polymethyl acrylate [402], magnesium metalorganic frameworks (MOF) [403,404], zeolite X [19] or silica hollow nanospheres [405].

The destabilization of AB is generally explained by two phenomena. The first one is the nanosizing of the hydride particles; it results in defect sites that initiate the dehydropolymerization of AB at lower temperatures. The second one is associated with $H^{\delta+}\cdots H^{\delta-}$ surface interactions, with $H^{\delta-}$ of the BH_3 moiety of AB and $H^{\delta+}$ belonging to surface/terminal hydroxyl groups (–O–H) generally found on carbonaceous or oxide nanoscaffolds. Such acid-base interactions enhance H_2 release but usually lead to an unstable material in room conditions [406].

One of the strategies is therefore focused on the use of nanoscaffolds free of reactive surface groups to be safe in room conditions. Boron nitride (BN), which can be produced as nanostructured and porous materials from precursors [248,407–410], has been used to confine AB. In particular, BN nanopolyhedrons with hollow core and mesoporous shell structure with a BET specific surface area of 200.5 $m^2 \cdot g^{-1}$, a total pore volume of 0.287 $cm^3 \cdot g^{-1}$ and a narrow pore size distribution centred at 3.5 nm have been used as nanoscaffolds of AB in order to improve its dehydrogenation properties [411,412]. The as-formed BN NPHs-confined AB is able to liberate H_2 at temperatures as low as 40 °C. Over the range 40–80 °C, the as-generated H_2 is pure, the only traces of by-product (*i.e.*, NH$_3$) being detected at >80 °C. Considering the effective regenerability of AB, the composite material represents a safe and practical hydrogen storage material which open the way to a very broad set of non-oxide materials including ceramics and nanocomposites using single-source molecules.

5.3.3. Heterogeneous Catalysis

In Section 2.2, we presented the chemical modification of preceramic polymers using coordination compounds. In this approach, a metal transfer from the complex to the polymer chain can occur, giving rise to metal modified polymers, which are cross-linked and pyrolyzed, forming the metal-containing PDCs namely nanocomposites. Depending on the nature of the metal, as-obtained nanocomposites can be used for heterogeneous catalysis for reactions in harsh conditions because of the robust nature of non-oxide ceramic such as SiC and SiCN, the difference in terms of polarity and acidic properties in comparison to oxide ceramics, and their inertness and chemical resistance. As a proof-of-principle experiments, the catalytic activity and thermal stability of highly porous and hierarchically ordered SiCN ceramics that integrate well-dispersed Pt nanoparticles was tested in the total oxidation (combustion) of methane as shown in Figure 21 [151].

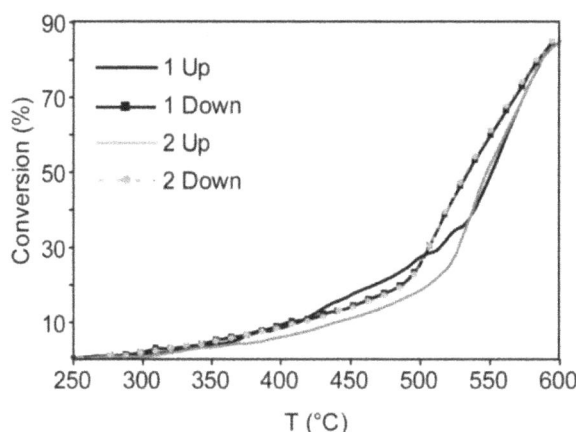

Figure 21. Methane conversion(activity) of highly porous and hierarchically ordered SiCN ceramics that integrate well-dispersed Pt nanoparticles as a function of the reaction temperature during two heating/cooling cycles (reprinted with permission from reference [151]; Copyright 2009 American Ceramic Society).

Current technologies for the conversion of natural gas to liquid products proceed by generation of carbon monoxide and hydrogen (syn-gas) that is then converted to higher products through Fischer-Tropsch chemistry. The direct, low-temperature, oxidative conversion of methane to an ester of methanol is best achieved with selected Pt complexes in high yield based on metal [413]. Platinum-supported SiCN ceramics showed the typical hysteresis behavior observed for platinum-catalyzed methane oxidation during heating and subsequent cooling cycles. The system combined excellent size control and high thermal stability of the catalytically active platinum with high structural flexibility rendering this system especially interesting for size-selective catalysis, monolith- and microreactor applications.

Interestingly, the use of coordination compounds with different metals allowed to generate a large range of metal-supported ceramic composition for various reactions. As an illustration, copper-containing silicon carbonitride ceramics (Cu@SiCN) by using silylaminopyridinato complexes showed catalytic activity towards the oxidation of cycloalkanes using air as oxidant [152]. As alkanes are inert and thermodynamically preferred total oxidation is highly competitive, there is a great need for catalysts that are able to convert simple alkanes into the corresponding mono-oxidation products. It was demonstrated that the selectivity issues can be addressed by tailoring the copper content as well as the nature of the metal loading in the polymer. Catalytic tests confirmed the applicability of these materials for the oxidation of cycloalkanes. Based on the same synthesis procedure, the same authors demonstrated that Ni@SiCN nanocatalysts were potential candidates for the hydrogenation of alkynes such as phenylacetylene into alkenes such as styrene [153]. For polymerization of styrene and obtain high-quality polymers, pure styrene feedstock is needed for a longer catalyst life. Even very small amount of phenylacetylene (which is used to produce styrene by dehydrogenation) in the styrene stream can deactivate the catalyst hence, styrene feed with very low concentration of phenylacetylene is mandatory. Monolithic microporous Ni@SiCN materials have been demonstrated as thermally robust until 500 °C in an oxidative environment and were selective hydrogenation catalysts for the conversion of phenylacetylene into styrene (selectivity ≥ 89% and conversion ≥ 99%). By changing the nature of the PDC phase (SiC), the resulting micro-, meso- and hierarchically porous Ni@SiC catalysts could be active and highly selective in the selective hydrogenolysis of the aromatic carbon-oxygen (C–O) bonds [155]. In particular, hierarchically structured Ni@SiC materials were the most active catalysts.

5.3.4 Anode Materials for Secondary Li Ion Batteries

As described above (see Section 3), the ceramization of preceramic polymers delivers amorphous ceramics as intrinsically complex nano-structured materials composed of nano-domains of 1–3 nm in size, which persist up to very high temperatures [91,414–416]. An important feature of the PDC-NCs is their possibility to accommodate significant amounts of segregated carbon in their microstructure [80]. The nature, distribution and amount of the segregated carbon phase are correlated with the macromolecular architecture and decomposition behavior of the preceramic polymer. Within this context, carbon-rich SiCN and SiOC ceramics were shown to be suitable anode materials for Li-ion batteries [417–420].

The electrochemical properties of silicon oxycarbides were intensively studied in the last two decades. Especially materials consisting of large amounts of segregated carbon present within an amorphous SiOC matrix were identified as promising anode materials for secondary lithium ion battery (LIB) applications [421–428]. Within this context nanocomposites based on Si or Sn nanoparticles dispersed within SiOC were also studied [218,429].

Thus, in a case study, silicon oxycarbide/tin nanocomposites (SiOC/Sn) were prepared by chemical modification of two different polysilsesquioxanes with Tin(II) acetate and subsequent pyrolysis at 1000 °C. The obtained samples consisted of an amorphous SiOC matrix, *in situ* precipitated metallic Sn nanoparticles and segregated carbon. Galvanostatic cycling of both composites revealed superior cycling stability and rate capability of the C/SiOC/Sn nanocomposite containing large carbon content and was attributed to the soft, carbon-rich SiOC matrix, which was able to accommodate the volume changes related to the Li uptake and release of the Sn phase. The poor cycling stability found for the C/SiOC/Sn nanocomposite containing rather low amount of segregated carbon relates to mechanical failure of the stiff and fragile, carbon-poor matrix generated from it. Moreover, incremental capacity measurements emphasized that the two nanocomposites can incorporated different amounts of Li within the Sn phase. Thus, in the carbon-rich nanocomposite Li_7Sn_2 was formed; whereas the nanocomposite with lower carbon content exhibits the formation of $Li_{22}Sn_5$ was observed. The suppression of the formation of $Li_{22}Sn_5$ in the carbon-rich sample is rationalized by a restriction of the expansion of the matrix and thus prevention of a higher Li content in the Sn-Li alloy. For carbon-poor material, the matrix is severely degrading (fracture, cracking), providing an unlimited free volume for expansion and thus the formation of the lithium-rich $Li_{22}Sn_5$ phase is favored.

6. Conclusions

The use of tailor-made preceramic polymers provides a unique preparative access to ceramic nanocomposites possessing adjustable phase compositions and microstructures. Thus, a deep understanding of the interrelations between the features of the preceramic polymers and the structure and properties of the resulting ceramic nanocomposites represents a key parameter for a knowledge-based development of multifunctional materials based on polymer-derived ceramic nanocomposites.

However, despite numerous studies trying to elucidate these correlations, there is no systematic understanding yet available.

Consequently, following aspects should be intensively addressed in the future in order to provide a rational design of PDC-NCs with tailored compositions, microstructures and properties: (i) a straight-forward and knowledge-based synthetic access to preceramic single-source precursors with tailored chemical compositions and molecular architectures; (ii) a fundamental understanding of the ceramization process of the preceramic precursor to deliver PDC-NCs with designed compositions and microstructures; (iii) extensive experimental and modeling data are necessary in order to assess the properties of the PDC-NCs and to understand how they are affected/determined by the microstructural features of the nanocomposites.

Acknowledgments

Gabriela Mera, Markus Gallei and Emanuel Ionescu acknowledge financial support from the German Research Foundation (DFG).

Additionally, financial support within the frame of the FP7 European Initial Training Network FUNEA (Functional Nitrides for Energy Applications) is gratefully acknowledged.

Author Contributions

All authors contributed equally to the compilation of the manuscript.

Conflicts of Interest

The authors declare no conflict of interest.

References

1. Matic, P. Overview of Multifunctional Materials. In Proceedings of the SPIE 5053, Smart Structures and Materials 2003: Active Materials: Behavior and Mechanics; San Diego, CA, USA, 2 March 2003.

2. Momoda, L.A. The future of engineering materials: Multifunction for performance-tailored structures. *Bridge.* **2004**, *34*, 18–21.

3. Christodoulou, L.; Venables, J. Multifunctional material systems: The first generation. *JOM* **2003**, *55*, 39–45.

4. Mera, G.; Ionescu, E. Silicon-Containing Preceramic Polymers. In *Encyclopedia of Polymer Science and Technology*; John Wiley & Sons, Inc.: Hoboken, NJ, USA, 2013.

5. Colombo, P.; Mera, G.; Riedel, R.; Sorarù, G.D. Polymer-derived ceramics: 40 years of research and innovation in advanced ceramics. *J. Am. Ceram. Soc.* **2010**, *93*, 1805–1837.

6. Ionescu, E.; Riedel, R. Polymer Processing of Ceramics. In *Ceramics and Composites Processing Methods*; Bansal, N., Boccaccini, A., Eds.; Wiley: Hoboken, NJ, USA, 2012; pp. 235–270.

7. Riedel, R.; Mera, G.; Hauser, R.; Klonczynski, A. Silicon-based polymer-derived ceramics: Synthesis properties and applications—A review. *J. Ceram. Soc. Jpn.* **2006**, *114*, 425–444.

8. Mera, G.; Riedel, R. Organosilicon-Based Polymers as Precursors for Ceramics. In *Polymer Derived Ceramics: From Nanostructure to Applications*; Colombo, P., Riedel, R., Soraru, G.D., Kleebe, H.-J., Eds.; DEStech Publications Inc.: Lancaster, PA, USA, 2010; pp. 51–89.

9. Ionescu, E.; Kleebe, H.-J.; Riedel, R. Silicon-containing polymer-derived ceramic nanocomposites (PDC-NCs): Preparative approaches and properties. *Chem. Soc. Rev.* **2012**, *41*, 5032–5052.

10. Ionescu, E.; Gervais, C.; Babonneau, F. Polymer-to-Ceramic Transformation. In *Polymer Derived Ceramics: From Nanostructure to Applications*; Colombo, P., Riedel, R., Soraru, G.D., Kleebe, H.-J., Eds.; DEStech Publications Inc.: Lancaster, PA, USA, 2010; pp. 108–127.

11. Miller, R.D.; Michl, J. Polysilane high polymers. *Chem. Rev.* **1989**, *89*, 1359–1410.

12. West, R. The polysilane high polymers. *J. Organomet. Chem.* **1986**, *300*, 327–346.

13. West, R.; Maxka, J.; Sinclair, R.; Cotts, P. Polysilane high polymers. *Abstr. Pap. Am. Chem. S* **1987**, *193*, 11–Inor.

14. West, R.; Maxka, J. Polysilane high polymers—An overview. *ACS Symp. Ser.* **1988**, *360*, 6–20.

15. Wynne, K.J. An introduction to inorganic and organometallic polymers. *ACS Symp. Ser.* **1988**, *360*, 1–4.

16. Kipping, F.S.; Sands, J.E. XCIII.—Organic derivatives of silicon. Part XXV. Saturated and unsaturated silicohydrocarbons, Si_4Ph_8. *J. Chem. Soc. Trans.* **1921**, *119*, 830–847.

17. Kipping, F.S. CCCVIII.—Organic derivatives of silicon. Part XXX. Complex silicohydrocarbons $[SiPh_2]_n$. *J. Chem. Soc. Trans.* **1924**, *125*, 2291–2297.

18. Wesson, J.P.; Williams, T.C. Organosilane polymers: I. Poly(dimethylsilylene). *J. Polym. Sci. Polym. Chem. Ed.* **1979**, *17*, 2833–2843.

19. West, R.; David, L.D.; Djurovich, P.I.; Stearley, K.L.; Srinivasan, K.S.V.; Yu, H. Phenylmethylpolysilanes: Formable silane copolymers with potential semiconducting properties. *J. Am. Chem. Soc.* **1981**, *103*, 7352–7354.

20. Trujillo, R.E. Preparation of long-chain poly(methylphenylsilane). *J. Organomet. Chem.* **1980**, *198*, C27–C28.

21. Kawabe, T.; Naito, M.; Fujiki, M. Multiblock polysilane copolymers: One-pot wurtz synthesis, fluoride anion-induced block-selective scission experiments, and spectroscopic characterization. *Macromolecules* **2008**, *41*, 1952–1960.

22. Jones, R.G.; Holder, S.J. High-yield controlled syntheses of polysilanes by the Wurtz-type reductive coupling reaction. *Polym. Int.* **2006**, *55*, 711–718.

23. Herzog, U.; West, R. Heterosubstituted polysilanes. *Macromolecules* **1999**, *32*, 2210–2214.

24. Lacave-Goffin, B.; Hevesi, L.; Devaux, J. Graphite? Potassium, a new reagent for the synthesis of polysilanes. *J. Chem. Soc. Chem. Commun.* **1995**, 769–770.

25. Jones, R.G.; Benfield, R.E.; Evans, P.J.; Swain, A.C. Poly(methylphenylsilane) with an enhanced isotactic content prepared using the graphite intercalation compound, C_8K. *J. Chem. Soc. Chem. Commun.* **1995**, 1465–1466.

26. Chang, L.S.; Corey, J.Y. Dehydrogenative coupling of diarylsilanes. *Organometallics* **1989**, *8*, 1885–1893.

27. Woo, H.G.; Walzer, J.F.; Tilley, T.D. .sigma.-Bond metathesis mechanism for dehydropolymerization of silanes to polysilanes by d0 metal-catalysts. *J. Am. Chem. Soc.* **1992**, *114*, 7047–7055.

28. Sakamoto, K.; Obata, K.; Hirata, H.; Nakajima, M.; Sakurai, H. Novel anionic-polymerization of masked disilenes to polysilylene high polymers and block copolymers. *J. Am. Chem. Soc.* **1989**, *111*, 7641–7643.

29. Cypryk, M.; Gupta, Y.; Matyjaszewski, K. Anionic ring-opening polymerization of 1,2,3,4-tetramethyl-1,2,3,4-tetraphenylcyclotetrasilane. *J. Am. Chem. Soc.* **1991**, *113*, 1046–1047.

30. Kashimura, S.; Tane, Y.; Ishifune, M.; Murai, Y.; Hashimoto, S.; Nakai, T.; Hirose, R.; Murase, H. Practical method for the synthesis of polysilanes using Mg and Lewis acid system. *Tetrahedron Lett.* **2008**, *49*, 269–271.

31. Yajima, S.; Hayashi, J. Continuous silicon carbide fiber of high tensile strength. *Chem. Lett.* **1975**, *4*, 931–934.

32. Shiina, K.; Kumada, M. Thermal rearrangement of hexamethyldisilane to trimethyl(dimethylsilylmethyl)-silane. *J. Org. Chem.* **1958**, *23*, 139.

33. Whitmarsh, C.K.; Interrante, L.V. Synthesis and structure of a highly branched polycarbosilane derived from (chloromethyl)trichlorosilane. *Organometallics* **1991**, *10*, 1336–1344.

34. Smith, J.; Troy, L. Process for the Production of Silicon Carbide by the Pyrolysis of a Polycarbosilane Polymer. *US Patent US4631179 A*, 23 December 1986.

35. Wu, H.J.; Interrante, L.V. Preparation of poly(dichlorosilaethylene) and poly(silaethylene) via ring-opening polymerization. *Macromolecules* **1992**, *25*, 1840–1841.

36. Nate, K.; Inoue, T.; Sugiyama, H.; Ishikawa, M. Organosilicon deep UV positive resist consisting of poly(P-disilanylenephenylene). *J. Appl. Polym. Sci.* **1987**, *34*, 2445–2455.

37. Iwahara, T.; Hayase, S.; West, R. Synthesis and properties of ethynylene disilanylene copolymers. *Macromolecules* **1990**, *23*, 1298–1301.

38. Corriu, R.J.P.; Guerin, C.; Henner, B.; Kuhlmann, T.; Jean, A.; Garnier, F.; Yassar, A. Organosilicon polymers—Synthesis of poly[(silanylene)diethynylene]s with conducting properties. *Chem. Mater.* **1990**, *2*, 351–352.

39. Ohshita, J.; Kanaya, D.; Ishikawa, M.; Yamanaka, T. Polymeric organo-silicon systems 6: Synthesis and properties of trans-poly[(disilanylene)ethenylene]. *J. Organomet. Chem.* **1989**, *369*, C18–C20.

40. Ohshita, J.; Kunai, A. Polymers with alternating organosilicon and Pi-conjugated units. *Acta Polym.* **1998**, *49*, 379–403.

41. Bacque, E.; Pillot, J.P.; Birot, M.; Dunogues, J.; Bourgeois, G. New model polycarbosilanes 3: Synthesis and characterization of functional linear carbosilanes. *J. Organomet. Chem.* **1988**, *346*, 147–160.

42. Kim, Y.H.; Gal, Y.S.; Kim, U.Y.; Choi, S.K. Cyclo-polymerization of dipropargylsilanes by transition-metal catalysts. *Macromolecules* **1988**, *21*, 1991–1995.

43. Shinar, J.; Ijadimaghsoodi, S.; Ni, Q.X.; Pang, Y.; Barton, T.J. Synthesis and study of a polysilole. *Synth. Met.* **1989**, *28*, C593–C598.

44. Zhang, X.H.; Zhou, Q.S.; Weber, W.P.; Horvath, R.F.; Chan, T.H.; Manuel, G. Anionic Ring-opening polymerization of sila-silacyclopent-3-enes and germacyclopent-3-enes. *Macromolecules* **1988**, *21*, 1563–1566.

45. Park, Y.T.; Zhou, Q.S.; Weber, W.P. Anionic ring-opening polymerization of 3,4-benzo-1,1-dimethyl-1-silacyclopentene—properties of poly(3,4-benzo-1,1-dimethyl-1-silapentene). *Polym. Bull.* **1989**, *22*, 349–353.

46. Park, Y.T.; Manuel, G.; Weber, W.P. Anionic ring-opening polymerization of 1,1,3-trimethyl-1-silacyclopent-3-ene—effect of temperature on poly(1,1,3-trimethyl-1-sila-cis-pent-3-ene) microstructure. *Macromolecules* **1990**, *23*, 1911–1915.

47. Zhou, S.Q.; Weber, W.P. Anionic-polymerization of 1-methyl-1-silacyclopent-3-ene—Characterization of poly(1-methyl-1-sila-cis-pent-3-ene) by H-1, C-13, and Si-29 NMR-spectroscopy and mechanism of polymerization. *Macromolecules* **1990**, *23*, 1915–1917.

48. Seyferth, D. Polycarbosilanes—An overview. *ACS Symp. Ser.* **1988**, *360*, 21–42.

49. Mera, G. Contributions to The Synthesis of Silicon-Rich Oligocarbosilanes and Their Use as Precursors for Electrically Conductive Films. Ph.D. Thesis, Ruhr-Universität Bochum, Bochum, Germany, 2005.

50. Jones, R.G.; Ando, W.; Chojnowski, J. *Silicon-Containing Polymers: The Science and Technology of Their Synthesis and Applications*; Kluwer: Dordrecht, The Netherland; Boston, MA, USA; London, UK, 2000.

51. Babonneau, F.; Thorne, K.; Mackenzie, J.D. Dimethyldiethoxysilane/tetraethoxysilane copolymers: Precursors for the silicon-carbon-oxygen system. *Chem. Mater.* **1989**, *1*, 554–558.

52. Soraru, G.D. Silicon oxycarbide glasses from gels. *J. Sol–Gel Sci. Technol.* **1994**, *2*, 843–848.

53. Soraru, G.D.; D'Andrea, G.; Campostrini, R.; Babonneau, F.; Mariotto, G. Structural characterization and high-temperature behavior of silicon oxycarbide glasses prepared from sol–gel precursors containing Si–H bonds. *J. Am. Ceram. Soc.* **1995**, *78*, 379–387.

54. Dire, S.; Ceccato, R.; Babonneau, F. Structural and microstructural evolution during pyrolysis of hybrid polydimethylsiloxane-titania nanocomposites. *J. Sol–Gel Sci. Technol.* **2005**, *34*, 53–62.

55. Dire, S.; Campostrini, R.; Ceccato, R. Pyrolysis chemistry of sol–gel-derived poly(dimethylsiloxane)-zirconia nanocomposites. Influence of zirconium on polymer-to-ceramic conversion. *Chem. Mater.* **1998**, *10*, 268–278.

56. Alonso, B.; Sanchez, C. Structural investigation of polydimethylsiloxane-vanadate hybrid materials. *J. Mater. Chem.* **2000**, *10*, 377–386.

57. Huang, H.H.; Orler, B.; Wilkes, G.L. Structure property behavior of new hybrid materials incorporating oligomeric species into sol–gel glasses 3: Effect of acid content, tetraethoxysilane content, and molecular-weight of poly(dimethylsiloxane). *Macromolecules* **1987**, *20*, 1322–1330.

58. Dire, S.; Babonneau, F.; Sanchez, C.; Livage, J. Sol–gel synthesis of siloxane oxide hybrid coatings $[Si(CH_3)_2O \cdot Mo_x$–M = Si, Ti, Zr, Al] with luminescent properties. *J. Mater. Chem.* **1992**, *2*, 239–244.

59. Yamada, N.; Yoshinaga, I.; Katayama, S. Synthesis and dynamic mechanical behaviour of inorganic-organic hybrids containing various inorganic components. *J. Mater. Chem.* **1997**, *7*, 1491–1495.

60. Fukushima, M.; Yasuda, E.; Nakamura, Y.; Tanabe, Y. Pyrolysis behavior of organic-inorganic hybrids with Si–O–Nb/Si–O–Ta oxygen bridged heterometallic bonds. *J. Ceram. Soc. Jpn.* **2003**, *111*, 857–859.

61. Saha, A.; Shah, S.R.; Raj, R. Amorphous silicon carbonitride fibers drawn from alkoxide modified Ceraset (TM). *J. Am. Ceram. Soc.* **2003**, *86*, 1443–1445.

62. Saha, A.; Shah, S.R.; Raj, R. Oxidation behavior of SiCN-ZrO_2 fiber prepared from alkoxide-modified silazane. *J. Am. Ceram. Soc.* **2004**, *87*, 1556–1558.

63. Fukushima, M.; Yasuda, E.; Nakamura, Y.; Shimizu, M.; Teranishi, Y.; Manocha, L.M.; Tanabe, Y. Oxidation behavior of Si–Nb–C–O ceramics prepared by the pyrolysis of methyltriethoxysilane based organic-inorganic hybrid gel. *J. Sol–Gel Sci. Technol.* **2005**, *34*, 15–21.

64. Kroke, E.; Li, Y.L.; Konetschny, C.; Lecomte, E.; Fasel, C.; Riedel, R. Silazane derived ceramics and related materials. *Mater. Sci. Eng. R* **2000**, *26*, 97–199.

65. Narsavage, D.M.; Interrante, L.V.; Marchetti, P.S.; Maciel, G.E. Condensation polymerization of tetrakis(ethylamino)silane and its thermal-decomposition to Si_3N_4/SiC ceramics. *Chem. Mater.* **1991**, *3*, 721–730.

66. Schutzenberger, H.; Colson, C.R. Silicon. *Rendus Hebd. Seances Acad. Sci.* **1885**, *92*, 1508–1511.

67. Glemser, O.; Naumann, P. Uber den thermischen abbau von siliciumdiimid $Si(NH)_2$. *Z. Anorg. Allg. Chem.* **1959**, *298*, 134–141.

68. Seyferth, D.; Wiseman, G.H.; Prudhomme, C. A liquid silazane precursor to silicon-nitride. *J. Am. Ceram. Soc.* **1983**, *66*, C13–C14.

69. Arai, M.; Sakurada, S.; Isoda, T.; Tomizawa, H. Preceramic polysilazane to silicon-nitride. *Abstr. Pap. Am. Chem. S* **1987**, *193*, 41–Inor.

70. Blanchard, C.R.; Schwab, S.T. X-ray-diffraction analysis of the pyrolytic conversion of perhydropolysilazane into silicon-nitride. *J. Am. Ceram. Soc.* **1994**, *77*, 1729–1739.

71. Seyferth, D.; Wiseman, G.H. High-yield synthesis of Si_3N_4/SiC ceramic materials by pyrolysis of a novel polyorganosilazane. *J. Am. Ceram. Soc.* **1984**, *67*, C132–C133.

72. Iwamoto, Y.; Kikuta, K.; Hirano, S. Crystallization and microstructure development of Si_2N_4-Ti(C,N)-Y_2O_3 ceramics derived from chemically modified perhydropolysilazane. *J. Ceram. Soc. Jpn.* **2000**, *108*, 1072–1078.

73. Iwamoto, Y.; Kikuta, K.; Hirano, S. Synthesis of poly-titanosilazanes and conversion into Si₃N₄–TiN ceramics. *J. Ceram. Soc. Jpn.* **2000**, *108*, 350–356.

74. Papendorf, B.; Nonnenmacher, K.; Ionescu, E.; Kleebe, H.-J.; Riedel, R. Strong influence of polymer architecture on the microstructural evolution of hafnium-alkoxide-modified silazanes upon ceramization. *Small* **2011**, *7*, 970–978.

75. Ionescu, E.; Papendorf, B.; Kleebe, H.-J.; Breitzke, H.; Nonnenmacher, K.; Buntkowsky, G.; Riedel, R. Phase separation of a hafnium alkoxide-modified polysilazane upon polymer-to-ceramic transformation—A case study. *J. Eur. Ceram. Soc.* **2012**, *32*, 1873–1881.

76. Iwamoto, Y.; Kikuta, K.; Hirano, S. Si₃N₄-TiN-Y₂O₃ ceramics derived from chemically modified perhydropolysilazane. *J. Mater. Res.* **1999**, *14*, 4294–4301.

77. Bechelany, M.C.; Proust, V.; Gervais, C.; Ghisleni, R.; Bernard, S.; Miele, P. *In situ* controlled growth of titanium nitride in amorphous silicon nitride: A general route toward bulk nitride nanocomposites with very high hardness. *Adv. Mater.* **2014**, *26*, 6548–6553.

78. Yuan, J.; Luan, X.; Riedel, R.; Ionescu, E. Preparation and hydrothermal corrosion behavior of Cf/SiCN and Cf/SiHfBCN ceramic matrix composites. *J. Eur. Ceram. Soc.* **2015**, in press.

79. Yuan, J.; Hapis, S.; Breitzke, H.; Xu, Y.P.; Fasel, C.; Kleebe, H.J.; Buntkowsky, G.; Riedel, R.; Ionescu, E. Single-source-precursor synthesis of hafnium-containing ultrahigh-temperature ceramic nanocomposites (UHTC-NCs). *Inorg. Chem.* **2014**, *53*, 10443–10455.

80. Mera, G.; Riedel, R.; Poli, F.; Muller, K. Carbon-rich SiCN ceramics derived from phenyl-containing poly(silylcarbodiimides). *J. Eur. Ceram. Soc.* **2009**, *29*, 2873–2883.

81. Ebsworth, E.A.; Mays, M.J. Zur Reaktion zwischen halogensilanen und silbercyanamid. *Angew. Chem. Int. Ed.* **1962**, *74*, doi:10.1002/ange.19620740307. (In German)

82. Ebsworth, E.A.; Mays, M.J. Preparation and properties of disilylcyanamide. *J. Chem. Soc.* **1961**, 4879–4882.

83. Pump, J.; Wannagat, U. Bis-(trimethylsilyl)-carbodiimid. *Angew. Chem. Int. Ed.* **1962**, *74*, 117.

84. Birkofer, L.; Ritter, A.; Richter, P. Uber siliciumorganische verbindungen 13: *N,N′*-bis-trimethylsilyl-carbodiimid. *Tetrahedron Lett.* **1962**, 195–198. (In German)

85. Klebe, J.F.; Murray, J.G. Organosiliconcarbodiimide Polymers and Process for Their Preparation. *US Patent 3,352,799*, 14 November 1967.

86. Razuvaev, G.A.; Gordetsov, A.S.; Kozina, A.P.; Brevnova, T.N.; Semenov, V.V.; Skobeleva, S.E.; Boxer, N.A.; Dergunov, Y.I. Synthesis of monomeric and oligomeric carbodiimides with polysilane and siloxane fragments. *J. Organomet. Chem.* **1987**, *327*, 303–309.

87. Gorbatenko, V.I.; Gertsyuk, M.N.; Samarai, L.I. Synthesis of 2,4-bis(trimethylsilyloxy)-6-chloro-1,3,5-triazine. *Zh. Org. Khim.* **1977**, *13*, 899.

88. Riedel, R.; Kroke, E.; Greiner, A.; Gabriel, A.O.; Ruwisch, L.; Nicolich, J.; Kroll, P. Inorganic solid-state chemistry with main group element carbodiimides. *Chem. Mater.* **1998**, *10*, 2964–2979.

89. Morcos, R.M.; Mera, G.; Navrotsky, A.; Varga, T.; Riedel, R.; Poli, F.; Muller, K. Enthalpy of formation of carbon-rich polymer-derived amorphous SiCN ceramics. *J. Am. Ceram. Soc.* **2008**, *91*, 3349–3354.

90. Mera, G.; Tamayo, A.; Nguyen, H.; Sen, S.; Riedel, R. Nanodomain structure of carbon-rich silicon carbonitride polymer-derived ceramics. *J. Am. Ceram. Soc.* **2010**, *93*, 1169–1175.

91. Widgeon, S.; Mera, G.; Gao, Y.; Stoyanov, E.; Sen, S.; Navrotsky, A.; Riedel, R. Nanostructure and energetics of carbon-rich SiCN ceramics derived from polysilylcarbodiimides: Role of the nanodomain interfaces. *Chem. Mater.* **2012**, *24*, 1181–1191.

92. Prasad, R.M.; Mera, G.; Morita, K.; Muller, M.; Kleebe, H.J.; Gurlo, A.; Fasel, C.; Riedel, R. Thermal decomposition of carbon-rich polymer-derived silicon carbonitrides leading to ceramics with high specific surface area and tunable micro- and mesoporosity. *J. Eur. Ceram. Soc.* **2012**, *32*, 477–484.

93. Graczyk-Zajac, M.; Mera, G.; Kaspar, J.; Riedel, R. Electrochemical studies of carbon-rich polymer-derived SiCN ceramics as anode materials for lithium-ion batteries. *J. Eur. Ceram. Soc.* **2010**, *30*, 3235–3243.

94. Pump, J.; Rochow, E.G. Silycarbodiimide 4: Sila-polycarbodiimide. *Z. Anorg. Allg. Chem.* **1964**, *330*, 101–106.

95. Gabriel, A.O.; Riedel, R.; Storck, S.; Maier, W.F. Synthesis and thermally induced ceramization of a non-oxidic poly(methylsilsesquicarbodiimide) gel. *Appl. Organomet. Chem.* **1997**, *11*, 833–841.

96. Kim, D.S.; Kroke, E.; Riedel, R.; Gabriel, A.O.; Shim, S.C. An anhydrous sol–gel system derived from methyldichlorosilane. *Appl. Organomet. Chem.* **1999**, *13*, 495–499.

97. Nahar-Borchart, S.; Kroke, E.; Riedel, R.; Boury, B.; Corriu, R.J.P. Synthesis and characterization of alkylene-bridged silsesquicarbodiimide hybrid xerogels. *J. Organomet. Chem.* **2003**, *686*, 127–133.

98. Iwamoto, Y.; Volger, W.; Kroke, E.; Riedel, R.; Saitou, T.; Matsunaga, K. Crystallization behavior of amorphous silicon carbonitride ceramics derived from organometallic precursors. *J. Am. Ceram. Soc.* **2001**, *84*, 2170–2178.

99. Kienzle, A.; Bill, J.; Aldinger, F.; Riedel, R. Nanosized Si–C–N-power by pyrolysis of highly crosslinked silylcarbodiimide. *Nanostruct. Mater.* **1995**, *6*, 349–352.

100. Riedel, R.; Greiner, A.; Miehe, G.; Dressier, W.; Fuess, H.; Bill, J.; Aldinger, F. The first crystalline solids in the ternary Si–C–N system. *Angew. Chem. Int. Ed. Engl.* **1997**, *36*, 603–606.

101. Dressler, W.; Riedel, R. Progress in silicon-based non-oxide structural ceramics. *Int. J. Refract. Met. H* **1997**, *15*, 13–47.

102. Schadler, H.D.; Jager, L.; Senf, I. Pseudoelement compounds 5: Pseudochalcogens—An attempt of an empirical and theoretical characterization. *Z. Anorg. Allg. Chem.* **1993**, *619*, 1115–1120.

103. Abd-El-Aziz, A.S.; Manners, I. *Frontiers in Transition Metal-Containing Polymers*; Wiley-Interscience: Hoboken, NJ, USA, 1997.

104. Wöhrle, D.; Pomogailo, A.D. *Metal Complexes and Metals in Macromolecules: Synthesis, Structure and Properties*; Wiley-VCH: Weinheim, Germany, 2003.

105. Carraher, C.E.; Abd-El-Aziz, A.S.; Pittman, C.; Sheats, J.; Zeldin, M. *A Half Century of Metal and Metalloid Containing Polymers*; Wiley: New York, NY, USA, 2003.

106. Manners, I. *Synthetic Metal-Containing Polymers*; VCH: Weinheim, Germany, 2004.

107. Whittell, G.R.; Manners, I. Metallopolymers: New multifunctional materials. *Adv. Mater.* **2007**, *19*, 3439–3468.

108. Whittell, G.R.; Hager, M.D.; Schubert, U.S.; Manners, I. Functional soft materials from metallopolymers and metallosupramolecular polymers. *Nat. Mater.* **2011**, *10*, 176–188.

109. Gallei, M. The renaissance of side-chain ferrocene-containing polymers: Scope and limitations of vinylferrocene and ferrocenyl methacrylates. *Macromol. Chem. Phys.* **2014**, *215*, 699–704.

110. Cai, T.; Qiu, W.F.; Liu, D.; Han, W.J.; Ye, L.; Zhao, A.J.; Zhao, T. Synthesis of soluble poly-yne polymers containing zirconium and silicon and corresponding conversion to nanosized ZrC/SiC composite ceramics. *Dalton Trans.* **2013**, *42*, 4285–4290.

111. Tao, X.Y.; Qiu, W.F.; Li, H.; Zhao, T.; Wei, X.Y. New route to synthesize preceramic polymers for zirconium carbide. *Chin. Chem. Lett.* **2012**, *23*, 1075–1078.

112. Wu, H.; Zhang, W.; Zhang, J. Pyrolysis synthesis and microstructure of zirconium carbide from new preceramic polymers. *Certif. Int.* **2014**, *40*, 5967–5972.

113. Wang, H.; Gao, B.; Chen, X.; Wang, J.; Chen, S.; Gou, Y. Synthesis and pyrolysis of a novel preceramic polymer PZMS from PMS to fabricate high-temperature-resistant ZrC/SiC ceramic composite. *Appl. Organomet. Chem.* **2013**, *27*, 166–173.

114. Demars, T.; Boltoeva, M.; Vigier, N.; Maynadié, J.; Ravaux, J.; Genre, C.; Meyer, D. From coordination polymers to doped rare-earth oxides. *Eur. J. Inorg. Chem.* **2012**, *2012*, 3875–3884.

115. Abd-El-Aziz, A.S.; Agatemor, C.; Etkin, N. Sandwich complex-containing macromolecules: Property tunability through versatile synthesis. *Macromol. Rapid Commun.* **2014**, *35*, 513–559.

116. Clendenning, S.B.; Han, S.; Coombs, N.; Paquet, C.; Rayat, M.S.; Grozea, D.; Brodersen, P.M.; Sodhi, R.N.S.; Yip, C.M.; Lu, Z.-H.; *et al.* Magnetic ceramic films from a metallopolymer resist using reactive ion etching in a secondary magnetic field. *Adv. Mater.* **2004**, *16*, 291–296.

117. Kulbaba, K.; Manners, I. Polyferrocenylsilanes: Metal-containing polymers for materials science, self-assembly and nanostructure applications. *Macromol. Rapid Commun.* **2001**, *22*, 711–724.

118. Elbert, J.; Didzoleit, H.; Fasel, C.; Ionescu, E.; Riedel, R.; Stühn, B.; Gallei, M. Surface-initiated anionic polymerization of [1]silaferrocenophanes for the preparation of colloidal preceramic materials. *Macromol. Rapid Commun.* **2015**, doi:10.1002/marc.201400581.

119. Manners, I. Poly(ferrocenylsilanes): Novel organometallic plastics. *Chem. Commun.* **1999**, 857–865.

120. MacLachlan, M.J.; Ginzburg, M.; Coombs, N.; Coyle, T.W.; Raju, N.P.; Greedan, J.E.; Ozin, G.A.; Manners, I. Shaped ceramics with tunable magnetic properties from metal-containing polymers. *Science* **2000**, *287*, 1460–1463.

121. Nguyen, P.; Gomez-Elipe, P.; Manners, I. Organometallic polymers with transition metals in the main chain. *Chem. Rev.* **1999**, *99*, 1515–1548.

122. Kumar, M.; Metta-Magana, A.J.; Pannell, K.H. Ferrocenylene- and carbosiloxane-bridged bis(sila[1]ferrocenophanes) E[SiMe2-X-SiMeFC](2) {E = (eta(5)-C5H4)Fe(eta(5)-C5H4) or O; X = (CH2)(n) (n = 2, 3, 6); CH=CH; FC = (eta(5)-C5H4)(2)Fe}. *Organometallics* **2008**, *27*, 6457–6463.

123. Sun, Q.H.; Lam, J.W.Y.; Xu, K.T.; Xu, H.Y.; Cha, J.A.K.; Wong, P.C.L.; Wen, G.H.; Zhang, X.X.; Jing, X.B.; Wang, F.S.; *et al.* Nanocluster-containing mesoporous magnetoceramics from hyperbranched organometallic polymer precursors. *Chem. Mater.* **2000**, *12*, 2617–2624.

124. Sun, Q.H.; Xu, K.T.; Peng, H.; Zheng, R.H.; Haussler, M.; Tang, B.Z. Hyperbranched organometallic polymers: Synthesis and properties of poly(ferrocenylenesilyne)s. *Macromolecules* **2003**, *36*, 2309–2320.

125. Shi, J.B.; Tong, B.; Li, Z.; Shen, J.B.; Zhao, W.; Fu, H.H.; Zhi, J.; Dong, Y.P.; Haussler, M.; Lam, J.W.Y.; *et al.* Hyperbranched poly(ferrocenylphenylenes): Synthesis, characterization, redox activity, metal complexation, pyrolytic ceramization, and soft ferromagnetism. *Macromolecules* **2007**, *40*, 8195–8204.

126. Gadt, T.; Ieong, N.S.; Cambridge, G.; Winnik, M.A.; Manners, I. Complex and hierarchical micelle architectures from diblock copolymers using living, crystallization-driven polymerizations. *Nat. Mater.* **2009**, *8*, 144–150.

127. Kulbaba, K.; Cheng, A.; Bartole, A.; Greenberg, S.; Resendes, R.; Coombs, N.; Safa-Sefat, A.; Greedan, J.E.; Stover, H.D.H.; Ozin, G.A.; *et al.* Polyferrocenylsilane microspheres: Synthesis, mechanism of formation, size and charge tunability, electrostatic self-assembly, and pyrolysis to spherical magnetic ceramic particles. *J. Am. Chem. Soc.* **2002**, *124*, 12522–12534.

128. Wang, X.S.; Guerin, G.; Wang, H.; Wang, Y.S.; Manners, I.; Winnik, M.A. Cylindrical block copolymer micelles and co-micelles of controlled length and architecture. *Science* **2007**, *317*, 644–647.

129. Rider, D.A.; Liu, K.; Eloi, J.-C.; Vanderark, L.; Yang, L.; Wang, J.-Y.; Grozea, D.; Lu, Z.-H.; Russell, T.P.; Manners, I.; *et al.* Nanostructured magnetic thin films from organometallic block copolymers: pyrolysis of self-assembled polystyrene-block-poly(ferrocenylethylmethylsilane). *ACS Nano* **2008**, *2*, 263–270.

130. Wurm, F.; Hilf, S.; Frey, H. Electroactive linear-hyperbranched block copolymers based on linear poly(ferrocenylsilane)s and hyperbranched poly(carbosilane)s. *Chem. Eur. J.* **2009**, *15*, 9068–9077.

131. Thomas, K.R.; Ionescu, A.; Gwyther, J.; Manners, I.; Barnes, C.H.W.; Steiner, U.; Sivaniah, E. Magnetic properties of ceramics from the pyrolysis of metallocene-based polymers doped with palladium. *J. Appl. Phys.* **2011**, *109*, 073904:1–073904:8.

132. Häußler, M.; Sun, Q.; Xu, K.; Lam, J.W.Y.; Dong, H.; Tang, B.Z. Hyperbranched poly(ferrocenylene)s containing groups 14 and 15 elements: Syntheses, optical and thermal properties, and pyrolytic transformations into nanostructured magnetoceramics. *J. Inorg. Organomet. Polym. Mater.* **2005**, *15*, 67–81.

133. Scheid, D.; Cherkashinin, G.; Ionescu, E.; Gallei, M. Single-source magnetic nanorattles by using convenient emulsion polymerization protocols. *Langmuir* **2014**, *30*, 1204–1209.

134. Scheid, D.; Lederle, C.; Vowinkel, S.; Schäfer, C.G.; Stühn, B.; Gallei, M. Redox- and mechano-chromic response of metallopolymer-based elastomeric colloidal crystal films. *J. Mater. Chem. C* **2014**, *2*, 2583–2590.

135. Hardy, C.G.; Ren, L.; Ma, S.; Tang, C. Self-assembly of well-defined ferrocene triblock copolymers and their template synthesis of ordered iron oxide nanoparticles. *Chem. Commun.* **2013**, *49*, 4373–4375.

136. Morsbach, J.; Natalello, A.; Elbert, J.; Winzen, S.; Kroeger, A.; Frey, H.; Gallei, M. Redox-responsive block copolymers: Poly(vinylferrocene)-*b*-poly(lactide) diblock and miktoarm star polymers and their behavior in solution. *Organometallics* **2013**, *32*, 6033–6039.

137. Mazurowski, M.; Gallei, M.; Li, J.; Didzoleit, H.; Stühn, B.; Rehahn, M. Redox-responsive polymer brushes grafted from polystyrene nanoparticles by means of surface initiated atom transfer radical polymerization. *Macromolecules* **2012**, *45*, 8970–8981.

138. Yu, Z.J.; Yang, L.; Min, H.; Zhang, P.; Zhou, C.; Riedel, R. Single-source-precursor synthesis of high temperature stable SiC/C/Fe nanocomposites from a processable hyperbranched polyferrocenylcarbosilane with high ceramic yield. *J. Mater. Chem. C* **2014**, *2*, 1057–1067.

139. Kenaree, A.R.; Berven, B.M.; Ragogna, P.J.; Gilroy, J.B. Highly-metallized phosphonium polyelectrolytes. *Chem. Commun.* **2014**, *50*, 10714–10717.

140. Puerta, A.R.; Remsen, E.E.; Bradley, M.G.; Sherwood, W.; Sneddon, L.G. Synthesis and ceramic conversion reactions of 9-BBN-modified allylhydridopolycarbosilane: A new single-source precursor to boron-modified silicon carbide. *Chem. Mater.* **2003**, *15*, 478–485.

141. Donaghy, K.J.; Carroll, P.J.; Sneddon, L.G. Reactions of 1,1'-bis(diphenylphosphino)ferrocene with boranes, thiaboranes, and carboranes. *Inorg. Chem.* **1997**, *36*, 547–553.

142. Boshra, R.; Venkatasubbaiah, K.; Doshi, A.; Jakle, F. Resolution of planar-chiral ferrocenylborane lewis acids: the impact of steric effects on the stereoselective binding of ephedrine derivatives. *Organometallics* **2009**, *28*, 4141–4149.

143. Eckensberger, U.D.; Kunz, K.; Bolte, M.; Lerner, H.W.; Wagner, M. Synthesis and structural characterization of the diborylated organometallics 1,3-bis(dibromoboryl)-1',2',3',4',5'-(pentamethyl)ferrocene and 1,3-bis(dibromoboryl)cymantrene. *Organometallics* **2008**, *27*, 764–768.

144. Bauer, F.; Braunschweig, H.; Schwab, K. 1,1-Diboration of isocyanides with [2]borametalloarenophanes. *Organometallics* **2010**, *29*, 934–938.

145. Scheibitz, M.; Li, H.Y.; Schnorr, J.; Perucha, A.S.; Bolte, M.; Lerner, H.W.; Jakle, F.; Wagner, M. Ferrocenylhydridoborates: Synthesis, structural characterization, and application to the preparation of ferrocenylborane polymers. *J. Am. Chem. Soc.* **2009**, *131*, 16319–16329.

146. Parab, K.; Jakle, F. Synthesis, characterization, and anion binding of redox-active triarylborane polymers. *Macromolecules* **2009**, *42*, 4002–4007.

147. Jakle, F.; Berenbaum, A.; Lough, A.J.; Manners, I. Selective ring-opening reactions of [1]ferrocenophanes with boron halides: A novel route to functionalized ferrocenylboranes and boron-containing oligo- and poly(ferrocene)s. *Chem. Eur. J.* **2000**, *6*, 2762–2771.

148. Heilmann, J.B.; Scheibitz, M.; Qin, Y.; Sundararaman, A.; Jakle, F.; Kretz, T.; Bolte, M.; Lerner, H.W.; Holthausen, M.C.; Wagner, M. A synthetic route to borylene-bridged poly(ferrocenylene)s. *Angew. Chem. Int. Ed.* **2006**, *45*, 920–925.

149. Heilmann, J.B.; Qin, Y.; Jakle, F.; Lerner, H.W.; Wagner, M. Boron-bridged poly(ferrocenylene)s as promising materials for nanoscale molecular wires. *Inorg. Chim. Acta* **2006**, *359*, 4802–4806.

150. Kong, J.; Schmalz, T.; Motz, G.; Muller, A.H.E. Novel hyperbranched ferrocene-containing poly(boro)carbosilanes synthesized via a convenient "A(2) + B-3" approach. *Macromolecules* **2011**, *44*, 1280–1291.

151. Kamperman, M.; Burns, A.; Weissgraeber, R.; van Vegten, N.; Warren, S.C.; Gruner, S.M.; Baiker, A.; Wiesner, U. Integrating structure control over multiple length scales in porous high temperature ceramics with functional platinum nanoparticles. *Nano Lett.* **2009**, *9*, 2756–2762.

152. Glatz, G.; Schmalz, T.; Kraus, T.; Haarmann, F.; Motz, G.; Kempe, R. Copper-containing SiCN precursor ceramics (Cu@SiCN) as selective hydrocarbon oxidation catalysts using air as an oxidant. *Chem. Eur. J.* **2010**, *16*, 4231–4238.

153. Zaheer, M.; Keenan, C.D.; Hermannsdorfer, J.; Roessler, E.; Motz, G.; Senker, J.; Kempe, R. Robust microporous monoliths with integrated catalytically active metal sites investigated by hyperpolarized Xe-129 NMR. *Chem. Mater.* **2012**, *24*, 3952–3963.

154. Zaheer, M.; Schmalz, T.; Motz, G.; Kempe, R. Polymer derived non-oxide ceramics modified with late transition metals. *Chem. Soc. Rev.* **2012**, *41*, 5102–5116.

155. Zaheer, M.; Hermannsdorfer, J.; Kretschmer, W.P.; Motz, G.; Kempe, R. Robust heterogeneous nickel catalysts with tailored porosity for the selective hydrogenolysis of aryl ethers. *ChemCatChem* **2014**, *6*, 91–95.

156. Bazarjani, M.S.; Kleebe, H.J.; Muller, M.M.; Fasel, C.; Baghaie Yazdi, M.; Gurlo, A.; Riedel, R. Nanoporous silicon oxycarbonitride ceramics derived from polysilazanes *in situ* modified with nickel nanoparticles. *Chem. Mater.* **2011**, *23*, 4112–4123.

157. Yajima, S.; Hasegawa, Y.; Hayashi, J.; Iimura, M. Synthesis of continuous silicon-carbide fiber with high-tensile strength and high youngs modulus 1: Synthesis of polycarbosilane as precursor. *J. Mater. Sci.* **1978**, *13*, 2569–2576.

158. Yajima, S.; Hasegawa, Y.; Okamura, K.; Matsuzawa, T. Development of high-tensile strength silicon-carbide fiber using an organosilicon polymer precursor. *Nature* **1978**, *273*, 525–527.

159. Laine, R.M.; Babonneau, F. Preceramic polymer routes to silicon-carbide. *Chem. Mater.* **1993**, *5*, 260–279.

160. Hasegawa, Y.; Iimura, M.; Yajima, S. Synthesis of continuous silicon-carbide fiber 2: Conversion of polycarbosilane fiber into silicon-carbide fibers. *J. Mater. Sci.* **1980**, *15*, 720–728.

161. Ichikawa, H.; Machino, F.; Mitsuno, S.; Ishikawa, T.; Okamura, K.; Hasegawa, Y. Synthesis of continuous silicon-carbide fiber 5: Factors affecting stability of polycarbosilane to oxidation. *J. Mater. Sci.* **1986**, *21*, 4352–4358.

162. Hasegawa, Y. Synthesis of continuous silicon-carbide fiber 6: Pyrolysis process of cured polycarbosilane fiber and structure of SiC fiber. *J. Mater. Sci.* **1989**, *24*, 1177–1190.

163. Taki, T.; Maeda, S.; Okamura, K.; Sato, M.; Matsuzawa, T. Oxidation curing mechanism of polycarbosilane fibers by solid-state Si-29 high-resolution NMR. *J. Mater. Sci. Lett.* **1987**, *6*, 826–828.

164. Hasegawa, Y.; Okamura, K. Synthesis of continuous silicon-carbide fiber 3: Pyrolysis process of polycarbosilane and structure of the products. *J. Mater. Sci.* **1983**, *18*, 3633–3648.

165. Babonneau, F.; Soraru, G.D.; Mackenzie, J.D. Si-29 Mas-NMR investigation of the conversion process of a polytitanocarbosilane into SiC-TiC ceramics. *J. Mater. Sci.* **1990**, *25*, 3664–3670.

166. Dunham, M.L.; Bailey, D.L.; Mixer, R.Y. New curing system for silicone rubber. *Ind. Eng. Chem.* **1957**, *49*, 1373–1376.

167. Valles, E.M.; Macosko, C.W. Properties of networks formed by end linking of poly(dimethylsiloxane). *Macromolecules* **1979**, *12*, 673–679.

168. Heidingsfeldova, M.; Capka, M. Rhodium complexes as catalysts for hydrosilylation crosslinking of silicone-rubber. *J. Appl. Polym. Sci.* **1985**, *30*, 1837–1846.

169. Grzelka, A.; Chojnowski, J.; Cypryk, M.; Fortuniak, W.; Hupfield, P.C.; Taylor, R.G. Polysiloxanol condensation and disproportionation in the presence of a superacid. *J. Organomet. Chem.* **2004**, *689*, 705–713.

170. Scheffler, M.; Bordia, R.; Travitzky, N.; Greil, P. Development of a rapid crosslinking preceramic polymer system. *J. Eur. Ceram. Soc.* **2005**, *25*, 175–180.

171. Yive, N.S.C.K.; Corriu, R.J.P.; Leclercq, D.; Mutin, P.H.; Vioux, A. Silicon carbonitride from polymeric precursors—Thermal cross-linking and pyrolysis of oligosilazane model compounds. *Chem. Mater.* **1992**, *4*, 141–146.

172. Lavedrine, A.; Bahloul, D.; Goursat, P.; Choong Kwet Yive, N.; Corriu, R.; Leclerq, D.; Mutin, H.; Vioux, A. Pyrolysis of polyvinylsilazane precursors to silicon carbonitride. *J. Eur. Ceram. Soc.* **1991**, *8*, 221–227.

173. Martinez-Crespiera, S.; Ionescu, E.; Kleebe, H.J.; Riedel, R. Pressureless synthesis of fully dense and crack-free SiOC bulk ceramics via photo-crosslinking and pyrolysis of a polysiloxane. *J. Eur. Ceram. Soc.* **2011**, *31*, 913–919.

174. Bill, J.; Aldinger, F. Precursor-derived covalent ceramics. *Adv. Mater.* **1995**, *7*, 775–787.

175. Greil, P.; Seibold, M. Modeling of dimensional changes during polymer ceramic conversion for bulk component fabrication. *J. Mater. Sci.* **1992**, *27*, 1053–1060.

176. Schwartz, K.B.; Rowcliffe, D.J. Modeling density contributions in preceramic polymer ceramic powder systems. *J. Am. Ceram. Soc.* **1986**, *69*, C106–C108.

177. Greil, P. Active-filler-controlled pyrolysis of preceramic polymers. *J. Am. Ceram. Soc.* **1995**, *78*, 835–848.

178. Greil, P. Near net shape manufacturing of polymer derived ceramics. *J. Eur. Ceram. Soc.* **1998**, *18*, 1905–1914.

179. Harshe, R.; Balan, C.; Riedel, R. Amorphous Si(Al)OC ceramic from polysiloxanes: Bulk ceramic processing, crystallization behavior and applications. *J. Eur. Ceram. Soc.* **2004**, *24*, 3471–3482.

180. Kaur, S.; Riedel, R.; Ionescu, E. Pressureless fabrication of dense monolithic SiC ceramics from a polycarbosilane. *J. Eur. Ceram. Soc.* **2014**, *34*, 3571–3578.

181. Linck, C.; Ionescu, E.; Papendorf, B.; Galuskova, D.; Galusek, D.; Sajgalik, P.; Riedel, R. Corrosion behavior of silicon oxycarbide-based ceramic nanocomposites under hydrothermal conditions. *Int. J. Mater. Res.* **2012**, *103*, 31–39.

182. Wen, Q.B.; Xu, Y.P.; Xu, B.B.; Fasel, C.; Guillon, O.; Buntkowsky, G.; Yu, Z.J.; Riedel, R.; Ionescu, E. Single-source-precursor synthesis of dense SiC/HfC_xN_{1-x}-based ultrahigh-temperature ceramic nanocomposites. *Nanoscale* **2014**, *6*, 13678–13689.

183. Ishihara, S.; Gu, H.; Bill, J.; Aldinger, F.; Wakai, F. Densification of precursor-derived Si–C–N ceramics by high-pressure hot isostatic pressing. *J. Am. Ceram. Soc.* **2002**, *85*, 1706–1712.

184. Esfehanian, M.; Oberacker, R.; Fett, T.; Hoffmann, M.J. Development of dense filler-free polymer-derived SiOC ceramics by field-assisted sintering. *J. Am. Ceram. Soc.* **2008**, *91*, 3803–3805.

185. Bechelany, M.C.; Salameh, C.; Viard, A.; Guichaoua, L.; Rossignol, F.; Chartier, T.; Bernard, S.; Miele, P. Preparation of polymer-derived Si–B–C–N monoliths by spark plasma sintering technique. *J. Eur. Ceram. Soc.* **2015**, *35*, 1361–1374.

186. Soraru, G.D.; Babonneau, F.; Mackenzie, J.D. Structural concepts on new amorphous covalent solids. *J. Non-Cryst. Solids* **1988**, *106*, 256–261.

187. Monthioux, M.; Oberlin, A.; Bouillon, E. Relationship between microtexture and electrical-properties during heat-treatment of SiC fiber precursor. *Compos. Sci. Technol.* **1990**, *37*, 21–35.

188. Corriu, R.J.P.; Leclercq, D.; Mutin, P.H.; Vioux, A. Preparation and structure of silicon oxycarbide glasses derived from polysiloxane precursors. *J. Sol–Gel Sci. Technol.* **1997**, *8*, 327–330.

189. Pantano, C.G.; Singh, A.K.; Zhang, H.X. Silicon oxycarbide glasses. *J. Sol–Gel Sci. Technol.* **1999**, *14*, 7–25.

190. Belot, V.; Corriu, R.J.P.; Leclercq, D.; Mutin, P.H.; Vioux, A. Silicon oxycarbide glasses with low O/Si ratio from organosilicon precursors. *J. Non-Cryst. Solids* **1994**, *176*, 33–44.

191. Brequel, H.; Enzo, S.; Walter, S.; Soraru, G.D.; Badheka, R.; Babonneau, F. Structure/property relationship in silicon oxycarbide glasses and ceramics obtained via the sol–gel method. *Mater. Sci. Forum.* **2002**, *386–388*, 359–364.

192. Dibandjo, P.; Dire, S.; Babonneau, F.; Soraru, G.D. Influence of the polymer architecture on the high temperature behavior of SiCO glasses: A comparison between linear- and cyclic-derived precursors. *J. Non-Cryst. Solids* **2010**, *356*, 132–140.

193. Zhang, H.; Pantano, C.G. Synthesis and characterization of silicon oxycarbide glasses. *J. Am. Ceram. Soc.* **1990**, *73*, 958–963.

194. Renlund, G.M.; Prochazka, S.; Doremus, R.H. Silicon oxycarbide glasses 1: Preparation and chemistry. *J. Mater. Res.* **1991**, *6*, 2716–2722.

195. Soraru, G.D.; Campostrini, R.; Maurina, S.; Babonneau, F. Gel precursor to silicon oxycarbide glasses with ultrahigh ceramic yield. *J. Am. Ceram. Soc.* **1997**, *80*, 999–1004.

196. Soraru, G.D.; Liu, Q.; Interrante, L.V.; Apple, T. Role of precursor molecular structure on the microstructure and high temperature stability of silicon oxycarbide glasses derived from methylene-bridged polycarbosilanes. *Chem. Mater.* **1998**, *10*, 4047–4054.

197. Belot, V.; Corriu, R.J.P.; Leclercq, D.; Mutin, P.H.; Vioux, A. Thermal redistribution reactions in cross-linked polysiloxanes. *J. Polym. Sci. Pol. Chem.* **1992**, *30*, 613–623.

198. Bois, L.; Maquet, J.; Babonneau, F.; Mutin, H.; Bahloul, D. Structural characterization of sol–gel derived oxycarbide glasses 1: Study of the pyrolysis process. *Chem. Mater.* **1994**, *6*, 796–802.

199. Kalfat, R.; Babonneau, F.; Gharbi, N.; Zarrouk, H. Si-29 MAS NMR investigation of the pyrolysis process of cross-linked polysiloxanes prepared from polymethylhydrosiloxane. *J. Mater. Chem.* **1996**, *6*, 1673–1678.

200. Gualandris, V.; Hourlier-Bahloul, D.; Babonneau, F. Structural investigation of the first stages of pyrolysis of Si–C–O preceramic polymers containing Si–H bonds. *J. Sol–Gel Sci. Technol.* **1999**, *14*, 39–48.

201. Ionescu, E.; Papendorf, B.; Kleebe, H.-J.; Poli, F.; Muller, K.; Riedel, R. Polymer-derived silicon oxycarbide/hafnia ceramic nanocomposites. Part I: Phase and microstructure evolution during the ceramization process. *J. Am. Ceram. Soc.* **2010**, *93*, 1774–1782.

202. Chomel, A.D.; Dempsey, P.; Latournerie, J.; Hourlier-Bahloul, D.; Jayasooriya, U.A. Gel to glass transformation of methyltriethoxysilane: A silicon oxycarbide glass precursor investigated using vibrational spectroscopy. *Chem. Mater.* **2005**, *17*, 4468–4473.

203. Ionescu, E.; Linck, C.; Fasel, C.; Mueller, M.; Kleebe, H.-J.; Riedel, R. Polymer-derived SiOC/ZrO$_2$ ceramic nanocomposites with excellent high-temperature stability. *J. Am. Ceram. Soc.* **2010**, *93*, 241–250.

204. Soraru, G.D.; Pederiva, L.; Latournerie, J.; Raj, R. Pyrolysis kinetics for the conversion of a polymer into an amorphous silicon oxycarbide ceramic. *J. Am. Ceram. Soc.* **2002**, *85*, 2181–2187.

205. Narisawa, M.; Watase, S.; Matsukawa, K.; Dohmaru, T.; Okamura, K. White Si–O–C(–H) particles with photoluminescence synthesized by decarbonization reaction on polymer precursor in a hydrogen atmosphere. *Bull. Chem. Soc. Jpn.* **2012**, *85*, 724–726.

206. Bill, J.; Seitz, J.; Thurn, G.; Durr, J.; Canel, J.; Janos, B.Z.; Jalowiecki, A.; Sauter, D.; Schempp, S.; Lamparter, H.P.; *et al.* Structure analysis and properties of Si–C–N ceramics derived from polysilazanes. *Phys. Status Solidi A* **1998**, *166*, 269–296.

207. Seitz, J.; Bill, J.; Egger, N.; Aldinger, F. Structural investigations of Si/C/N-ceramics from polysilazane precursors by nuclear magnetic resonance. *J. Eur. Ceram. Soc.* **1996**, *16*, 885–891.

208. Li, Y.L.; Kroke, E.; Riedel, R.; Fasel, C.; Gervais, C.; Babonneau, F. Thermal cross-linking and pyrolytic conversion of poly(ureamethylvinyl)silazanes to silicon-based ceramics. *Appl. Organomet. Chem.* **2001**, *15*, 820–832.

209. Laine, R.M.; Babonneau, F.; Blowhowiak, K.Y.; Kennish, R.A.; Rahn, J.A.; Exarhos, G.J.; Waldner, K. The evolutionary process during pyrolytic transformation of poly(N-methylsilazane) from a preceramic polymer into an amorphous-silicon nitride carbon composite. *J. Am. Ceram. Soc.* **1995**, *78*, 137–145.

210. Schuhmacher, J.; Weinmann, M.; Bill, J.; Aldinger, F.; Muller, K. Solid-state NMR studies of the preparation of Si–C–N ceramics from polysilylcarbodiimide polymers. *Chem. Mater.* **1998**, *10*, 3913–3922.

211. Riedel, R.; Toma, L.; Fasel, C.; Miehe, G. Polymer-derived mullite-SiC-based nanocomposites. *J. Eur. Ceram. Soc.* **2009**, *29*, 3079–3090.

212. Dire, S.; Ceccato, R.; Gialanella, S.; Babonneau, F. Thermal evolution and crystallisation of polydimethylsiloxane-zirconia nanocomposites prepared by the sol–gel method. *J. Eur. Ceram. Soc.* **1999**, *19*, 2849–2858.

213. Ionescu, E.; Papendorf, B.; Kleebe, H.J.; Riedel, R. Polymer-derived silicon oxycarbide/hafnia ceramic nanocomposites. Part II: Stability toward decomposition and microstructure evolution at T1000 degrees C. *J. Am. Ceram. Soc.* **2010**, *93*, 1783–1789.

214. Bazarjani, M.S.; Kleebe, H.J.; Muller, M.M.; Fasel, C.; Yazdi, M.B.; Gurlo, A.; Riedel, R. Nanoporous silicon oxycarbonitride ceramics derived from polysilazanes *in situ* modified with nickel nanoparticles. *Chem. Mater.* **2011**, *23*, 4112–4123.

215. Hauser, R.; Francis, A.; Theismann, R.; Riedel, R. Processing and magnetic properties of metal-containing SiCN ceramic micro- and nano-composites. *J. Mater. Sci.* **2008**, *43*, 4042–4049.

216. Francis, A.; Ionescu, E.; Fasel, C.; Riedel, R. Crystallization behavior and controlling mechanism of iron-containing Si–C–N ceramics. *Inorg. Chem.* **2009**, *48*, 10078–10083.

217. Zaheer, M.; Motz, G.; Kempe, R. The generation of palladium silicide nanoalloy particles in a SiCN matrix and their catalytic applications. *J. Mater. Chem.* **2011**, *21*, 18825–18831.

218. Kaspar, J.; Terzioglu, C.; Ionescu, E.; Graczyk-Zajac, M.; Hapis, S.; Kleebe, H.J.; Riedel, R. Stable SiOC/Sn nanocomposite anodes for lithium-ion batteries with outstanding cycling stability. *Adv. Funct. Mater.* **2014**, *24*, 4097–4104.

219. Hojamberdiev, M.; Prasad, R.M.; Fasel, C.; Riedel, R.; Ionescu, E. Single-source-precursor synthesis of soft magnetic Fe_3Si- and Fe_5Si_3-containing SiOC ceramic nanocomposites. *J. Eur. Ceram. Soc.* **2013**, *33*, 2465–2472.

220. Umicevic, A.B.; Cekic, B.D.; Belosevic-Cavor, J.N.; Koteski, V.J.; Papendorf, B.; Riedel, R.; Ionescu, E. Evolution of the local structure at Hf sites in SiHfOC upon ceramization of a hafnium-alkoxide-modified polysilsesquioxane: A perturbed angular correlation study. *J. Eur. Ceram. Soc.* **2015**, *35*, 29–35.

221. Ionescu, E.; Terzioglu, C.; Linck, C.; Kaspar, J.; Navrotsky, A.; Riedel, R. Thermodynamic control of phase composition and crystallization of metal-modified silicon oxycarbides. *J. Am. Ceram. Soc.* **2013**, *96*, 1899–1903.

222. Sing, K.S.W.; Everett, D.H.; Haul, R.A.W.; Moscou, L.; Pierotti, R.A.; Rouquerol, J.; Siemieniewska, T. Reporting physisorption data for gas solid systems with special reference to the determination of surface-area and porosity (Recommendations 1984). *Pure Appl. Chem.* **1985**, *57*, 603–619.

223. Kresge, C.T.; Leonowicz, M.E.; Roth, W.J.; Vartuli, J.C.; Beck, J.S. Ordered mesoporous molecular-sieves synthesized by a liquid-crystal template mechanism. *Nature* **1992**, *359*, 710–712.

224. Beck, J.S.; Vartuli, J.C.; Roth, W.J.; Leonowicz, M.E.; Kresge, C.T.; Schmitt, K.D.; Chu, C.T.W.; Olson, D.H.; Sheppard, E.W.; Mccullen, S.B.; *et al.* A New family of mesoporous molecular-sieves prepared with liquid-crystal templates. *J. Am. Chem. Soc.* **1992**, *114*, 10834–10843.

225. Kresge, C.T.; Vartuli, J.C.; Roth, W.J.; Leonowicz, M.E.; Beck, J.S.; Schmitt, K.D.; Chu, C.T.W.; Olson, D.H.; Sheppard, E.W.; McCullen, S.B.; *et al.* M41S: A new family of mesoporous molecular sieves prepared with liquid crystal templates. *Stud. Surf. Sci. Catal.* **1995**, *92*, 11–19.

226. Yanagisawa, T.; Shimizu, T.; Kuroda, K.; Kato, C. The preparation of alkyltrimethylammonium-kanemite complexes and their conversion to microporous materials. *Bull. Chem. Soc. Jpn.* **1990**, *63*, 988–992.

227. Shi, Y.F.; Wan, Y.; Zhao, D.Y. Ordered mesoporous non-oxide materials. *Chem. Soc. Rev.* **2011**, *40*, 3854–3878.

228. Wan, Y.; Zhao, D.Y. On the controllable soft-templating approach to mesoporous silicates. *Chem. Rev.* **2007**, *107*, 2821–2860.

229. Kamperman, M.; Garcia, C.B.W.; Du, P.; Ow, H.S.; Wiesner, U. Ordered mesoporous ceramics stable up to 1500 degrees C from diblock copolymer mesophases. *J. Am. Chem. Soc.* **2004**, *126*, 14708–14709.

230. Malenfant, P.R.L.; Wan, J.L.; Taylor, S.T.; Manoharan, M. Self-assembly of an organic-inorganic block copolymer for nano-ordered ceramics. *Nat. Nanotechnol.* **2007**, *2*, 43–46.

231. Nghiem, Q.D.; Kim, D.; Kim, D.P. Synthesis of inorganic-organic diblock copolymers as a precursor of ordered mesoporous SiCN ceramic. *Adv. Mater.* **2007**, *19*, 2351–2354.

232. Ryoo, R.; Joo, S.H.; Jun, S. Synthesis of highly ordered carbon molecular sieves via template-mediated structural transformation. *J. Phys. Chem. B* **1999**, *103*, 7743–7746.

233. Shi, Y.F.; Meng, Y.; Chen, D.H.; Cheng, S.J.; Chen, P.; Yang, T.F.; Wan, Y.; Zhao, D.Y. Highly ordered mesoporous silicon carbide ceramics with large surface areas and high stability. *Adv. Funct. Mater.* **2006**, *16*, 561–567.

234. Shi, Y.; Wan, Y.; Tu, B.; Zhao, D. Nanocasting synthesis of ordered mesoporous silicon nitrides with a high nitrogen content. *J. Phys. Chem. C* **2008**, *112*, 112–116.

235. Yan, J.; Wang, A.; Kim, D.-P. Preparation of ordered mesoporous SiC from preceramic polymer templated by nanoporous silica. *J. Phys. Chem. B* **2006**, *110*, 5429–5433

236. Monthioux, M.; Delverdier, O. Thermal behavior of (organosilicon) polymer-derived ceramics 5: Main facts and trends. *J. Eur. Ceram. Soc.* **1996**, *16*, 721–737.

237. Yan, J.; Wang, A.J.; Kim, D.P. Preparation of ordered mesoporous SiCN ceramics with large surface area and high thermal stability. *Microporous Mesoporous Mater.* **2007**, *100*, 128–133.

238. Shi, Y.F.; Wan, Y.; Zhai, Y.P.; Liu, R.L.; Meng, Y.; Tu, B.; Zhao, D.Y. Ordered mesoporous SiOC and SiCN ceramics from atmosphere-assisted *in situ* transformation. *Chem. Mater.* **2007**, *19*, 1761–1771.

239. Bill, J.; Kamphowe, T.W.; Muller, A.; Wichmann, T.; Zern, A.; Jalowieki, A.; Mayer, J.; Weinmann, M.; Schuhmacher, J.; Muller, K.; *et al.* Precursor-derived Si–(B–)C–N ceramics: Thermolysis, amorphous state and crystallization. *Appl. Organomet. Chem.* **2001**, *15*, 777–793.

240. Yan, X.B.; Gottardo, L.; Bernard, S.; Dibandjo, P.; Brioude, A.; Moutaabbid, H.; Miele, P. Ordered mesoporous silicoboron carbonitride materials via preceramic polymer nanocasting. *Chem. Mater.* **2008**, *20*, 6325–6334.

241. Majoulet, O.; Alauzun, J.G.; Gottardo, L.; Gervais, C.; Schuster, M.E.; Bernard, S.; Miele, P. Ordered mesoporous silicoboron carbonitride ceramics from boron-modified polysilazanes: Polymer synthesis, processing and properties. *Microporous Microporous Mater.* **2011**, *140*, 40–50.

242. Bernard, S.; Majoulet, O.; Sandra, F.; Malchere, A.; Miele, P. Direct synthesis of periodic mesoporous silicoboron carbonitride frameworks via the nanocasting from ordered mesoporous silica with boron-modified polycarbosilazane. *Adv. Eng. Mater.* **2013**, *15*, 134–140.

243. Lewis, M.H.; Barnard, P. Oxidation mechanisms in Si–Al–O–N ceramics. *J. Mater. Sci.* **1980**, *15*, 443–448.

244. MacKenzie, K.J.D.; Shimada, S.; Aoki, T. Thermal oxidation of carbothermal β'-sialon powder: Reaction sequence and kinetics. *J. Mater. Chem.* **1997**, *7*, 527–530.

245. Nordberg, L.O.; Nygren, M.; Kall, P.O.; Shen, Z.J. Stability and oxidation properties of RE-alpha-sialon ceramics (RE = Y, Nd, Sm, Yb). *J. Am. Ceram. Soc.* **1998**, *81*, 1461–1470.

246. An, L.N.; Wang, Y.G.; Bharadwaj, L.; Zhang, L.G.; Fan, Y.; Jiang, D.P.; Sohn, Y.H.; Desai, V.H.; Kapat, J.; Chow, L.C.; *et al.* Silicoaluminum carbonitride with anomalously high resistance to oxidation and hot corrosion. *Adv. Eng. Mater.* **2004**, *6*, 337–340.

247. Majoulet, O.; Salameh, C.; Schuster, M.E.; Demirci, U.B.; Sugahara, Y.; Bernard, S.; Miele, P. Preparation, characterization, and surface modification of periodic mesoporous silicon-aluminum-carbon-nitrogen frameworks. *Chem. Mater.* **2013**, *25*, 3957–3970.

248. Alauzun, J.G.; Ungureanu, S.; Brun, N.; Bernard, S.; Miele, P.; Backov, R.; Sanchez, C. Novel monolith-type boron nitride hierarchical foams obtained through integrative chemistry. *J. Mater. Chem.* **2011**, *21*, 14025–14030.

249. Majoulet, O.; Sandra, F.; Bechelany, M.C.; Bonnefont, G.; Fantozzi, G.; Joly-Pottuz, L.; Malchere, A.; Bernard, S.; Miele, P. Silicon-boron-carbon-nitrogen monoliths with high, interconnected and hierarchical porosity. *J. Mater. Chem. A* **2013**, *1*, 10991–11000.

250. Bernard, S.; Miele, P. Ordered mesoporous polymer-derived ceramics and their processing into hierarchically porous boron nitride and silicoboron carbonitride monoliths. *New J. Chem.* **2014**, *38*, 1923–1931.

251. Bates, F.S.; Fredrickson, G.H. Block copolymers—designer soft materials. *Phys. Today* **1999**, *52*, 32–38.

252. Hamley, I.W. *The Physics of Block Copolymers*; Oxford University Press: Oxford, UK, 1998.

253. Hillmyer, M.A. Nanoporous materials fromblock copolymer precursors. *Adv. Polym. Sci.* **2005**, *190*, 137–181.

254. Darling, S.B. Directing the self-assembly of block copolymers. *Prog. Polym. Sci.* **2007**, *32*, 1152–1204.

255. Aissou, K.; Shaver, J.; Fleury, G.; Pecastaings, G.; Brochon, C.; Navarro, C.; Grauby, S.; Rampnoux, J.M.; Dilhaire, S.; Hadziioannou, G.; *et al.* Nanoscale block copolymer ordering induced by visible interferometric micropatterning: A route towards large scale block copolymer 2D crystals. *Adv. Mater.* **2013**, *25*, 213–217.

256. Koo, K.; Ahn, H.; Kim, S.-W.; Ryu, D.Y.; Russell, T.P. Directed self-assembly of block copolymers in the extreme: Guiding microdomains from the small to the large. *Soft Matter* **2013**, *9*, 9059–9071.

257. Orilall, M.C.; Wiesner, U. Block copolymer based composition and morphology control in nanostructured hybrid materials for energy conversion and storage: Solar cells, batteries, and fuel cells. *Chem. Soc. Rev.* **2011**, *40*, 520–535.

258. She, M.-S.; Lo, T.-Y.; Hsueh, H.-Y.; Ho, R.-M. Nanostructured thin films of degradable block copolymers and their applications. *NPG Asia Mater.* **2013**, *5*, e42:1–e42:9.

259. Ren, Y.; Ma, Z.; Bruce, P.G. Ordered mesoporous metal oxides: Synthesis and applications. *Chem. Soc. Rev.* **2012**, *41*, 4909–4927.

260. Innocenzi, P.; Malfatti, L. Mesoporous thin films: Properties and applications. *Chem. Soc. Rev.* **2013**, *42*, 4198–4216.

261. Petkovich, N.D.; Stein, A. Controlling macro- and mesostructures with hierarchical porosity through combined hard and soft templating. *Chem. Soc. Rev.* **2013**, *42*, 3721–3739.

262. Rawolle, M.; Niedermeier, M.A.; Kaune, G.; Perlich, J.; Lellig, P.; Memesa, M.; Cheng, Y.J.; Gutmann, J.S.; Muller-Buschbaum, P. Fabrication and characterization of nanostructured titania films with integrated function from inorganic-organic hybrid materials. *Chem. Soc. Rev.* **2012**, *41*, 5131–5142.

263. Hsueh, H.Y.; Ho, R.M. Bicontinuous ceramics with high surface area from block copolymer templates. *Langmuir* **2012**, *28*, 8518–8529.

264. Hsueh, H.Y.; Chen, H.Y.; She, M.S.; Chen, C.K.; Ho, R.M.; Gwo, S.; Hasegawa, H.; Thomas, E.L. Inorganic gyroid with exceptionally low refractive index from block copolymer templating. *Nano Lett.* **2010**, *10*, 4994–5000.

265. Jones, B.H.; Lodge, T.P. High-temperature nanoporous ceramic monolith prepared from a polymeric bicontinuous microemulsion template. *J. Am. Chem. Soc.* **2009**, *131*, 1676–1677.

266. Reitz, C.; Haetge, J.; Suchomski, C.; Brezesinski, T. Facile and general synthesis of thermally stable ordered mesoporous rare-earth oxide ceramic thin films with uniform mid-size to large-size pores and strong crystalline texture. *Chem. Mater.* **2013**, *25*, 4633–4642.

267. Rauda, I.E.; Buonsanti, R.; Saldarriaga-Lopez, L.C.; Benjauthrit, K.; Schelhas, L.T.; Stefik, M.; Augustyn, V.; Ko, J.; Dunn, B.; Wiesner, U.; *et al.* General method for the synthesis of hierarchical nanocrystal-based mesoporous materials. *ACS Nano* **2012**, *6*, 6386–6399.

268. Liu, Y.; Zhang, J.; Tu, M. Synthesis of porous methylphenylsiloxane/poly(dimethylsiloxane) composite using poly(dimethylsiloxane)-poly(ethylene oxide) (PDMS-PEO) as template. *J. Mater. Sci.* **2012**, *47*, 3350–3353.

269. Kamperman, M.; Du, P.; Scarlat, R.O.; Herz, E.; Werner-Zwanziger, U.; Graf, R.; Zwanziger, J.W.; Spiess, H.W.; Wiesner, U. Composition and morphology control in ordered mesostructured high-temperature ceramics from block copolymer mesophases. *Macromol. Chem. Phys.* **2007**, *208*, 2096–2108.

270. Wan, J.; Malenfant, P.R.L.; Taylor, S.T.; Loureiro, S.M.; Manoharan, M. Microstructure of block copolymer/precursor assembly for Si–C–N based nano-ordered ceramics. *Mater. Sci. Eng. A* **2007**, *463*, 78–88.

271. Nghiem, Q.D.; Kim, D.-P. Direct preparation of high surface area mesoporous SiC-based ceramic by pyrolysis of a self-assembled polycarbosilane-block-polystyrene diblock copolymer. *Chem. Mater.* **2008**, *20*, 3735–3739.

272. Matsumoto, K.; Matsuoka, H. Synthesis of core-crosslinked carbosilane block copolymer micelles and their thermal transformation to silicon-based ceramics nanoparticles. *J. Polym. Sci. A* **2005**, *43*, 3778–3787.

273. Nguyen, C.T.; Hoang, P.H.; Perumal, J.; Kim, D.P. An inorganic-organic diblock copolymer photoresist for direct mesoporous SiCN ceramic patterns via photolithography. *Chem. Commun.* **2011**, *47*, 3484–3486.

274. Chan, V.Z.-H.; Hoffman, J.; Lee, V.Y.; Iatrou, H.; Avgeropoulos, A.; Hadjichristidis, N.; Miller, R.D.; Thomas, E.L. Ordered bicontinuous nanoporous and nanorelief ceramic films from self assembling polymer precursors. *Science* **1999**, *286*, 1716–1719.

275. Li, Z.; Sai, H.; Tan, K.W.; Hoheisel, T.N.; Gruner, S.M.; Wiesner, U. Ordered nanostructured ceramic-metal composites through multifunctional block copolymer-metal nanoparticle self-assembly. *J. Sol–Gel Sci. Technol.* **2014**, *70*, 286–291.

276. Chao, C.-C.; Ho, R.-M.; Georgopanos, P.; Avgeropoulos, A.; Thomas, E.L. Silicon oxy carbide nanorings from polystyrene-b-polydimethylsiloxane diblock copolymer thin films. *Soft Matter* **2010**, *6*, 3582–3587.

277. Pillai, S.K.; Kretschmer, W.P.; Denner, C.; Motz, G.; Hund, M.; Fery, A.; Trebbin, M.; Forster, S.; Kempe, R. SiCN nanofibers with a diameter below 100 nm synthesized via concerted block copolymer formation, microphase separation, and crosslinking. *Small* **2013**, *9*, 984–989.

278. Ahmed, R.; Priimagi, A.; Faul, C.F.; Manners, I. Redox-active, organometallic surface-relief gratings from azobenzene-containing polyferrocenylsilane block copolymers. *Adv. Mater.* **2012**, *24*, 926–931.

279. Cao, L.; Massey, J.A.; Winnik, M.A.; Manners, I.; Riethmüller, S.; Banhart, F.; Spatz, J.P.; Möller, M. Reactive ion etching of cylindrical polyferrocenylsilane block copolymer micelles: Fabrication of ceramic nanolines on semiconducting substrates. *Adv. Funct. Mater.* **2003**, *13*, 271–276.

280. Temple, K.; Kulbaba, K.; Power-Billard, K.N.; Manners, I.; Leach, K.A.; Xu, T.; Russell, T.P.; Hawker, C.J. Spontaneous vertical ordering and pyrolytic formation of nanoscopic ceramic patterns from poly(styrene-*b*-ferrocenylsilane). *Adv. Mater.* **2003**, *15*, 297–300.

281. Wang, X.-S.; Arsenault, A.; Ozin, G.A.; Winnik, M.A.; Manners, I. Shell cross-linked cylinders of polyisoprene-*b*-ferrocenyldimethylsilane: Formation of magnetic ceramic replicas and microfluidic channel alignment and patterning. *J. Am. Chem. Soc.* **2003**, *125*, 12686–12687.

282. Cheng, J.Y.; Ross, C.A.; Chan, V.Z.-H.; Thomas, E.L.; Lammertink, R.G.H.; Vancso, G.J. Formation of a cobalt magnetic dot array via block copolymer lithography. *Adv. Mater.* **2001**, *13*, 1174–1178.

283. Gallei, M.; Tockner, S.; Klein, R.; Rehahn, M. Silacyclobutane-based diblock copolymers with vinylferrocene, ferrocenylmethyl methacrylate, and [1]dimethylsilaferrocenophane. *Macromol. Rapid Commun.* **2010**, *31*, 889–896.

284. Gallei, M.; Schmidt, B.V.K.J.; Klein, R.; Rehahn, M. Defined poly[styrene-block-(ferrocenylmethyl methacrylate)] diblock copolymers via living anionic polymerization. *Macromol. Rapid Commun.* **2009**, *30*, 1463–1469.

285. Rittscher, V.; Gallei, M. A convenient synthesis strategy for microphase-separating functional copolymers: The cyclohydrocarbosilane tool box. *Polym. Chem.* **2015**, doi:10.1039/C1035PY00065C.

286. Ge, J.; Yin, Y. Responsive photonic crystals. *Angew. Chem. Int. Ed. Engl.* **2011**, *50*, 1492–1522.

287. Schäfer, C.G.; Lederle, C.; Zentel, K.; Stühn, B.; Gallei, M. Utilising stretch-tunable thermochromic elastomeric opal films as novel reversible switchable photonic materials. *Macromol. Rapid Commun.* **2014**, *35*, 1852–1860.

288. Schäfer, C.G.; Viel, B.; Hellmann, G.P.; Rehahn, M.; Gallei, M. Thermo-cross-linked elastomeric opal films. *ACS Appl. Mater. Interfaces* **2013**, *5*, 10623–10632.

289. Schäfer, C.G.; Smolin, D.A.; Hellmann, G.P.; Gallei, M. Fully reversible shape transition of soft spheres in elastomeric polymer opal films. *Langmuir* **2013**, *29*, 11275–11283.

290. Schäfer, C.G.; Gallei, M.; Zahn, J.T.; Engelhardt, J.; Hellmann, G.P.; Rehahn, M. Reversible light-, thermo-, and mechano-responsive elastomeric polymer opal films. *Chem. Mater.* **2013**, *25*, 2309–2318.

291. Galisteo-López, J.F.; Ibisate, M.; Sapienza, R.; Froufe-Pérez, L.S.; Blanco, Á.; López, C. Self-assembled photonic structures. *Adv. Mater.* **2011**, *23*, 30–69.

292. Von Freymann, G.; Kitaev, V.; Lotsch, B.V.; Ozin, G.A. Bottom-up assembly of photonic crystals. *Chem. Soc. Rev.* **2013**, *42*, 2528–2554.

293. Schäfer, C.G.; Vowinkel, S.; Hellmann, G.P.; Herdt, T.; Contiu, C.; Schneider, J.J.; Gallei, M. A polymer based and template-directed approach towards functional multidimensional microstructured organic/inorganic hybrid materials. *J. Mater. Chem. C* **2014**, *2*, 7960–7975.

294. Galloro, J.; Ginzburg, M.; Miguez, H.; Yang, S.M.; Coombs, N.; Safa-Sefat, A.; Greedan, J.E.; Manners, I.; Ozin, G.A. Replicating the structure of a crosslinked polyferrocenylsilane inverse opal in the form of a magnetic ceramic. *Adv. Funct. Mater.* **2002**, *12*, 382–388.

295. Zhou, J.; Li, H.; Ye, L.; Liu, J.; Wang, J.; Zhao, T.; Jiang, L.; Song, Y. Facile fabrication of tough SiC inverse opal photonic crystals. *J. Phys. Chem. C* **2010**, *114*, 22303–22308.

296. Brequel, H.; Soraru, G.D.; Schiffini, L.; Enzo, S. Radial distribution function of amorphous silicon oxycarbide compounds. *Metastable Mech. Alloy. Nanocryst. Mater.* **2000**, *8*, 677–682.

297. Brequel, H.; Parmentier, J.; Walter, S.; Badheka, R.; Trimmel, G.; Masse, S.; Latournerie, J.; Dempsey, P.; Turquat, C.; Desmartin-Chomel, A.; *et al*. Systematic structural characterization of the high-temperature behavior of nearly stoichiometric silicon oxycarbide glasses. *Chem. Mater.* **2004**, *16*, 2585–2598.

298. Belot, V.; Corriu, R.; Leclercq, D.; Mutin, P.H.; Vioux, A. Thermal reactivity of hydrogenosilsesquioxane gels. *Chem. Mater.* **1991**, *3*, 127–131.

299. Kleebe, H.J.; Turquat, C.; Soraru, G.D. Phase separation in an SiCO glass studied by transmission electron microscopy and electron energy-loss spectroscopy. *J. Am. Ceram. Soc.* **2001**, *84*, 1073–1080.

300. Burns, G.T.; Taylor, R.B.; Xu, Y.R.; Zangvil, A.; Zank, G.A. High-temperature chemistry of the conversion of siloxanes to silicon carbide. *Chem. Mater.* **1992**, *4*, 1313–1323.

301. Saha, A.; Raj, R. Crystallization maps for SiCO amorphous ceramics. *J. Am. Ceram. Soc.* **2007**, *90*, 578–583.

302. Nonnenmacher, K.; Kleebe, H.J.; Rohrer, J.; Ionescu, E.; Riedel, R. Carbon mobility in SiOC/HfO$_2$ ceramic nanocomposites. *J. Am. Ceram. Soc.* **2013**, *96*, 2058–2060.

303. Schiavon, M.A.; Gervais, C.; Babonneau, F.; Soraru, G.D. Crystallization behavior of novel silicon boron oxycarbide glasses. *J. Am. Ceram. Soc.* **2004**, *87*, 203–208.

304. Klonczynski, A.; Schneider, G.; Riedel, R.; Theissmann, R. Influence of boron on the microstructure of polymer derived SiCO ceramics. *Adv. Eng. Mater.* **2004**, *6*, 64–68.

305. Ngoumeni-Yappi, R.; Fasel, C.; Riedel, R.; Ischenko, V.; Pippel, E.; Woltersdorf, J.; Clade, J. Tuning of the rheological properties and thermal behavior of boron-containing polysiloxanes. *Chem. Mater.* **2008**, *20*, 3601–3608.

306. Soraru, G.D.; Pena-Alonso, R.; Kleebe, H.J. The effect of annealing at 1400 °C on the structural evolution of porous C-rich silicon (boron)oxycarbide glass. *J. Eur. Ceram. Soc.* **2012**, *32*, 1751–1757.

307. Chavez, R.; Ionescu, E.; Fasel, C.; Riedel, R. Silicon-containing polyimide-based polymers with high temperature stability. *Chem. Mater.* **2010**, *22*, 3823–3825.

308. Riedel, R.; Seher, M. Crystallization behaviour of amorphous silicon nitride. *J. Eur. Ceram. Soc.* **1991**, *7*, 21–25.

309. Weinmann, M.; Zern, A.; Aldinger, F. Stoichiometric silicon nitride/silicon carbide composites from polymeric precursors. *Adv. Mater.* **2001**, *13*, 1704–1708.

310. Riedel, R.; Kienzle, A.; Dressler, W.; Ruwisch, L.; Bill, J.; Aldinger, F. A silicoboron carbonitride ceramic stable to 2000 °C. *Nature* **1996**, *382*, 796–798.

311. Wang, Z.C.; Aldinger, F.; Riedel, R. Novel silicon-boron-carbon-nitrogen materials thermally stable up to 2200 °C. *J. Am. Ceram. Soc.* **2001**, *84*, 2179–2183.

312. Matsunaga, K.; Iwamoto, Y.; Fisher, C.A.J.; Matsubara, H. Molecular dynamics study of atomic structures in amorphous Si–C–N ceramics. *J. Ceram. Soc. Jpn.* **1999**, *107*, 1025–1031.

313. Matsunaga, K.; Iwamoto, Y. Molecular dynamics study of atomic structure and diffusion behavior in amorphous silicon nitride containing boron. *J. Am. Ceram. Soc.* **2001**, *84*, 2213–2219.

314. Tavakoli, A.H.; Gerstel, P.; Golczewski, J.A.; Bill, J. Crystallization kinetics of Si$_3$N$_4$ in Si–B–C–N polymer-derived ceramics. *J. Mater. Res.* **2010**, *25*, 2150–2158.

315. Tavakoli, A.H.; Gerstel, P.; Golczewski, J.A.; Bill, J. Kinetic effect of boron on the crystallization of Si_3N_4 in Si–B–C–N polymer-derived ceramics. *J. Mater. Res.* **2011**, *26*, 600–608.

316. Tavakoli, A.H.; Golczewski, J.A.; Bill, J.; Navrotsky, A. Effect of boron on the thermodynamic stability of amorphous polymer-derived Si(B)CN ceramics. *Acta Mater.* **2012**, *60*, 4514–4522.

317. Wang, Y.G.; Fei, W.F.; An, L.N. Oxidation/corrosion of polymer-derived SiAlCN ceramics in water vapor. *J. Am. Ceram. Soc.* **2006**, *89*, 1079–1082.

318. Sujith, R.; Kousaalya, A.B.; Kumar, R. Coarsening induced phase transformation of hafnia in polymer-derived Si–Hf–C–N–O ceramics. *J. Am. Ceram. Soc.* **2011**, *94*, 2788–2791.

319. Gurlo, A. Electrical Properties of Polymer-Derived Ceramics. In *Polymer Derived Ceramics: From Nanostructure to Applications*; Colombo, P., Riedel, R., Kleebe, H.-J., Soraru, G.D., Eds.; DesTech Publications, Inc.: Lancaster, PA, USA, 2010; pp. 261–274.

320. Chollon, G. Oxidation behaviour of ceramic fibres from the Si–C–N–O system and related sub-systems. *J. Eur. Ceram. Soc.* **2000**, *20*, 1959–1974.

321. Naslain, R.; Guette, A.; Rebillat, F.; le Gallet, S.; Lamouroux, F.; Filipuzzi, L.; Louchet, C. Oxidation mechanisms and kinetics of SiC-matrix composites and their constituents. *J. Mater. Sci.* **2004**, *39*, 7303–7316.

322. Rodriguezviejo, J.; Sibieude, F.; Clavagueramora, M.T.; Monty, C. O-18 diffusion through amorphous SiO_2 and cristobalite. *Appl. Phys. Lett.* **1993**, *63*, 1906–1908.

323. Modena, S.; Soraru, G.D.; Blum, Y.; Raj, R. Passive oxidation of an effluent system: The case of polymer-derived SiCO. *J. Am. Ceram. Soc.* **2005**, *88*, 339–345.

324. Butchereit, E.; Nickel, K.G. Oxidation behaviour of precursor derived Si–(B)–C–N ceramics. *J. Mater. Process. Manuf.* **1998**, *7*, 15–21.

325. Butchereit, E.; Nickel, K.G. Oxidation behaviour of precursor derived ceramics in the system Si–(B)–C–N. *Key Eng. Mater.* **2000**, *175–176*, 69–77.

326. Muller, A.; Gerstel, P.; Butchereit, E.; Nickel, K.G.; Aldinger, F. Si/B/C/N/Al precursor-derived ceramics: Synthesis, high temperature behaviour and oxidation resistance. *J. Eur. Ceram. Soc.* **2004**, *24*, 3409–3417.

327. Nguyen, V.L. Molecular Approaches to Novel SiHfBCN Ceramic Nanocomposites: Synthesis and High-Temperature Behavior. Master Thesis, Technische Universität Darmstadt, Darmstadt, Germany, 2011.

328. Wang, Y.G.; An, L.N.; Fan, Y.; Zhang, L.G.; Burton, S.; Gan, Z.H. Oxidation of polymer-derived SiAlCN ceramics. *J. Am. Ceram. Soc.* **2005**, *88*, 3075–3080.

329. Wang, Y.G.; Fan, Y.; Zhang, L.G.; Zhang, W.G.; An, L.A. Polymer-derived SiAlCN ceramics resist oxidation at 1400 °C. *Scr. Mater.* **2006**, *55*, 295–297.

330. Wang, Y.G.; Sohn, Y.H.; Fan, Y.; Zhang, L.G.; An, L.N. Oxygen diffusion through Al-doped amorphous SiO_2. *J. Phase Equilib. Diffus.* **2006**, *27*, 671–675.

331. Renlund, G.M.; Prochazka, S.; Doremus, R.H. Silicon oxycarbide glasses 2: Structure and properties. *J. Mater. Res.* **1991**, *6*, 2723–2734.

332. Soraru, G.D.; Dallapiccola, E.; DAndrea, G. Mechanical characterization of sol–gel-derived silicon oxycarbide glasses. *J. Am. Ceram. Soc.* **1996**, *79*, 2074–2080.

333. Rouxel, T.; Massouras, G.; Soraru, G.D. High temperature behavior of a gel-derived SiOC glass: Elasticity and viscosity. *J. Sol–Gel Sci. Technol.* **1999**, *14*, 87–94.

334. Rouxel, T.; Soraru, G.D.; Vicens, J. Creep viscosity and stress relaxation of gel-derived silicon oxycarbide glasses. *J. Am. Ceram. Soc.* **2001**, *84*, 1052–1058.

335. Papendorf, B.; Ionescu, E.; Kleebe, H.J.; Guillon, O.; Nonnenmacher, K.; Riedel, R. High-temperature creep behavior of dense SiOC-based ceramic nanocomposites: Microstructural and phase composition effects. *J. Am. Ceram. Soc.* **2013**, *96*, 272–280.

336. Hampshire, S.; Drew, R.A.L.; Jack, K.H. Viscosities, glass-transition temperatures, and microhardness of Y–Si–Al–O–N glasses. *J. Am. Ceram. Soc.* **1984**, *67*, C46–C47.

337. Rouxel, T.; Huger, M.; Besson, J.L. Rheological properties of Y–Si–Al–O–N glasses—Elastic moduli, viscosity and creep. *J. Mater. Sci.* **1992**, *27*, 279–284.

338. Ionescu, E.; Balan, C.; Kleebe, H.J.; Muller, M.M.; Guillon, O.; Schliephake, D.; Heilmaier, M.; Riedel, R. High-temperature creep behavior of SiOC glass-ceramics: Influence of network carbon versus segregated carbon. *J. Am. Ceram. Soc.* **2014**, *97*, 3935–3942.

339. An, L.A.; Riedel, R.; Konetschny, C.; Kleebe, H.J.; Raj, R. Newtonian viscosity of amorphous silicon carbonitride at high temperature. *J. Am. Ceram. Soc.* **1998**, *81*, 1349–1352.

340. Shah, S.R.; Raj, R. Nanoscale densification creep in polymer-derived silicon carbonitrides at 1350 °C. *J. Am. Ceram. Soc.* **2001**, *84*, 2208–2212.

341. Zimmermann, A.; Bauer, A.; Christ, M.; Cai, Y.; Aldinger, F. High-temperature deformation of amorphous Si–C–N and Si–B–C–N ceramics derived from polymers. *Acta Mater.* **2002**, *50*, 1187–1196.

342. Riedel, R.; Ruwisch, L.M.; An, L.N.; Raj, R. Amorphous silicoboron carbonitride ceramic with very high viscosity at temperatures above 1500 °C. *J. Am. Ceram. Soc.* **1998**, *81*, 3341–3344.

343. Kumar, N.V.R.; Prinz, S.; Cai, Y.; Zimmermann, A.; Aldinger, F.; Berger, F.; Muller, K. Crystallization and creep behavior of Si–B–C–N ceramics. *Acta Mater.* **2005**, *53*, 4567–4578.

344. Kim, K.J.; Eom, J.-H.; Kim, Y.-W.; Seo, W.-S. Electrical conductivity of dense, bulk silicon-oxycarbide ceramics. *J. Eur. Ceram. Soc.* **2015**, *35*, 1355–1360.

345. Cordelair, J.; Greil, P. Electrical conductivity measurements as a microprobe for structure transitions in polysiloxane derived Si–O–C ceramics. *J. Eur. Ceram. Soc.* **2000**, *20*, 1947–1957.

346. Trassl, S.; Motz, G.; Rossler, E.; Ziegler, G. Characterisation of the free-carbon phase in precursor-derived SiCN ceramics. *J. Non-Cryst. Solids* **2001**, *293*, 261–267.

347. Trassl, S.; Kleebe, H.J.; Stormer, H.; Motz, G.; Rossler, E.; Ziegler, G. Characterization of the free-carbon phase in Si–C–N ceramics: Part II, comparison of different polysilazane precursors. *J. Am. Ceram. Soc.* **2002**, *85*, 1268–1274.

348. Ionescu, E.; Francis, A.; Riedel, R. Dispersion assessment and studies on AC percolative conductivity in polymer-derived Si–C–N/CNT ceramic nanocomposites. *J. Mater. Sci.* **2009**, *44*, 2055–2062.

349. Hermann, A.M.; Wang, Y.T.; Ramakrishnan, P.A.; Balzar, D.; An, L.N.; Haluschka, C.; Riedel, R. Structure and electronic transport properties of Si–(B)–C–N ceramics. *J. Am. Ceram. Soc.* **2001**, *84*, 2260–2264.

350. Ramakrishnan, P.A.; Wang, Y.T.; Balzar, D.; An, L.A.; Haluschka, C.; Riedel, R.; Hermann, A.M. Silicoboron-carbonitride ceramics: A class of high-temperature, dopable electronic materials. *Appl. Phys. Lett.* **2001**, *78*, 3076–3078.

351. Zhang, L.G.; Wang, Y.S.; Wei, Y.; Xu, W.X.; Fang, D.J.; Zhai, L.; Lin, K.C.; An, L.N. A silicon carbonitride ceramic with anomalously high piezoresistivity. *J. Am. Ceram. Soc.* **2008**, *91*, 1346–1349.

352. Wang, Y.; Zhang, L.; Fan, Y.; Jiang, D.; An, L. Stress-dependent piezoresistivity of tunneling-percolation systems. *J. Mater. Sci.* **2009**, *44*, 2814–2819.

353. Riedel, R.; Toma, L.; Janssen, E.; Nuffer, J.; Melz, T.; Hanselka, H. Piezoresistive Effect in SiOC Ceramics for Integrated Pressure Sensors. *J. Am. Ceram. Soc.* **2010**, *93*, 920–924.

354. Toma, L.; Kleebe, H.J.; Muller, M.M.; Janssen, E.; Riedel, R.; Melz, T.; Hanselka, H. Correlation between intrinsic microstructure and piezoresistivity in a SiOC polymer-derived ceramic. *J. Am. Ceram. Soc.* **2012**, *95*, 1056–1061.

355. Terauds, K.; Sanchez-Jimenez, P.E.; Raj, R.; Vakifahmetoglu, C.; Colombo, P. Giant piezoresistivity of polymer-derived ceramics at high temperatures. *J. Eur. Ceram. Soc.* **2010**, *30*, 2203–2207.

356. Roth, F.; Schmerbauch, C.; Nicoloso, N.; Guillon, O.; Riedel, R.; Ionescu, E. High-temperature piezoresistive C/SiOC sensors. *J. Sens. Sens. Struct.* **2015**, in press.

357. Kolel-Veetil, M.K.; Keller, T.M. Organometallic routes into the nanorealms of binary Fe–Si phases. *Materials* **2010**, *3*, 1049–1088.

358. Yu, Y.X.; An, L.N.; Chen, Y.H.; Yang, D.X. Synthesis of SiFeC magnetoceramics from reverse polycarbosilane-based microemulsions. *J. Am. Ceram. Soc.* **2010**, *93*, 3324–3329.

359. Mishra, R.; Tiwari, R.K.; Saxena, A.K. Synthesis of Fe-SiC Nanowires via precursor route. *J. Inorg. Organomet. Polym. Mater.* **2009**, *19*, 223–227.

360. Chen, X.J.; Su, Z.M.; Zhang, L.; Tang, M.; Yu, Y.X.; Zhang, L.T.; Chen, L.F. Iron Nanoparticle-containing silicon carbide fibers prepared by pyrolysis of Fe(CO)(5)-doped polycarbosilane fibers. *J. Am. Ceram. Soc.* **2010**, *93*, 89–95.

361. Saha, A.; Shah, S.R.; Raj, R.; Russek, S.E. Polymer-derived SiCN composites with magnetic properties. *J. Mater. Res.* **2003**, *18*, 2549–2551.

362. Park, J.H.; Park, K.H.; Kim, D.P. Superparamagnetic Si_3N_4-Fe-containing ceramics prepared from a polymer-metal complex. *J. Ind. Eng. Chem.* **2007**, *13*, 27–32.

363. Dumitru, A.; Ciupina, V.; Stamatin, I.; Prodan, G.; Morozan, A.; Mirea, C. Plasma polymerization of ferrocene with silane and silazane monomers for design of nanostructured magnetic ceramics. *J. Optoelectron. Adv. Mater.* **2006**, *8*, 50–54.

364. Vakifahmetoglu, C.; Pippel, E.; Woltersdorf, J.; Colombo, P. Growth of one-dimensional nanostructures in porous polymer-derived ceramics by catalyst-assisted pyrolysis. Part I: Iron catalyst. *J. Am. Ceram. Soc.* **2010**, *93*, 959–968.

365. Hornig, S.; Manners, I.; Newkome, G.R.; Schubert, U.S. Metal-containing and metallo-supramolecular polymers and materials. *Macromol. Rapid Commun.* **2010**, *31*, 771.

366. Kolar, F.; Machovic, V.; Svitilova, J. Cobalt-containing silicon oxycarbide glasses derived from poly[methyl(phenyl)]siloxane and cobalt phthalate. *J. Non-Cryst. Solids* **2006**, *352*, 2892–2896.

367. Weisbarth, R.; Jansen, M. SiBN3C Ceramic workpieces by pressureless pyrolysis without sintering aids: Preparation, characterization and electrical properties. *J. Mater. Chem.* **2003**, *13*, 2975–2978.

368. Haug, R.; Weinmann, M.; Bill, J.; Aldinger, F. Plastic forming of preceramic polymers. *J. Eur. Ceram. Soc.* **1999**, *19*, 1–6.

369. Riedel, R.; Passing, G.; Schonfelder, H.; Brook, R.J. Synthesis of dense silicon-based ceramics at low-temperatures. *Nature* **1992**, *355*, 714–717.

370. Wilfert, J.; Meier, K.; Hahn, K.; Grin, Y.; Jansen, M. SiC/BN composites by spark plasma sintering (SPS) of precursor-derived SiBNC powders. *J. Ceram. Sci. Technol.* **2010**, *1*, 1–6.

371. Janakiraman, N.; Aldinger, F. Indentation analysis of elastic and plastic deformation of precursor-derived Si–C–N ceramics. *J. Eur. Ceram. Soc.* **2010**, *30*, 775–785.

372. Janakiraman, N.; Aldinger, F. Yielding, strain hardening, and creep under nanoindentation of precursor-derived Si–C–N ceramics. *J. Am. Ceram. Soc.* **2010**, *93*, 821–829.

373. Moraes, K.V.; Interrante, L.V. Processing, fracture toughness, and vickers hardness of allylhyd ridopolycarbosilane-derived silicon carbide. *J. Am. Ceram. Soc.* **2003**, *86*, 342–346.

374. Rouxel, T.; Sangleboeuf, J.C.; Guin, J.P.; Keryvin, V.; Soraru, G.D. Surface damage resistance of gel-derived oxycarbide glasses: Hardness, toughness, and scratchability. *J. Am. Ceram. Soc.* **2001**, *84*, 2220–2224.

375. Galusek, D.; Riley, F.L.; Riedel, R. Nanoindentation of a polymer-derived amorphous silicon carbonitride ceramic. *J. Am. Ceram. Soc.* **2001**, *84*, 1164–1166.

376. Shah, S.R.; Raj, R. Mechanical properties of a fully dense polymer derived ceramic made by a novel pressure casting process. *Acta Mater.* **2002**, *50*, 4093–4103.

377. Nishimura, T.; Haug, R.; Bill, J.; Thurn, G.; Aldinger, F. Mechanical and thermal properties of Si–C–N material from polyvinylsilazane. *J. Mater. Sci.* **1998**, *33*, 5237–5241.

378. Sujith, R.; Kumar, R. Experimental investigation on the indentation hardness of precursor derived Si–B–C–N ceramics. *J. Eur. Ceram. Soc.* **2013**, *33*, 2399–2405.

379. Cross, T.; Raj, R. Mechanical and tribological behavior of polymer-derived ceramics constituted from SiCxOyNz. *J. Am. Ceram. Soc.* **2006**, *89*, 3706–3714.

380. Cross, T.J.; Raj, R.; Cross, T.J.; Prasad, S.V.; Tallant, D.R. Synthesis and tribological behavior of silicon oxycarbonitride thin films derived from poly(urea)methyl vinyl silazane. *Int. J. Appl. Ceram. Technol.* **2006**, *3*, 113–126.

381. Iwamoto, Y. Precursors-derived ceramic membranes for high-temperature separation of hydrogen. *J. Ceram. Soc. Jpn.* **2007**, *115*, 947–954.

382. Iwamoto, Y. Membranes. In *Polymer Derived Ceramics: From Nanostructure to Applications*; Colombo, P., Riedel, R., Kleebe, H.-J., Soraru, G.D., Eds.; DesTech Publications, Inc.: Lancaster, PA, USA, 2010; pp. 397–402.

383. Yoshino, Y.; Suzuki, T.; Nair, B.N.; Taguchi, H.; Itoh, N. Development of tubular substrates, silica based membranes and membrane modules for hydrogen separation at high temperature. *J. Membr. Sci.* **2005**, *267*, 8–17.

384. Nagano, T.; Sato, K.; Saitoh, T.; Iwamoto, Y. Gas permeation properties of amorphous SiC membranes synthesized from polycarbosilane without oxygen-curing process. *J. Ceram. Soc. Jpn.* **2006**, *114*, 533–538.

385. Nagano, T.; Uno, N.; Saitoh, T.; Yamazaki, S.; Iwamoto, Y. Gas permeance behavior at elevated temperature in mesoporous anodic oxidized alumina synthesized by pulse-sequential voltage method. *Chem. Eng. Commun.* **2007**, *194*, 158–169.

386. Iwamoto, Y.; Sato, K.; Kato, T.; Inada, T.; Kubo, Y. A hydrogen-permselective amorphous silica membrane derived from polysilazane. *J. Eur. Ceram. Soc.* **2005**, *25*, 257–264.

387. Shelekhin, A.B.; Grosgogeat, E.J.; Hwang, S.T. Gas separation properties of a new polymer inorganic composite membrane. *J. Membr. Sci.* **1992**, *66*, 129–141.

388. Wach, R.A.; Sugimoto, M.; Yoshikawa, M. Formation of silicone carbide membrane by radiation curing of polycarbosilane and polyvinylsilane and its gas separation up to 250 °C. *J. Am. Ceram. Soc.* **2007**, *90*, 275–278.

389. Suda, H.; Yamauchi, H.; Uchimaru, Y.; Fujiwara, I.; Haraya, K. Structural evolution during conversion of polycarbosilane precursor into silicon carbide-based microporous membranes. *J. Ceram. Soc. Jpn.* **2006**, *114*, 539–544.

390. Kusakabe, K.; Li, Z.Y.; Maeda, H.; Morooka, S. Preparation of supported composite membrane by pyrolysis of polycarbosilane for gas separation at high-temperature. *J. Membr. Sci.* **1995**, *103*, 175–180.

391. Li, Z.Y.; Kusakabe, K.; Morooka, S. Preparation of thermostable amorphous Si–C–O membrane and its application to gas separation at elevated temperature. *J. Membr. Sci* **1996**, *118*, 159–168.

392. Volger, K.W.; Hauser, R.; Kroke, E.; Riedel, R.; Ikuhara, Y.H.; Iwamoto, Y. Synthesis and characterization of novel non-oxide sol–gel derived mesoporous amorphous Si–C–N membranes. *J. Ceram. Soc. Jpn.* **2006**, *114*, 567–570.

393. Hauser, R.; Nahar-Borchard, S.; Riedel, R.; Ikuhara, Y.H.; Iwamoto, Y. Polymer-derived SiBCN ceramic and their potential application for high temperature membranes. *J. Ceram. Soc. Jpn.* **2006**, *114*, 524–528.

394. Prasad, R.M.; Iwamoto, Y.; Riedel, R.; Gurlo, A. Multilayer amorphous-Si-B-C-N/γ-Al$_2$O$_3$/α-Al$_2$O$_3$ membranes for hydrogen purification. *Adv. Eng. Mater.* **2010**, *12*, 522–528.

395. Bazarjani, M.S.; Muller, M.M.; Kleebe, H.-J.; Fasel, C.; Riedel, R.; Gurlo, A. *In situ* formation of tungsten oxycarbide, tungsten carbide and tungsten nitride nanoparticles in micro- and mesoporous polymer-derived ceramics. *J. Mater. Chem. A* **2014**, *2*, 10454–10464.

396. Seifollahi Bazarjani, M.; Müller, M.M.; Kleebe, H.-J.; Jüttke, Y.; Voigt, I.; Baghaie Yazdi, M.; Alff, L.; Riedel, R.; Gurlo, A. High-temperature stability and saturation magnetization of superparamagnetic nickel nanoparticles in microporous polysilazane-derived ceramics and their gas permeation properties. *ACS Appl. Mater. Interfaces* **2014**, *6*, 12270–12278.

397. Schitco, C.; Bazarjani, M.S.; Riedel, R.; Gurlo, A. NH$_3$-assisted synthesis of microporous silicon oxycarbonitride ceramics from preceramic polymers: A combined N$_2$ and CO$_2$ adsorption and small angle X-ray scattering study. *J. Mater. Chem. A* **2015**, *3*, 805–818.

398. De Jongh, P.E.; Adelhelm, P. Nanosizing and nanoconfinement: New strategies towards meeting hydrogen storage goals. *Chemsuschem* **2010**, *3*, 1332–1348.

399. Gutowska, A.; Li, L.Y.; Shin, Y.S.; Wang, C.M.M.; Li, X.H.S.; Linehan, J.C.; Smith, R.S.; Kay, B.D.; Schmid, B.; Shaw, W.; *et al.* Nanoscaffold mediates hydrogen release and the reactivity of ammonia borane. *Angew. Chem. Int. Ed.* **2005**, *44*, 3578–3582.

400. Li, L.; Yao, X.; Sun, C.H.; Du, A.J.; Cheng, L.N.; Zhu, Z.H.; Yu, C.Z.; Zou, J.; Smith, S.C.; Wang, P.; *et al.* Lithium-catalyzed dehydrogenation of ammonia borane within mesoporous carbon framework for chemical hydrogen storage. *Adv. Funct. Mater.* **2009**, *19*, 265–271.

401. Li, S.F.; Guo, Y.H.; Sun, W.W.; Sun, D.L.; Yu, X.B. Platinum nanoparticle functionalized CNTs as nanoscaffolds and catalysts to enhance the dehydrogenation of ammonia-borane. *J. Phys. Chem. C* **2010**, *114*, 21885–21890.

402. Zhao, J.Z.; Shi, J.F.; Zhang, X.W.; Cheng, F.Y.; Liang, J.; Tao, Z.L.; Chen, J. A soft hydrogen storage material: poly(methyl acrylate)-confined ammonia borane with controllable dehydrogenation. *Adv. Mater.* **2010**, *22*, 394–397.

403. Gadipelli, S.; Ford, J.; Zhou, W.; Wu, H.; Udovic, T.J.; Yildirim, T. Nanoconfinement and catalytic dehydrogenation of ammonia borane by magnesium-metal-organic-framework-74. *Chem. Eur. J.* **2011**, *17*, 6043–6047.

404. Srinivas, G.; Travis, W.; Ford, J.; Wu, H.; Guo, Z.X.; Yildirim, T. Nanoconfined ammonia borane in a flexible metal-organic framework Fe-MIL-53: Clean hydrogen release with fast kinetics. *J. Mater. Chem. A* **2013**, *1*, 4167–4172.

405. Zhang, T.R.; Yang, X.J.; Yang, S.Q.; Li, D.X.; Cheng, F.Y.; Tao, Z.L.; Chen, J. Silica hollow nanospheres as new nanoscaffold materials to enhance hydrogen releasing from ammonia borane. *Phys. Chem. Chem. Phys.* **2011**, *13*, 18592–18599.

406. Moussa, G.; Bernard, S.; Demirci, U.B.; Chiriac, R.; Miele, P. Room-temperature hydrogen release from activated carbon-confined ammonia borane. *Int. J. Hydrog. Energy* **2012**, *37*, 13437–13445.

407. Bernard, S.; Miele, P. Nanostructured and architectured boron nitride from boron, nitrogen and hydrogen-containing molecular and polymeric precursors. *Mater. Today* **2014**, *17*, 443–450.

408. Schlienger, S.; Alauzun, J.; Michaux, F.; Vidal, L.; Parmentier, J.; Gervais, C.; Babonneau, F.; Bernard, S.; Miele, P.; Parra, J.B.; *et al.* Micro-, mesoporous boron nitride-based materials templated from zeolites. *Chem. Mater.* **2012**, *24*, 88–96.

409. Bernard, S.; Salles, V.; Li, J.P.; Brioude, A.; Bechelany, M.; Demirci, U.B.; Miele, P. High-yield synthesis of hollow boron nitride nano-polyhedrons. *J. Mater. Chem.* **2011**, *21*, 8694–8699.

410. Salles, V.; Bernard, S.; Li, J.P.; Brioude, A.; Chehaidi, S.; Foucaud, S.; Miele, P. Design of highly dense boron nitride by the combination of spray-pyrolysis of borazine and additive-free sintering of derived ultrafine powders. *Chem. Mater.* **2009**, *21*, 2920–2929.

411. Moussa, G.; Demirci, U.B.; Malo, S.; Bernard, S.; Miele, P. Hollow core@mesoporous shell boron nitride nanopolyhedron-confined ammonia borane: A pure B–N–H composite for chemical hydrogen storage. *J. Mater. Chem. A* **2014**, *2*, 7717–7722.

412. Moussa, G.; Salameh, C.; Bruma, A.; Malo, S.; Demirci, U.; Bernard, S.; Miele, P. Nanostructured boron nitride: From molecular design to hydrogen storage application. *Inorganics* **2014**, *2*, 396–409.

413. Periana, R.A.; Taube, D.J.; Gamble, S.; Taube, H.; Satoh, T.; Fujii, H. Platinum catalysts for the high-yield oxidation of methane to a methanol derivative. *Science* **1998**, *280*, 560–564.

414. Saha, A.; Raj, R.; Williamson, D.L. Characterization of nanodomains in polymer-derived SiCN ceramics employing multiple techniques. *J. Am. Ceram. Soc.* **2005**, *88*, 232–234.

415. Saha, A.; Raj, R.; Williamson, D.L. A model for the nanodomains in polymer-derived SiCO. *J. Am. Ceram. Soc.* **2006**, *89*, 2188–2195.

416. Widgeon, S.J.; Sen, S.; Mera, G.; Ionescu, E.; Riedel, R.; Navrotsky, A. Si-29 and C-13 solid-state NMR spectroscopic study of nanometer-scale structure and mass fractal characteristics of amorphous polymer derived silicon oxycarbide ceramics. *Chem. Mater.* **2010**, *22*, 6221–6228.

417. Dibandjo, P.; Graczyk-Zajac, M.; Riedel, R.; Pradeep, V.S.; Soraru, G.D. Lithium insertion into dense and porous carbon-rich polymer-derived SiOC ceramics. *J. Eur. Ceram. Soc.* **2012**, *32*, 2495–2503.

418. Graczyk-Zajac, M.; Toma, L.; Fasel, C.; Riedel, R. Carbon-rich SiOC anodes for lithium-ion batteries: Part I. Influence of material UV-pre-treatment on high power properties. *Solid State Ion.* **2012**, *225*, 522–526.

419. Kaspar, J.; Graczyk-Zajac, M.; Riedel, R. Carbon-rich SiOC anodes for lithium-ion batteries: Part II. Role of thermal cross-linking. *Solid State Ion.* **2012**, *225*, 527–531.

420. Graczyk-Zajac, M.; Reinold, L.M.; Kaspar, J.; Pradeep, V.S.; Soraru, G.D.; Riedel, R. New insights into understanding irreversible and reversible lithium storage within SiOC and SiCN ceramics. *Nanomaterials* **2015**, *5*, 233–245.

421. Wilson, A.M.; Reimers, J.N.; Fuller, E.W.; Dahn, J.R. Lithium insertion in pyrolyzed siloxane polymers. *Solid State Ion.* **1994**, *74*, 249–254.

422. Wilson, A.M.; Zank, G.; Eguchi, K.; Xing, W.; Dahn, J.R. Pyrolysed silicon-containing polymers as high capacity anodes for lithium-ion batteries. *J. Power Sources* **1997**, *68*, 195–200.

423. Wilson, A.M.; Xing, W.B.; Zank, G.; Yates, B.; Dahn, J.R. Pyrolysed pitch-polysilane blends for use as anode materials in lithium ion batteries 2: The effect of oxygen. *Solid State Ion.* **1997**, *100*, 259–266.

424. Xing, W.B.; Wilson, A.M.; Eguchi, K.; Zank, G.; Dahn, J.R. Pyrolyzed polysiloxanes for use as anode materials in lithium-ion batteries. *J. Electrochem. Soc.* **1997**, *144*, 2410–2416.

425. Xing, W.B.; Wilson, A.M.; Zank, G.; Dahn, J.R. Pyrolysed pitch-polysilane blends for use as anode materials in lithium ion batteries. *Solid State Ion.* **1997**, *93*, 239–244.

426. Ahn, D.; Raj, R. Thermodynamic measurements pertaining to the hysteretic intercalation of lithium in polymer-derived silicon oxycarbide. *J. Power Sources* **2010**, *195*, 3900–3906.

427. Ahn, D.; Raj, R. Cyclic stability and C-rate performance of amorphous silicon and carbon based anodes for electrochemical storage of lithium. *J. Power Sources* **2011**, *196*, 2179–2186.

428. Weidman, P.D.; Ahn, D.; Raj, R. Diffusive relaxation of Li in particles of silicon oxycarbide measured by galvanostatic titrations. *J. Power Sources* **2014**, *249*, 219–230.

429. Kaspar, J.; Graczyk-Zajac, M.; Lauterbach, S.; Kleebe, H.J.; Riedel, R. Silicon oxycarbide/nano-silicon composite anodes for Li-ion batteries: Considerable influence of nano-crystalline *vs.* nano-amorphous silicon embedment on the electrochemical properties. *J. Power Sources* **2014**, *269*, 164–172.

Alumina Matrix Composites with Non-Oxide Nanoparticle Addition and Enhanced Functionalities

Dušan Galusek [1,2,*] and Dagmar Galusková [1,2]

[1] Joint Glass Centre of the Institute of Inorganic Chemistry, Slovak Academy of Sciences, Alexander Dubček University of Trenčín, Študentská 2, 91150 Trenčín, Slovak Republic; E-Mail: dagmar.galuskova@tnuni.sk

[2] Faculty of Chemical and Food Technology, Slovak University of Technology in Bratislava, Vazovova 5, 81243 Bratislava, Slovak Republic

* Author to whom correspondence should be addressed; E-Mail: dusan.galusek@tnuni.sk

Academic Editor: Emanuel Ionescu

Abstract: The addition of SiC or TiC nanoparticles to polycrystalline alumina matrix has long been known as an efficient way of improving the mechanical properties of alumina-based ceramics, especially strength, creep, and wear resistance. Recently, new types of nano-additives, such as carbon nanotubes (CNT), carbon nanofibers (CNF), and graphene sheets have been studied in order not only to improve the mechanical properties, but also to prepare materials with added functionalities, such as thermal and electrical conductivity. This paper provides a concise review of several types of alumina-based nanocomposites, evaluating the efficiency of various preparation methods and additives in terms of their influence on the properties of composites.

Keywords: alumina-based nanocomposites; SiC; CNT; CNF; mechanical properties; functional properties

1. Introduction

Many ceramic materials used in engineering suffer from inherent brittleness and generally inferior mechanical properties compared to metals. Concentrated efforts in the last couple of decades have therefore been aimed at identification of methods that would result in better ceramics, ceramic-matrix composites (CMCs) chief among them. Since the pioneering work of Niihara in the 1990s [1], the addition of nano-particles or whiskers of a second phase was considered to be one of the most promising ways of improving the mechanical properties of polycrystalline alumina-based ceramics. Extensive literature published on the topic indicates that the addition of silicon carbide particles (SiC$_p$) or whiskers (SiC$_w$) to polycrystalline alumina improves strength [1–6], fracture toughness [6–8], wear resistance [9–11], and creep resistance [12–14], compared to monolithic polycrystalline alumina. However, despite the tremendous efforts documented by thousands of research papers published on the topic, the so-called "nanocomposites" have generally remained a topic of academic research, failing to make their breakthrough to large-scale production. The problems with homogeneous distribution of nanoparticles, and related problems with reproducible preparation of materials with improved mechanical properties, can be considered as a chief obstacle. Renewed interest in CMCs with alumina matrix was observed with the invention, and commercial availability, of carbon nanotubes (CNT) and nanofibers (CNF). The intrinsic properties of the CNTs are impressive. The theoretical mechanical strength of 30 GPa calculated for both the single-wall (SWCNT) and multi-wall carbon nanotubes (MWCNT), together with an extremely high Young's modulus (1 and 1–1.8 TPa, respectively) [15–18], immediately made them first-class candidates for preparation of CMCs with significantly improved mechanical properties. Other properties of the CNTs are even more impressive: the "metallic" character of SWCNT with armchair structure results in high electrical conductivity, which combines with extremely high thermal conductivity. The theoretically calculated values of the latter range between 2800 and 6000 W·m^{-1}·K^{-1}, making them the best heat conductors known [19]. Attempts to use the CNTs for preparation of CMCs with added functionalities, such as high thermal and electrical conductivity, then came as no surprise, providing added value to expected improvement of mechanical properties. This paper is an attempt to summarize the preparation of CMCs with polycrystalline alumina matrix, with improved mechanical properties and added functionalities, supplemented by critical evaluation of reported achievements.

2. Preparation of Nanocomposites

2.1. Homogeneous Distribution of Nanoparticles

Homogeneous mixing of both the matrix and reinforcing phases and even distribution of the reinforcing phase in ceramic matrix are prerequisites for achievement of the desired properties in CMCs. This is of particular importance if nanoparticles are used as the reinforcing phase, due to their extremely high specific surface and, hence, intrinsically high tendency to agglomeration. With the use of nanotubes or nanofibers, the problem is even more aggravated by their shape, leading to formation of severely entangled bundles, which are extremely difficult to de-agglomerate. The following section therefore deals with the ways devised to ensure de-agglomeration and/or homogeneous distribution of nanoparticles in a ceramic matrix.

2.1.1. Al$_2$O$_3$-SiC

The traditional (so called powder) route consists of mixing the alumina and SiC nano-powders in a suitable aqueous or non-aqueous media, drying, and green body shaping. This method has several serious drawbacks. It is extremely difficult to prevent agglomeration of the SiC nanoparticles and to ensure homogeneous mixing of SiC with Al$_2$O$_3$. Drying of suspensions is another source of agglomeration, which results in uneven sintering and void and crack formation in the course of densification. Walker *et al.* [20] investigated different drying methods of composite suspensions, and concluded that a freeze-drying technique can be successfully applied in order to avoid the formation of hard agglomerates. A pH adjustment can also be used to induce flocculation of the slurry, preventing segregation and agglomeration of the silicon carbide particles [21].

In order to prevent the problems caused by mixing of SiC and Al_2O_3 powders, composite powders can be synthesized by carbothermal reduction of a mixture of silica and alumina or of natural aluminosilicates like kaolinite [22], kyanite [23], and andalusite [24]. Suitable adjustment of reaction conditions (chemical composition of starting mixture or mineral, reaction temperature, source and partial pressure of C, dwell time at reaction temperature, or presence of impurities or catalysts) allows the control of the content and morphology of SiC fraction in the mixed powder. Sol-gel synthesis can also be applied for preparation of alumina-SiC nanopowders: The SiC nanopowder is dispersed in a suitable liquid medium to create a stable suspension and mixed with a liquid alumina precursor, such as Al_2O_3 [25] or $AlCl_3$ solution [26,27]. After gelation and drying, the xerogel containing SiC nanoparticles is calcined, crushed, sieved, and finally used for preparation of nanocomposites. Other attempts to prepare the nanocomposite powders use more exotic techniques, such as a Teflon-activated self-propagating aluminothermic reaction (Equation (1)) [28]:

$$4Al + SiO_2 + 3C \rightarrow 3SiC + 2Al_2O_3 \tag{1}$$

or thermal-gradient chemical vapor infiltration of SiC porous preforms with a gaseous mixture of $AlCl_3$, H_2, and CO_2, resulting in inhomogeneous distribution of constituent phases [29].

The most promising among the non-traditional routes of preparation of nanocomposite powders is the so-called "hybrid" route, in other words the route utilizing the ceramization of organosilicon polymeric precursors of SiC, typically polycarbosilanes [30–34]. This is usually based on coating the alumina particles with a dissolved polymer, followed by drying, cross-linking of the polymer, pyrolysis, and densification. The method allows preparation of alumina-based nanocomposites with ultrafine particles of SiC (~12 nm) evenly distributed in the alumina matrix both at intra- [30,31] and inter-granular [32] positions. The agglomeration in the course of drying is usually avoided by application of advanced drying techniques like freeze-drying and freeze granulation, or by the use of wet shaping techniques like slip casting, tape casting, and pressure filtration. An alternative shaping route adapted from processing of polymers is represented by axial pressing of polymer-coated powders at elevated temperatures of 300–400 °C (called also warm pressing), yielding a dense green body with alumina particles embedded in a matrix of a highly cross-linked preceramic polymer, which is transformed, upon heating in inert atmosphere (Ar), to SiC [34].

The preceramic polymers can be also used for infiltration of pre-sintered porous alumina matrix. For that purpose either liquid polymers are used (e.g., poly-allylcarobosilanes), or solid polymers are dissolved in a suitable solvent (e.g., cyclohexane). After infiltration the solvent is evaporated, and the polymer transformed to SiC *in situ*, entirely avoiding the problems associated with de-agglomeration. The size and distribution of SiC nanoparticles are then readily controlled by the size and distribution of open pores in the alumina matrix, which is adjusted by selection of the alumina powder, the shaping technique used, and the conditions of pre-sintering [35].

2.1.2. Al$_2$O$_3$-CNT, CNF

In principle, there exist two different attitudes aimed at achievement of even and homogeneous distribution of CNTs in the composites with polycrystalline alumina matrix. The first one relies on the ability to disentangle the nanotube bundles, separate the individual nanotubes by chemical or mechanical means (or their combination), and then mix the nanotubes with alumina powder or suspension. Mechanical de-agglomeration of CNTs requires an energetic method of separation, such as shear mixing, ball milling, ultra-sonication [36], gas purging sonication (combination of sonication with simultaneous purging with nitrogen gas), or others in a suitable liquid medium, ethanol being the most frequently used [36,37]. A suitable dispersant, such as sodium dodecyl sulfate (SDS), is often used to stabilize the suspension [37]. The attitude often requires previous chemical "functionalization" of nanotubes, *i.e.*, their pre-treatment in a mixture of inorganic acids at elevated temperature. The process involves heating nanotubes for several hours in a mixture of nitric and sulfuric acid at temperatures often exceeding 100 °C [38]. This rather harsh treatment has two outcomes: first, the metallic catalyst, as the impurity resulting from synthesis of the nanotubes, is thoroughly removed. Second, polar functional groups (such as C=O, and C–O–H) are formed at the surface of the nanotubes, changing their intrinsically hydrophobic nature to hydrophilic, and creating reaction centers facilitating attachment of dispersant molecules [39]. However, as an undesired byproduct, the treatment often leads to deterioration and defect formation at the surface of nanotubes. A typical procedure then involves dropwise addition of alumina suspension into a stabilized and vigorously stirred suspension of the nanotubes, and its homogenization by ball milling or sonication [36]. The composite suspension is then dried in a way that prevents re-agglomeration of the nanotubes; among the possible ways, freeze granulation has been recently reported as the most successful [40]. The composite powder is then consolidated by a suitable forming method and densified, as will be discussed in more detail in Section 2.2. One of the alternatives is represented by preparation of alumina-coated nanotubes by hydrothermal crystallization, thus achieving better compatibility and stronger bonding of the nanotubes with the alumina matrix [41].

The second approach is based on the ability of nanotubes to grow from a suitable gas atmosphere on a metallic (Ni, Co, or Fe) precursor *in situ*, *i.e.*, directly on alumina powder particles or within a porous alumina matrix. Various solutions of salts of the metal catalysts are used to disperse Fe, Ni, or Co onto ceramic supports [42–44]. A typical procedure includes preparation of a solution of suitable nitrates with alumina powder or an alumina precursor, such as Al(NO$_3$)$_3$·9H$_2$O with a mixture of citric acid and urea used as fuel. The mixture is heated until ignition of the fuel: in this way alumina powder homogenously doped with the metallic catalyst is prepared by combustion synthesis [45,46]. After calcination the powder is placed in a reaction furnace or a CCVD chamber and heated up to 1000 °C in a mixed H$_2$/CH$_4$ atmosphere. Carbon required for the growth of nanotubes originates from catalytic decomposition of methane on transition metal nanoparticles formed by combustion synthesis [47]. In this way the difficulties related to de-agglomeration and homogeneous dispersion of the nanotubes in the alumina matrix are avoided, and composite green bodies can be readily prepared.

2.2. Densification

2.2.1. Al$_2$O$_3$-SiC

All available consolidation techniques yield composite green bodies with relatively high porosity (usually around 50%) and more or less homogeneously distributed reinforcing SiC particles, which ideally contain no defects or agglomerates. However, the nature of the composite powders' preparation results in a microstructural arrangement where the SiC nanoparticles are located at the interfaces between alumina grains, acting as efficient obstacles to densification, and impeding grain boundary motion through a pinning mechanism. The use of pressure-assisted sintering techniques, such as hot pressing, spark plasma sintering, or application of high sintering temperatures (typically between 1700 and 1850 °C) in the case of pressureless sintering, are usually required for preparation of fully dense Al$_2$O$_3$-SiC nanocomposites. However, the high temperatures accelerate grain boundary motion, resulting in coarse grained microstructure. In addition, the SiC particles at intergranular positions are swallowed by fast moving grain boundaries, creating inclusions inside the alumina matrix grains (Figure 1). The resulting microstructure is coarse grained, with the size of alumina matrix grains in the range of about 5 μm or more [48]. Problems encountered during the pressureless sintering can be, at least partially, solved by the use of sintering additives. Fully dense alumina–SiC nanocomposites were prepared by free sintering followed by gas pressure sintering of the mixture of submicrometer alumina and SiC powder doped with 0.1 wt% MgO and/or Y$_2$O$_3$ [49]. The addition of other liquid-forming additives, such as MnO$_2$.SiO$_2$ and CaO.ZnO.SiO$_2$, allows densification of Al$_2$O$_3$-SiC green compacts by free sintering at temperatures as low as 1300 °C, but the resulting microstructures are usually coarse-grained. Abnormally large alumina grains are present, and the distribution of SiC particles is uneven [50]. The ability to densify the nanocomposites by pressureless sintering is limited by the content of SiC nanoparticles, which usually does not exceed 10 vol%. Fully dense composites with higher SiC content (>20 vol%) are prepared by combination of pressureless sintering with hot isostatic pressing (sinter-HIP) [51]. In this case a pre-sintered alumina compact is multiple pressure-infiltrated by a polymer SiC precursor. The precursor is then pyrolyzed in Ar, and sintered without pressure in inert gas or vacuum until the porosity closes, *i.e.*, after the relative density of about 92%–95% is achieved. Then the temperature is increased up to 1750 °C, and a high pressure (up to 150 MPa) of inert gas (Ar) is applied to achieve complete densification. Such a consolidation method benefits from the advantages of both pressureless and high-pressure processes and is a suitable alternative for mass production of ceramic nanocomposite components with complex shapes.

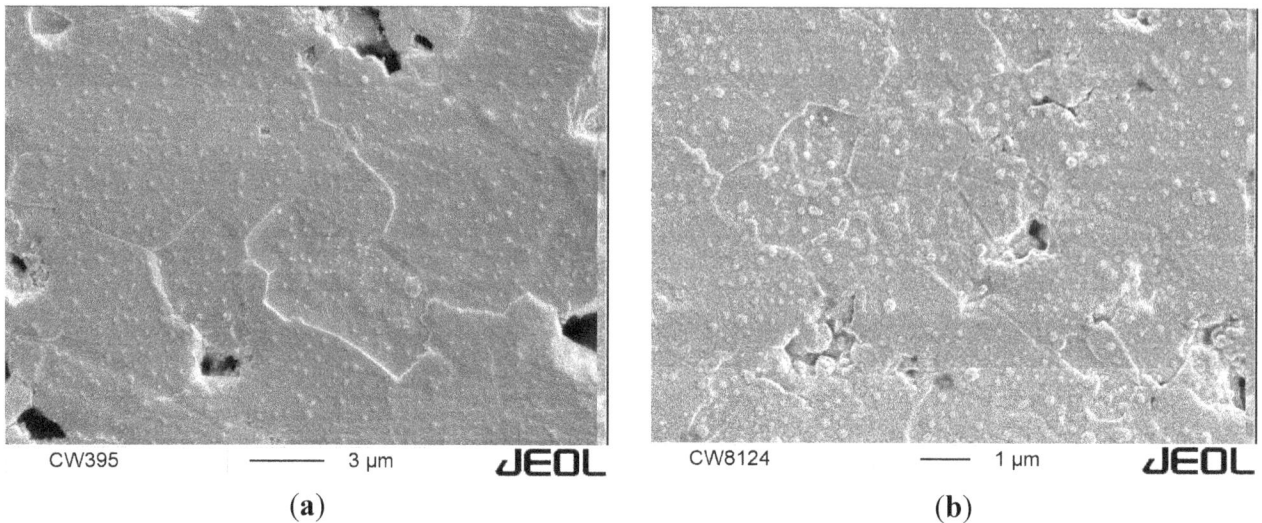

Figure 1. Microstructure refinement observed in Al$_2$O$_3$-SiC nanocomposites with increasing volume fraction of SiC nanoparticles: (**a**) 3 vol% SiC; (**b**) 8 vol% SiC. The nanocomposites were sintered without pressure at 1750 °C from a green body prepared by warm pressing of poly(allyl)carbosilane-coated alumina powder [34].

2.2.2. Al$_2$O$_3$-CNT, CNF

If the pinning effect of intergranular SiC particles is a serious obstacle for achieving high relative density in Al$_2$O$_3$-SiC nano-composites, the same applies, in an even more serious way, to the composites with added carbon nanotubes or carbon nanofibers. Due to their high aspect ratio, high specific surface, and chemical incompatibility with the surrounding alumina matrix, the CNTs act in two different ways: (a) they impair densification, with resulting decrease of sintered densities; and (b) they reduce the size of alumina matrix grains by sharply decreasing the grain boundary mobility through highly efficient grain boundary pinning [52–54]. Another factor contributing to low sintered densities is the presence of CNT clusters at the grain boundaries, as a result of imperfect de-agglomeration. The agglomerates act both as solid obstacles at grain boundary interfaces, which impair densification and grain boundary mobility even more efficiently than individual nanotubes, but also as solid un-sinterable porous objects, which decrease the sintered density [55,56]. It therefore stands to reason that the densification of carbon nanotubes containing nanocomposites is difficult, with difficulties growing with increasing content of CNT and CNFs. Although numerous attempts have been reported on pressureless sintering of CNT-reinforced alumina composites, they were seldom successful. The sintering temperatures usually range between 1200 and 1800 °C in air [36] or, more typically, in inert atmosphere (usually Ar), which is used to avoid oxidation damage and burning out of the CNTs from ceramic matrix. The difficulties related to densification of CNT-containing composites are demonstrated for example by the works of Zhang *et al.* [55] and Rice *et al.* [56], who found a sharp drop of relative density from 98.5% in the composite containing 1 wt% of CNTs to less than 95% when the CNT content increased above 3 vol%. However, in order to avoid the reactions between the alumina matrix and the CNTs, which could lead to CNT loss, the maximum temperature of sintering has to be kept below 1550 °C [56]. Similar results were obtained by Yamamoto *et al.* [57], who sintered Al_2O_3-MWCNT composites with up to 3 vol% of MWCNTs with various mechanical properties at 1400 °C in a flowing 95% Ar/5% H_2 atmosphere. The authors observed a monotonous decrease of relative density with increasing content of the MWCNT in all studied composites, from 99% at 0.5 vol% of the MWCNT to approximately 94% when the MWCNT contents approached 3 vol%.

More attention has therefore been paid to pressure assisted techniques, including hot pressing (HP), hot isostatic pressing (HIP), and spark plasma sintering (SPS). Although the pressure-assisted techniques can inflict some mechanical damage on CNTs, they are usually considered more suitable due to the following reasons: (1) they allow the use of lower temperatures, and shorter times of densification, thus reducing the thermal and oxidation damage of the nanotubes; and (2) they facilitate the achievement of higher relative densities, and through reducing the time and temperature of sintering also produce a finer grained final microstructure in the composite. Several authors successfully densified the Al_2O_3-CNT nanocomposites by hot pressing in a graphite die at an applied mechanical pressure up to 40 MPa, and temperatures ranging from 1600 to 1800 °C [39,52,58,59]. The method, if properly conducted, facilitates the preparation of the Al_2O_3-MWCNT nanocomposites containing up to 3 vol% of the nanotubes, and with the relative density at the level, or exceeding 99%. Spark plasma sintering has been favored in the last few years due to its ability to apply very high heating rates and achieve complete densification in minutes, rather than hours as in HP [60,61]. However, for this technique high residual porosity is also typical in the composites with high CNT contents. Ahmad et al. [62], spark plasma sintered CNT–alumina composites with a CNT content ranging between 1.1 and 10.4 vol% for 3 min at 1400 °C and an axial pressure of 50 MPa, and observed a marked drop of relative density from 98.5% (1.1% MWCNT) to 92% (10.4% MWCNT). Inability to achieve complete densification, together with imperfect de-agglomeration and poor control over the interfaces between CNTs and Al_2O_3, are thus major obstacles to unambiguous evaluation of the influence of CNT addition on functional and mechanical properties of CNT-containing CMCs.

3. Room and High-Temperature Mechanical Properties

3.1. Al_2O_3-SiC

Significant attention has been attracted to Al_2O_3-SiC composites by a pioneering work of Niihara, whose concept of nanocomposites (addition of nano-sized particles of SiC to microcrystalline alumina matrix) allowed preparation of the Al_2O_3-SiC_p materials with the flexural strength exceeding 1 GPa and increased fracture toughness [1]. Despite tremendous effort, the reason for such an improvement remains unclear. Niihara himself suggests that the strengthening arises due to the refinement of the microstructural scale from the order of the alumina matrix grain size to the order of the SiC interparticle spacing, thus reducing the critical flaw size. Strengthening can also be explained by the toughening effect caused by crack deflection due to the tensile stresses developed in alumina grains around the SiC particles as a result of thermo-elastic mismatch [1]. However, the observed toughness increase is not sufficient to account for observed strengthening. Many authors also failed to reproduce the results reported by Niihara. As a result, there exists no general agreement on the existence of the so-called "nanocomposite" effect, and alternative explanations of the observed strengthening, often related to processing or machining effects, are provided. These include: (1) elimination of processing flaws and suppression of grain growth in nanocomposites [63,64]; (2) elimination of grain pull-out during surface machining with resulting enhanced resistance to surface defect nucleation [10]; (3) generation of high level surface compressive residual stresses during machining [4,5,65]; or (4) increased tendency of nanocomposites to crack during annealing [4,66].

Even though the increase of fracture toughness in nanocomposites was never reported to be high, the results achieved by various authors are still more controversial. While several papers report modest increase of toughness of nanocomposites over that of unreinforced ceramics [67], others do not find any appreciable change [5,63,68], or even report a reduction in the fracture toughness depending on the measurement technique. One of the reasons may be that the mechanical properties of nanocomposites are strongly influenced by slight changes of the processing route [3]. Another factor, whose role is not clear, is the role of intra- and intergranular SiC particles in defining the mechanical properties, especially fracture toughness. Unlike monolithic aluminas, which usually fail by grain boundary fracture, cracks in Al_2O_3-SiC nanocomposites follow an almost entirely transgranular path. Some authors suggest that the change of the fracture mode is caused by tensile tangential stresses in alumina matrix grains around intragranular SiC inclusions. Combined with radial, grain boundary strengthening compressive stresses, the cracks are turned into alumina matrix grains so that they follow a transgranular path, being attracted by intragranular SiC inclusions. The increase of fracture energy is not observed, as the increase of toughness resulting from the change of fracture mode from intergranular to transgranular is compensated by the crack passing through tensile stress fields between second phase and matrix particles [5]. However, Jiao and Jenkins, who performed a detailed analysis of crack propagation in nanocomposites, observed no such attraction, not even in a crack moving very close to an intragranular SiC particle [69]. Other authors consider the ratio of volume fractions of intra- and intergranular SiC as an important parameter, which influences the fracture toughness of nanocomposites. The cracks are attracted to intergranular particles due to the formation of tensile residual stress fields around particles, and perpendicular to adjacent boundaries. This mechanism is expected to increase crack deflection length, at least to a certain extent, and thus to contribute to toughening of the nanocomposite [67]. Significant toughening with a steep R-curve is achieved only by the addition of SiC whiskers [6,8]. High fracture toughness also results in markedly improved thermal shock resistance of whisker-reinforced alumina composites [7,70].

Wear resistance is probably the only room temperature mechanical property where unambiguous improvement in comparison to monolithic alumina is observed [71]. The published data on wear of Al_2O_3-SiC nanocomposites and the monolithic alumina report that erosion resistance is more than three times higher [9,72,73], and note a reduction of the dry sliding wear rate [74] of composites with respect to the monolithic alumina with comparable grain size. Addition of SiC nanoparticles into polycrystalline alumina also produces a noticeable improvement in surface quality during lapping and polishing [10,75]. This is considered to be the result of a reduction of grain pullout during grinding and polishing, which, in turn, is believed to be the consequence of an altered method of fracture—from intergranular in monolithic alumina to transgranular in nanocomposites [1]. There exist various explanations for the observed change of the fracture mode, ranging from the strengthening of grain boundaries [76,77] and crack deflection from grain boundaries into the interior of alumina grains by thermal residual stresses around intragranular SiC particles [67], through changes in surface flaw population, to the presence of surface residual stresses [78]. Todd and Limpichaipanit suggest that the role of SiC in nanocomposites with high SiC volume fractions (10 vol%) is in suppression of brittle fracture of alumina by blocking the formation of long twins and dislocation pileups, which are thought to be responsible for crack initiation by intragranular SiC particles (*i.e.*, a form of slip homogenization). They also suggest that the reason for the observed change of fracture mode from intergranular in monolithic alumina to transgranular in SiC-containing composites (including those with added micrometer-sized SiC particles) can be sought in the change of the system's chemistry, rather than in purely mechanical interactions between alumina and SiC [79]. However, there exists no general agreement on which mechanism is responsible for the observed changes in mechanical and wear properties of "nanocomposites", and it remains unclear whether the SiC particles inside the alumina grains or those at the grain boundaries are primarily responsible for these changes.

As to the high temperature mechanical properties, several observations suggest that the addition of SiC into the Al_2O_3 matrix generally increases the creep resistance of Al_2O_3-SiC nanocomposites by one to two orders of magnitude in comparison to the monolithic Al_2O_3 [80,81]. The mechanisms responsible for the improvement of creep resistance of the Al_2O_3-SiC nanocomposites are still under investigation [14,82]. The improvement of creep resistance is generally attributed to the presence of residual stresses, which are created around SiC inclusions in the course of cooling from the temperature of sintering, due to different thermal expansion coefficients of alumina and SiC. The inherent stresses at the alumina–SiC interfaces are compressive, resulting in stronger particle/matrix bonding and inhibition of grain boundary diffusion by intergranular SiC particles and, hence, improved creep resistance [83,84]. In other words, the Al_2O_3-SiC interface is much stronger than the alumina–alumina interface, the interfacial fracture energy of an Al_2O_3-SiC interface being two times higher than the interfacial fracture energy of an alumina–alumina boundary. Another mechanism contributing to creep resistance is the grain boundary pinning by intergranular SiC particles. As they are engaged with the Al_2O_3 grains, the SiC particles rotate, inhibiting the grain boundary sliding and reducing the strain rate in the composite [85,86]. The Al_2O_3-SiC nanocomposites with 17 vol% of SiC nanoparticles tested at temperatures up to 1300 °C and at mechanical loads ranging from 50 to 150 MPa, exhibit a creep rate about three orders of magnitude lower and a creep life 10 times longer, than that of the monolithic Al_2O_3 under the same conditions. The addition of 5 vol% of SiC to the Al_2O_3-based nanocomposites leads to results similar to those mentioned above [87]. However, some authors suggest that higher SiC contents actually decrease the creep resistance due to surface oxidation of the SiC particles. Silica forms an amorphous silicate grain boundary film, which reduces the strength of the Al_2O_3-SiC interface bonding. In our previous work we studied the influence of the volume fraction of SiC particles with a mean size of 200 nm on the microstructure and creep behavior of the composites at temperatures up to 1450 °C and mechanical load up to 200 MPa (Figure 2) [88]. The composite with 10 vol% of SiC can withstand stress of 200 MPa at 1350 °C and 1400 °C for 150 h, while the monolithic Al_2O_3 reference fails already after 0.8 h at 1350 °C and a load of 75 MPa. The creep resistance of the composites increases with increasing volume fraction of SiC in the concentration range between 3 and 10 vol%. The improvement is attributed to the pinning effect of intergranular SiC particles. At higher SiC contents (15 and 20 vol%), the creep resistance is impaired significantly as the result of microstructure refinement.

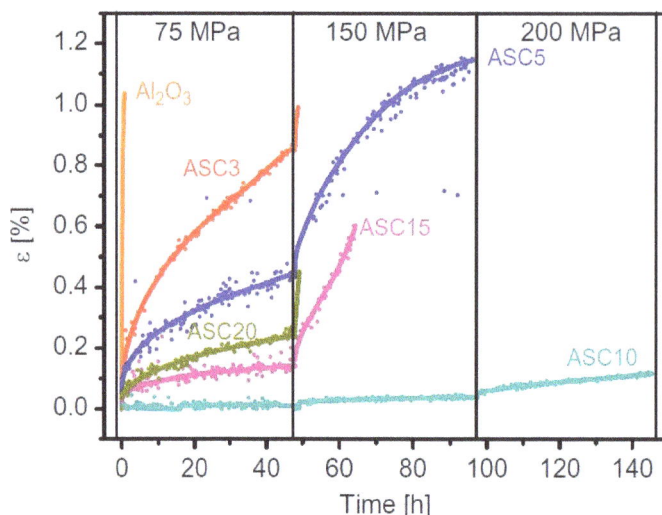

Figure 2. Creep deformation of Al_2O_3-SiC (AS) nanocomposites measured at 1350 °C and mechanical load of 75, 150, and 200 MPa. The increase of load is reflected as a break at the stress-strain curve. The number in the sample denomination represents the volume fraction of SiC in the material, *i.e.*, AS5c represents the Al_2O_3-SiC nanocomposite with 5 vol% of SiC [88].

3.2. Al_2O_3-CNT, CNF

Tremendous effort in the last couple of years focused on the research into CNT- or CNF-containing alumina-based nanocomposites was motivated by the extraordinary mechanical properties of both the single- and multi-walled carbon nanotubes, initiating large expectations concerning the improvement of the mechanical properties of polycrystalline alumina. As yet, the results remain controversial. Due to the high aspect ratio of CNT, some improvement of mechanical properties of alumina-based ceramics can be expected at a CNT content as low as 0.01 wt%. However, in many cases much higher CNT contents do not result in any observable improvement. On the contrary, many authors report deterioration of mechanical properties through addition of the CNTs. In order to understand such discrepancies the nature of CNTs, as well as the processing conditions, must be considered. The main problem is a great variety of choice of carbon nanotubes, which differ significantly due to the conditions of their preparation, and subsequent treatment. Different types of CNTs exhibit various levels of mechanical strength, density, and affinity to the ceramic matrix due to differences in their tubular structure, numbers and crystallinity of rolled graphene sheets, number and nature of surface defects, and surface chemistry, just to name only the most important ones [89]. Inappropriate choice of carbon nanotubes or treatment then impairs the mechanical properties of the nanocomposite [36]. This is further complicated by the fact that the understanding of nanostructured composites, and especially the nature of interfacial phenomena between CNTs and the Al_2O_3 matrix, which is of crucial importance for mechanical behavior, is far from satisfactory. Recent reports indicate that the grain boundary structure at the CNT–alumina interfaces has a strong influence on mechanical properties [53]. In order to exploit the exceptional elastic properties of CNTs, a strong interfacial bonding at the Al_2O_3-CNTs interface is considered vital. An Al_2OC phase possibly formed by carbothermal reduction of Al_2O_3, with good chemical compatibility with both the CNTs and the alumina matrix, has been found to increase the pullout resistance of the CNTs from alumina matrix. This way the high elastic modulus of the CNTs is exploited, bridging the cracks, hindering the crack propagation, and leading to improved fracture toughness [52].

There are also other factors affecting the mechanical properties of CNT-reinforced composites. The first and most crucial one is uneven distribution of the CNTs when an unsuitable dispersion technique is used. Suitability of the dispersion method is defined by the nature of the matrix phase and its surface charges and particle size distribution in the composite suspension, as well as the diameter and length of the used CNTs. The second reason is related to problems with densification, sometimes also due to uneven dispersion of the CNTs. In this case, ceramics with low relative density and poor mechanical properties are obtained despite a high CNT dispersion, irrespective of whether MWCNT or SWCNT is used [90]. The influence of CNT addition on various mechanical properties will be discussed in more detail in the following sections.

3.2.1. Elastic Modulus

The effective elastic modulus of the alumina composites containing up to 1 vol% of MWCNT is usually comparable or slightly lower than that of the monolithic alumina. With increasing volume fractions of the MWCNT, the modulus further decreases [38,55]. The main reason for the observed decrease is the presence of a significant amount of residual porosity: the pores are known to act as a second phase with zero modulus [91]. Another reason cited is the low elastic modulus of the MWCNT. Yu *et al.* [92], report the elastic modulus of the MWCNT at the level of 270 GPa due to imperfections in its structure, and structural defects originating from its treatment. Such damaged nanotubes can then be expected to reduce the elastic modulus of the composites.

3.2.2. Hardness

Several counteracting influences must be taken into account when evaluating the hardness of CNT-reinforced alumina composites. However, in most reported cases the addition of CNTs results in decrease of microhardness. These results are most often related to the presence of residual porosity, which increases with the CNT content. The trend is further aggravated by uneven dispersion of the CNTs in the matrix, the presence of porous bundles of nanotubes, which act as defects with no load-bearing capacity, and poor cohesion between CNTs and the matrix [93,94]. As a result, a decrease of hardness from 17 GPa in monolithic alumina to 12.5 GPa was observed in the nanocomposite with the addition of a mixture of MWCNT and SWCNT [36]. Other works report monotonous decrease of Vickers hardness with increasing volume fraction of carbon nanotubes, sometimes with small positive deviation from the trend at 1 vol% of MWCNTs (Figure 3a) [95]. The result is attributed to weak interfacial bonding between the MWCNT and the alumina matrix grains. The positive deviation observed at 1 vol% of the MWCNT is attributed to the slightly higher relative density of the composite. The hardness decrease at higher volume fractions of MWCNT is due to the presence of poorly dispersed bundles of MWCNTs, acting as residual porosity in the matrix. These effects can be to a certain extent counteracted by refinement of alumina matrix grains in the final stage of sintering due to pinning action of CNTs at grain boundaries. In fine (and especially submicron) grains, the slip of dislocations is blocked by grain boundaries, resulting in increased hardness of the material. In general, decrease of hardness in the nanocomposites with increasing CNT content is usually attributed to several factors, including the presence of soft phases at the alumina grain boundaries (hardness of MWNTs in the radial direction is 6–10 GPa [89]), poor adherence between the CNTs and the ceramic matrix grains, the lubricating nature of the CNTs, and poor dispersion of CNTs in the alumina matrix, which counteract the influence of microstructure refinement due to the pinning effect of the CNTs at grain boundaries.

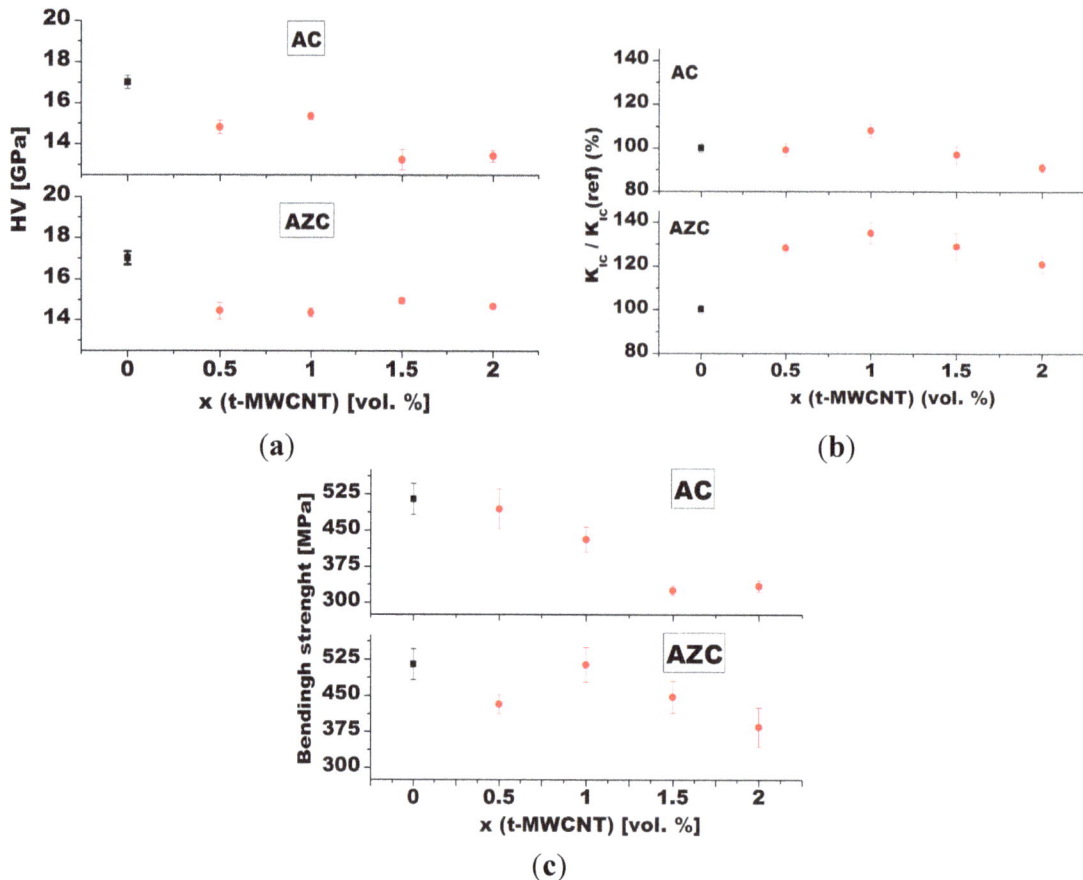

Figure 3. Composition dependence of (**a**) hardness; (**b**) fracture toughness; and (**c**) fracture strength of Al_2O_3-MWCNT (denoted AC) and Al_2O_3-ZrO_2-MWCNT (denoted AZC) nanocomposites. Comparison to monolithic alumina reference [95]. Red circles represent the respective properties of nanocomposite with various volume fractions of MWCNT. Black squares represent the same property of the monolithic alumina reference.

3.2.3. Fracture Toughness and Strength

Large interest in the use of carbon nanotubes as a strengthening and toughening agent has been spurred by the work of Zhan *et al.* [96], who reported a fracture toughness of 9.7 MPa·m$^{1/2}$ in a composite with alumina matrix containing 10 vol% of SWCNTs manufactured by spark-plasma sintering. This represents a nearly 300% increase in comparison to monolithic alumina. Since then many attempts have been made to use carbon nanotubes as a toughening agent (Figure 3b) [95,97–99]. Most investigations focused on alumina-based composites, using either SWCNTs or MWCNTs, but conclusive demonstration of toughening has not been achieved [100]. The results are generally (but not unambiguously) disappointing for the composites containing MWCNTs, where most authors observed only a marginal increase, or even a decrease of fracture toughness in comparison to monolithic alumina [98,101–104]. However, several authors reported a significant improvement of fracture toughness in MWCNT–alumina composites [105,106], similar to that achieved by Zhan [96]. Concerning SWCNTs, a few conflicting reports exist [97,107]. Contrary to the results of Zhan [96], several authors report no toughening effect in the Al₂O₃ nanocomposites with 10 vol% of SWCNTs, which were as brittle as the monolithic alumina [93]. Despite the contradictions, most authors believe SWCNTs to be more efficient toughening agents than MWCNTs. This is sometimes explained by the fact that MWCNTs exhibit easy sliding between individual graphene sheets in a "sword and sheath" manner, with small ability to carry or transfer loads [108]. Moreover, the MWCNTs have much lower bending strength and stiffness than the SWCNTs, which makes them less efficient reinforcement aids [109]. Yamamoto et al. [110], observed initiation of cracks at defect sites in the outer wall of the MWCNTs in external tensile stress field. The cracks propagate through the MWCNTs without any interwall sliding, i.e., the nanotubes simply break. The results suggest that defects detrimental to the load bearing capacity of MWCNTs can by created through thermal damage in the course of high temperature treatment of the nanotubes.

As to the mechanisms responsible for anticipated toughening, even more controversy exists. Most authors agree on the ability of CNTs to induce conventional toughening mechanisms, such as crack bridging and CNT pullout [111–113], in some cases combined with crack deflection at CNT–alumina interfaces [96,114]. Some authors also report the weakening of grain boundaries with CNTs, most likely due to formation of CNT agglomerates, difference of thermal expansion coefficients of the CNT and the alumina matrix, and resulting tensile thermal residual stresses across the grain interfaces [107,115,116]. Intergranular fractures characteristic of the CNT-reinforced composites are considered as evidence for such grain boundary weakening [96,105]. However, due to the fact that the addition of CNT usually results in marked refinement of alumina matrix grains (often to sub-micron level) the contribution of crack deflection to fracture toughness is negligible. If the load-bearing capacity of CNT is to be fully utilized, a strong interfacial bonding must exist between the CNTs and the matrix. The friction force between CNTs and matrix is thus the most important factor mediating the energy absorbing role of the CNTs during the crack bridging and pull-out process [105,112]. Tailoring the interfacial strength thus seems an efficient way to increase toughness of the composites. Song et al. [37], increased the adhesion between MWCNTs and the alumina matrix by coating the nanotubes with a layer of Al₂O₃ nanocrystals, significantly increasing the friction force between MWCNTs and the Al₂O₃ matrix grains. Strong interfacial bonding then makes full use of the high elastic modulus and tensile strength of the CNTs, leading to increased toughness of the composite. The toughening action of the CNTs can be then in principle described as follows: The alumina matrix breaks intergranularly, the crack preferentially following the grain boundaries, where weak matrix–CNT interfaces are created due to the CNTs accumulation. The cracked surfaces are then bridged by the CNTs, oriented perpendicularly to the direction of propagating crack, their ends firmly attached in the matrix. The energy dissipation is attributed to the work done by elastic extension of CNTs [105], combined with the work of friction related to pulling out either end of the CNTs from the matrix.

Marginal improvement, or even deterioration, of flexural strength of CNT-reinforced nanocomposites is usually associated with the failure to achieve required dispersion of the CNTs and problems with densification, which result in the presence of residual porosity. Both residual pores and residual bundles of agglomerated nanotubes act as strength defining defects and origins of fracture (Figure 3c) [95].

From the point of view of improving the mechanical properties of alumina ceramics, an interesting option is represented by the so-called hybrid microstructure design, *i.e.*, preparation of complex Al_2O_3-ZrO_2-CNT or Al_2O_3-SiC-CNT microstructures [60,90]. The addition of as little as 0.01 wt% of MWCNTs to conventionally sintered zirconia toughened alumina (ZTA) ceramics is reported to result in an increase of fracture toughness, with the MWCNT acting as an efficient toughening agent. Similar results observed in ZTA ceramics reinforced by the addition of up to 2 vol% of MWCNT are attributed to synergy effect of transformation toughening with small contribution of crack bridging and CNT pull-out (Figure 4) [95].

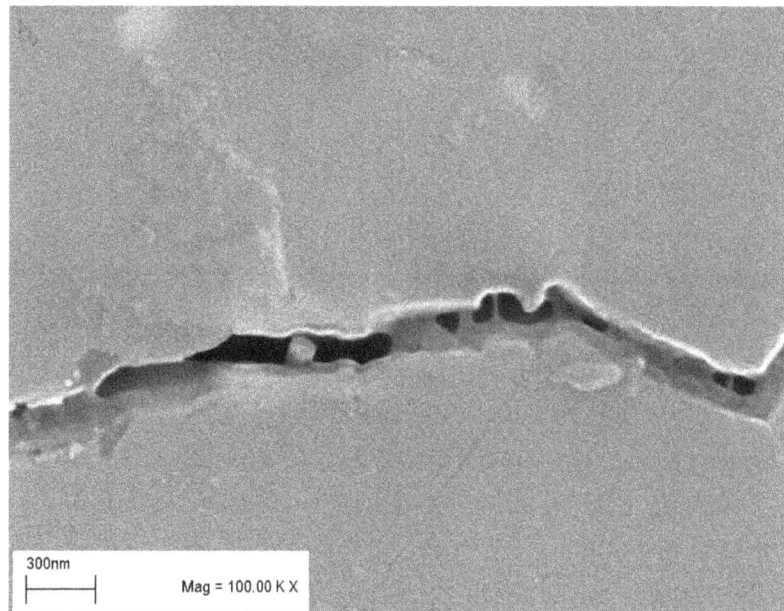

Figure 4. Indication of toughening mechanisms observed in Al_2O_3-ZrO_2-MWCNT nanocomposites [95].

In the Al_2O_3-SiC-CNT composites, the observed toughening is attributed to the strengthening of grain boundaries and toughening of the alumina matrix by nanosized SiC particles combined with fiber toughening mechanisms from MWCNTs. The incorporation of SiC nanoparticles is also believed to remove residual stresses at the alumina–alumina boundaries, and in matrix grains by generating dislocations around the particles [115,116]. Elimination of tensile stresses strengthens the grain boundaries and impedes the intergranular fracture observed in alumina with added CNTs.

In order to employ excellent elastic properties of carbon nanotubes, and to achieve toughening in alumina matrix composites, the following points are thus crucial: (1) homogeneous dispersion and de-agglomeration of carbon nanotubes in the matrix must be achieved; (2) the composite must be sintered to a high density and the residual porosity eliminated; (3) strong interfacial bonding between CNT and the matrix must be achieved, and (4) high temperature or oxidation damage to the CNTs must be avoided [117].

3.2.4. Tribological Properties

The trends observed in tribological characteristics of CNT-containing alumina-based composites are ambiguous. Some improvement of sliding wear resistance is observed at lower CNT contents, attributed to the grain size effect (refinement of alumina matrix grains with the addition of CNTs) combined with lubrication action of the CNTs (carbon) resulting in marked decrease of friction coefficient. However, despite a sharp decrease of friction coefficient at high CNT volume fractions, the wear losses increase significantly at a CNT content above 10 vol%. This trend is usually explained by deterioration of mechanical properties and increased level of residual porosity due to already discussed difficulties with even dispersion of CNTs and poor cohesion of CNTs with the alumina matrix [58].

3.2.5. Carbon Nanofibers

Carbon nanofibers are often considered as a more readily available and cheaper substitute for carbon nanotubes, with similar capacity for improving both the mechanical and the functional (electric conductivity) properties of alumina-based composites. However, similarly to CNTs, the results are often confusing and contradicting. Generally speaking, the addition of carbon nanofibers leads either to marginal improvement [118] or significant deterioration—by about 40% in comparison to monolithic alumina [119]—of fracture strength. Hardness and fracture toughness are degraded [118]. The disappointing results are usually attributed to the high affinity of the CNFs to form aggregates due to strong van der Waals interactions among them, and, as a consequence, uneven dispersion of CNFs in the composites. The addition of CNFs results in an increase of wear resistance (lower wear rates, decrease of friction coefficient) measured under the conditions of sliding wear using the ball-on-disk technique [120]. The improvement is attributed to the lubricating effect of the CNFs, making the Al_2O_3-CNF composites promising candidates for unlubricated tribological applications.

4. Functional Properties

4.1. Al_2O_3-SiC

Only a few research papers on the functional properties (thermal and electrical conductivity) of Al_2O_3-SiC nanocomposites have been published so far [121,122], despite the fact that thermal conductivity is an important parameter in many applications of alumina-based ceramics, including high temperature structural components, refractories for glass and metal production industries, gas radiant burners, wear parts, and cutting tools. In all these applications, the thermal conductivity has to be as high as possible, in order to reduce thermal shock-related failure of the components. Addition of SiC particles can be expected to improve thermal conductivity of the Al_2O_3-based nanocomposites due to intrinsically high thermal conductivity of SiC. However, only moderate increase of thermal conductivity has been achieved so far in the alumina matrix composites with SiC inclusions (Figure 5) [122,123]. The relations between the SiC addition and the thermal conductivity of alumina-based composites are complicated, and the influence of interfacial barriers, impurities, and various defects must be considered. The existence of interfacial barriers impairs the conduction of heat by scattering phonons with related increase of the interfacial thermal resistance [124]. It is therefore assumed that the presence of a thermal barrier at the matrix/dispersion boundaries is responsible for the relatively low values of thermal conductivity of this type of composite [125]. In our previous work we investigated the thermal conductivity of Al_2O_3-SiC nanocomposites containing 3–20 vol% of SiC particles of two different sizes, 40 and 200 nm. The maximum room temperature thermal conductivity is achieved in the samples containing 20 vol% of SiC particles (38 $W \cdot m^{-1} \cdot K^{-1}$), irrespective of the size of the SiC particles, which represents a 35% increase in comparison to the monolithic Al_2O_3 reference (28 $W \cdot m^{-1} \cdot K^{-1}$). The thermal conductivity decreased with increasing temperature, falling down to 10–15 $W \cdot m^{-1} \cdot K^{-1}$ at 1000 °C (Figure 5) [126].

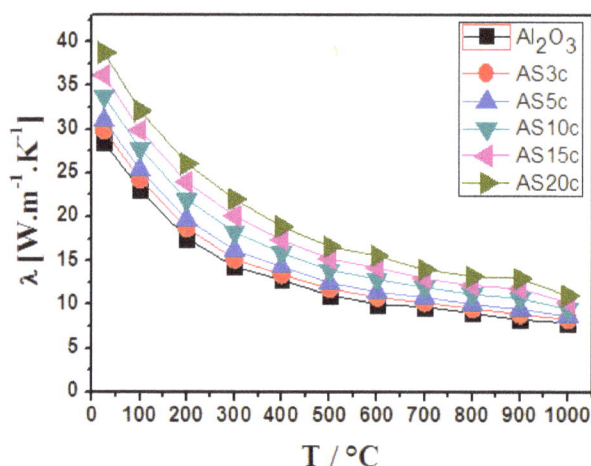

Figure 5. Temperature dependence of thermal conductivity of Al_2O_3-SiC (AS) nanocomposites with various volume fractions of SiC. Comparison to monolithic alumina reference [126]. The number in the sample denomination represents the volume fraction of SiC in the material (*i.e.*, AS3c represents the Al_2O_3-SiC nanocomposite containing 3 vol% SiC).

Unlike the moderate improvement of thermal conductivity, the electrical conductivity of alumina-based composites can be tailored in a much wider range [121]. The composites reinforced with conductive or semiconductive phases (such as silicon carbide), added in the amount at which they percolate the insulating alumina matrix, have received particular attention. Such electro-conductive or semiconductive ceramics are of special interest in a wide range of industrial applications. In the nanocomposites the electric properties are determined by many critical factors, such as the composition of powder mixtures (the volume fraction of SiC, content of silica as the product of surface oxidation of submicron SiC particles), content of other impurities, and the parameters of the final microstructure of the composite, including the size of alumina matrix grains and the size (micrometer or nanosized) and distribution (intergranular, intragranular or both) of SiC inclusions. The addition of SiC improves DC electrical conductivity, which increases with the volume fraction of SiC [126]. In the composite with 20 vol% of SiC, conductivity of 4.05×10^{-2} $S \cdot m^{-1}$ was measured, which represents an increase of four orders of magnitude in comparison to the monolithic alumina reference (7.80×10^{-6} $S \cdot m^{-1}$) (Figure 6). The electrical conductivity of the Al_2O_3-SiC nanocomposites with the same volume fraction but different size of SiC particles is comparable.

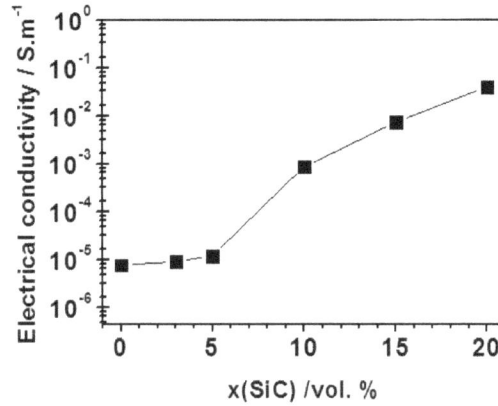

Figure 6. Composition dependence of DC electric conductivity of Al$_2$O$_3$-SiC nanocomposites with various volume fractions of SiC. Comparison to monolithic alumina reference [126].

4.2. Al$_2$O$_3$-CNT, CNF

4.2.1. Electric Properties

Due to their intrinsically high electric conductivity, carbon nanotubes (SWCNT or MWCNT) are considered as an ideal candidate for enhancement of the electric conductivity of ceramic materials [127,128] without impairing their mechanical properties. Indeed, the electric conductivity is the only physical property that is, beyond any doubt, markedly improved through the addition of CNTs. The addition of CNFs has am influence on electrical conductivity similar to that of the more expensive MWCNTs [119]. Apart from the high electric conductivity of CNTs or CNFs, an important factor contributing to the electric conductivity of a composite with electrically insulating matrix is the ability of a conductive phase to achieve percolation threshold. The CNTs and CNFs facilitate achievement of the percolation threshold at very low volume fractions, 20 times lower than the percolation threshold achieved in random two-phase composites with micrometer-scale microstructure and isometric morphology of individual components. This low value is attributed to the enormous aspect ratio of MWNTs [128,129]. The DC conductivity near the percolation threshold can be described by power law (Equations (2) and (3)) [130]:

$$\sigma_m = \sigma_c(f_{MWNT} - f_c)^t \text{ for } f_{MWNT} > f_c \tag{2}$$

$$\sigma_m = \sigma_i(f_c - f_{MWNTc})^{-s'} \text{ for } f_{MWNT} < f_c, \tag{3}$$

where σ_m is the total DC conductivity of the composite; and σ_c and σ_i are the DC conductivities of the conductive phase and the insulating ceramic matrix, respectively. The symbol f_{MWNT} stands for the volume fraction of the conductive phase, and f_c is the volume fraction of the conductive phase at which the percolation threshold is achieved. The exponent t is the conductivity exponent reflecting the dimensionality of the system, assuming the values 1.33 and 2 for two- and three-dimensional conductivity, respectively, and usually varying between 1.33 and 1.94. However, t values lower than 1.94 reflect thermally induced charge hopping transport between loosely connected parts of CNTs rather than the existence of a less than three-dimensional conductive network [131]. Exponent s' is the critical exponent in the insulating region, usually assuming universal values between 0.8 and 1.0.

Published experimental data generally indicate an increase of electric conductivity of CNT-containing composites with increasing content of the CNTs. Moreover, the composites exhibit a typical insulator-conductor transition around the percolation threshold. The volume fraction of the CNTs at which the percolation threshold is achieved depends on several factors, including aspect ratio of the used nanotubes, the level of de-agglomeration, and homogenous distribution of the CNTs in insulating matrix. Various authors report the f_c values ranging from 0.094 to 2.5 wt% of the CNT [61,128,132,133]. The composites typically exhibit a dramatic increase of electric conductivity (at the level of eight orders of magnitude from 10^{-12} S·m^{-1} characteristic for insulating alumina matrix to 10^{-4} S·m^{-1} or more in nanocomposites) when the percolation threshold is achieved. In our previous work we achieved the maximum value of the electrical conductivity at the level of 10^{-1} S·m^{-1} at 2 vol% addition of MWCNTs, which represents an improvement of 11 orders of magnitude with regard to the monolithic alumina reference (Figure 7) [95]. The results exceed by far those reported by Zhou et al. [134], who achieved a conductivity of 6.2×10^{-2} S·m^{-1} in the composite with the same content of MWCNTs. Such high electrical conductivity is attributed to simultaneous action of two conductive mechanisms: (1) formation of a conductive path through interconnected and percolated network of carbon nanotubes; and (2) evaporation of carbon from MWCNTs, which takes place at the temperature of hot pressing: The carbon deposits at grain boundaries, increasing their electrical conductivity. Further increase in the amount of CNTs beyond the percolation threshold results in only a marginal increase in conductivity, which tends to level off at higher concentrations of carbon nanotubes.

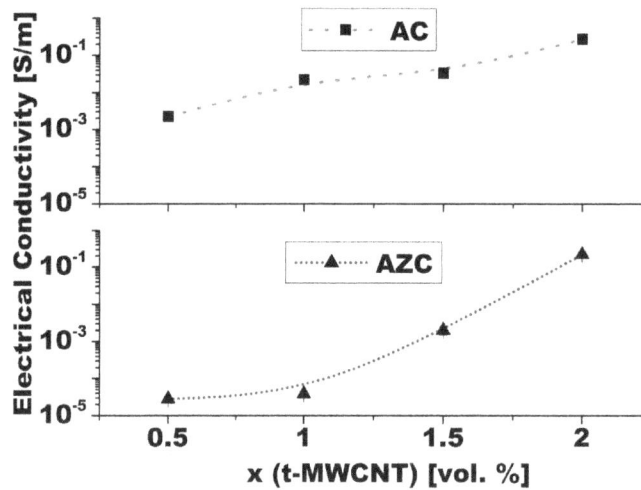

Figure 7. Composition dependence of DC electric conductivity of Al$_2$O$_3$-MWCNT (material AC) and Al$_2$O$_3$-ZrO$_2$-MWCNT (material AZC) nanocomposites [95].

Similarly to DC electric conductivity, the variation of dielectric constant of the composites also follows, in the area close to the percolation threshold, the power law (Equation (4)) [135]:

$$\varepsilon_c = \varepsilon_0 (f_c - f_{MWNT})^{-s} \tag{4}$$

where ε_c and ε_0 are the dielectric constants of the composite and the matrix, respectively; f_{MWNT} is the volume fraction of carbon nanotubes; f_c is the percolation threshold, and s is the critical exponent.

According to Ahmad et al. [128], the dielectric behavior of the CNT-containing nanocomposites can be divided into two categories. In the composites with less than 0.7 vol% of MWNTs, the dielectric constant at room temperature is frequency independent and is defined by dielectric properties of the alumina matrix. As the concentration of MWCNTs approaches the percolation threshold, the dielectric constant increases markedly, assuming a value of about 4600 at a frequency of 1 kHz and MWCNT content of 1.74 vol%. The behavior is attributed to the presence of large number of conducting clusters isolated by thin dielectric layers. Each cluster acts as a minicapacitor: Polarization between the clusters improves electric charge storage. The combination of these factors then contributes to the increase of dielectric constant [136].

4.2.2. Thermal Conductivity

Although there are some works that report increased thermal conductivity in CNT- or CNF-doped alumina-based composites (Figure 8) [95,119,137], the majority of the published data indicates a decrease of thermal conductivity in comparison to monolithic alumina matrix when CNTs are added [129,138]. Such a decrease is understandable if the heat transport mechanism is considered. In ceramic composites with added carbon nanotubes, heat is conducted through propagation of phonon waves [139,140]. Thermal conductivity is then influenced by the sound speed in the composite (related to its elastic modulus), phonon mean free path, and thermal resistance at CNT-alumina interfaces [62]. The thermal conductivity K_e of CNT-alumina composites can be described using Equation (5) [141]:

$$K_e = \frac{1}{3}C_v ml \qquad (5)$$

where C_v is the heat capacity per unit volume; m is the speed of sound; and l is the phonon mean free path. At temperatures above the Debye temperature, C_v changes very little in comparison to the other two quantities.

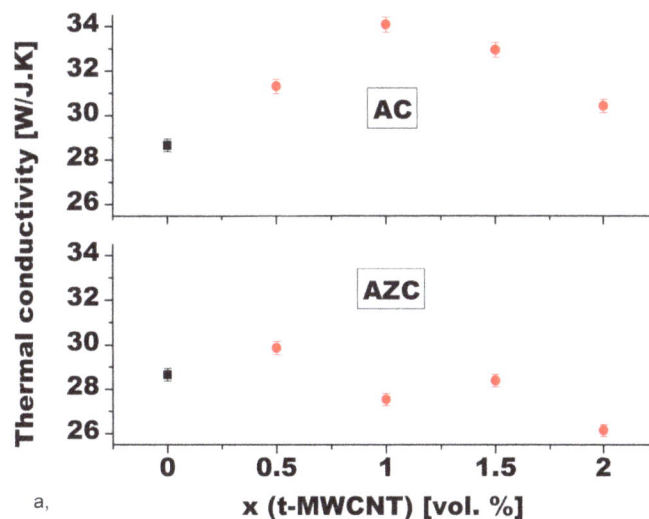

Figure 8. Composition dependence of thermal conductivity of Al_2O_3-MWCNT (material AC) and Al_2O_3-ZrO_2-MWCNT (material AZC) nanocomposites [95].

The thermal conductivity is therefore controlled by the changes in the phonon mean free path and sound speed. The elastic modulus, and hence the sound speed, is known to decrease with increasing content of CNTs in the composite [62]. The phonon mean free path is influenced by scattering processes related to the arrangement of CNTs and CNT-Al_2O_3 interfaces in the composite. The mean free phonon path is reported to decrease with an increasing content of MWCNTs, reaching the minimum values as low as 1 nm, which is much less than the size of matrix grains. Hence, the probability of grain boundary phonon scattering is low [62]. Other authors report much higher values of the phonon mean free path, ranging from 20 to 500 nm [142]. However, even if the grain boundary scattering can be neglected, intertube and intergraphene layer coupling together with scattering at matrix defects markedly contribute to reduction of the phonon mean free path [19,142]. In addition, the MWCNTs dispersed inside the alumina matrix induce interface scattering and further reduce the phonon mean free path.

Apart from reduction of the phonon mean free path, additional mechanisms responsible for decrease of thermal conductivity must also be taken into account. These include high thermal resistance at alumina-CNT interfaces [129] and agglomeration of the CNTs with much lower thermal conductivity than the conductivity of individual CNTs due to intense intertube scattering [143].

5. Conclusions

The paper provides a review of alumina-based nanocomposites with added SiC, CNTs, and CNFs, evaluating the efficiency of the additives in terms of their influence on the mechanical and functional properties of the composites. Despite tremendous effort in the last decades focused on improvement of mechanical properties of alumina ceramics, the results remain controversial. Although in some cases significant improvement of mechanical properties is reported (e.g., fracture strength as high as 1000 MPa in Al_2O_3-SiC, and fracture toughness of 9.7 $MPa \cdot m^{1/2}$ in Al_2O_3-CNT nanocomposites), subsequent work encountered serious problems reproducing the results, achieved more modest improvements, failed to achieve any improvement entirely, or reported deterioration of mechanical properties. The failure to achieve any significant improvement is attributed to a poor understanding of physical and chemical interactions between alumina matrix and the reinforcing phases, problems with de-agglomeration of reinforcing phases and their homogenous distribution in the matrix, and problems related to complete elimination of residual porosity during sintering. Among the mechanical properties, wear resistance is the only one where significant improvement is achieved through the addition of SiC nanoparticles into a polycrystalline alumina matrix. Among the functional properties, electrical conductivity is markedly increased through the addition of highly conductive secondary phases, such as CNTs and CNFs, at concentrations where percolation threshold is achieved, while the thermal conductivity of the nanocomposites is usually impaired by the addition of the second phase.

Acknowledgment

The financial support of this work by the grants APVV 0108-12 and VEGA 2/0058/14 is gratefully acknowledged. This publication was created in the framework of the project "Centre of excellence for ceramics, glass, and silicate materials" ITMS code 262 201 20056, based on the Operational Program Research and Development funded by the European Regional Development Fund.

Author Contributions

The authors contribution to creation of the manuscript is as follows: Dušan Galusek: literature review on processing and functional properties, preparation of introduction and conclusions, as well as the chapters on processing and functional properties, final editing of the manuscript; Dagmar Galusková: literature review on mechanical properties of nanocomposites, preparation of the chapters on mechanical properties.

Conflicts of Interest

The authors declare no conflict of interest.

References

1. Niihara, K. New design concept of structural ceramics—Ceramic nanocomposites. *J. Ceram. Soc. Jpn.* **1991**, *99*, 974–982.

2. Davidge, R.W.; Brook, R.J.; Cambier, F.; Poorteman, M.; Leriche, A.; O'Sullivan, D.; Hampshire, S.; Kennedy, T. Fabrication, properties, and modelling of engineering ceramics reinforced with nanoparticles of silicon carbide. *Br. Ceram. Trans.* **1997**, *96*, 121–127.

3. Perez-Rigueiro, J.; Pastor, J.Y.; Llorca, J.; Elices, M.; Miranzo, P.; Moya, J.S. Revisiting the mechanical behavior of alumina silicon carbide nanocomposites. *Acta Mater.* **1998**, *46*, 5399–5411.

4. Wu, H.Z.; Lawrence, C.W.; Roberts, S.G.; Derby, B. The strength of Al_2O_3/SiC nanocomposites after grinding and annealing. *Acta Mater.* **1998**, *46*, 3839–3848.

5. Zhao, J.; Stearns, L.C.; Harmer, M.P.; Chan, H.M.; Miller, G.A. Mechanical behavior of alumina silicon-carbide nanocomposites. *J. Am. Ceram. Soc.* **1993**, *76*, 503–510.

6. Jiang, D.L.; Huang, Z.R. SiC whiskers and particles reinforced Al_2O_3 matrix composites and N_2-HIP post-treatment. *Key Eng. Mater.* **1999**, *159*, 379–386.

7. Collin, M.I.K.; Rowcliffe, D.J.; Influence of thermal conductivity and fracture toughness on the thermal shock resistance of alumina-silicon-carbide-whisker composites. *J. Am. Ceram. Soc.* **2001**, *84*, 1334–1340.

8. Akatsu, T.; Suzuki, M.; Tanabe, Y.; Yasuda, E. Effects of whisker content and dimensions on the R-curve behavior of an alumina matrix composite reinforced with silicon carbide whiskers. *J. Mater. Res.* **2001**, *16*, 1919–1927.

9. Davidge, R.W.; Twigg, P.C.; Riley, F.L. Effects of silicon carbide nano-phase on the wet erosive wear of polycrystalline alumina. *J. Eur. Ceram. Soc.* **1996**, *16*, 799–802.

10. Sternitzke, M.; Dupas, E.; Twigg, P.C.; Derby, B. Surface mechanical properties of alumina matrix nanocomposites. *Acta Mater.* **1997**, *45*, 3963–3973.

11. Chen, H.J.; Rainforth, W.M.; Lee, W.E. The wear behaviour of Al_2O_3-SiC ceramic nanocomposites. *Scripta Mat.* **2000**, *42*, 555–560.

12. De Arellano-Lopez, A.R.; Dominguez-Rodriguez, A.; Routbort, L. Microstructural constraints for creep in SiC-whisker-reinforced Al_2O_3. *Acta Mater.* **1998**, *46*, 6361–6373.

13. Deng, Z.-Y.; Shi, J.-L.; Zhang, Y.-F.; Lai, T.-R.; Guo, J.-K. Creep and creep-recovery behavior in silicon-carbide-particle-reinforced alumina. *J. Am. Ceram. Soc.* **1999**, *82*, 944–952.

14. Tai, Q.; Mocellin, A. Review: High temperature deformation of Al_2O_3-based ceramic particle or whisker composites. *Ceram. Int.* **1999**, *25*, 395–408.

15. Gao, G.; Cagin, T.; Goddard, W.A., III. Energetics, structure, thermodynamic and mechanical properties of nanotubes. *Nanotechnology* **1998**, *9*, 183–191.

16. Krishnan, A.; Dujardin, E.; Ebbesen, T.W.; Yianilos, P.N.; Treacy, M.M.J. Young's modulus of single-walled nanotubes. *Phys. Rev. B* **1998**, *58*, 14013–14019.

17. Hernández, E.; Goze, C.; Bernier, P.; Rubio, A. Elastic properties of single-wall nanotubes. *Appl. Phys. A* **1999**, *68*, 287–292.

18. Treacy, M.M.J.; Ebbesen, T.W.; Gibson, J.M. Exceptionally high Young's modulus observed for individual carbon nanotubes. *Nature* **1996**, *381*, 678–680.

19. Berber, S.; Kwon, Y.-K.; Tománek, D. Unusually high thermal conductivity of carbon nanotubes. *Phys. Rev. Lett.* **2000**, *84*, 4613–4616.

20. Walker, C.N.; Borsa, C.E.; Todd, R.I.; Davidge, R.W.; Brook, R.J. Fabrication, Characterisation and Properties of Alumina Matrix Nanocomposites. In *British Ceramic Proceedings, No. 53: Novel Synthesis and Processing of Ceramics*; Sale, F.R., Ed.; The Institute of Materials: London, UK, 1994; pp. 249–264.

21. Timms, L.A.; Ponton, C.B. Processing of Al_2O_3/SiC nanocomposites—Part 1: Aqueous colloidal processing. *J. Eur. Ceram. Soc.* **2002**, *22*, 1553–1567.

22. Panda, P.K.; Mariappan, L.; Kannan, T.S. The effect of various reaction parameters on carbothermal reduction of kaolinite. *Ceram. Int.* **1999**, *25*, 467–473.

23. Amroune, A.; Fantozzi, G. Synthesis of Al_2O_3-SiC from kyanite precursor. *J. Mater. Res.* **2001**, *16*, 1609–1613.

24. Amroune, A.; Fantozzi, G.; Dubois, J.; Deloume, J.-P.; Durand, B.; Halimi, R. Formation of Al_2O_3-SiC powder from andalusite and carbon. *Mater. Sci. Eng.* **2000**, *A290*, 11–15.

25. Xu, Y.; Nakahira, A.; Niihara, K. Characteristics of Al_2O_3-SiC nanocomposite prepared by sol-gel processing. *J. Ceram. Soc. Jap.* **1994**, *102*, 312–315.

26. Gao, L.; Wang, H.Z.; Hong, J.S.; Miyamoto, H.; Miyamoto, K.; Nishikawa, Y.; Torre, S.D.D.L. Mechanical properties and microstructure of nano-SiC-Al_2O_3 composites densified by spark plasma sintering. *J. Eur. Ceram. Soc.* **1999**, *19*, 609–613.

27. Wang, H.Z.; Gao, L.; Guo, J.K. The effect of nanoscale SiC particles on the microstructure of Al_2O_3 ceramics. *Ceram. Int.* **2000**, *26*, 391–396.

28. Abovyan, L.S.; Nersisyan, H.H.; Kharatyan, S.L.; Orru, R.; Saiu, R.; Cao, G.; Zedda, D. Synthesis of alumina-silicon carbide composites by chemically activated self-propagating reactions. *Ceram. Int.* **2001**, *27*, 163–169.

29. Tago, T.; Kawase, M.; Morita, K.; Hashimoto, K. Fabrication of silicon carbide whisker/alumina composite by thermal-gradient chemical vapor infiltration. *J. Am. Ceram. Soc.* **1999**, *82*, 3393–3400.

30. Su, B.; Sternitzke, M. *Fourth Euro-Ceramics; Basic Science and Trends in Emerging Materials and Applications*; Bellosi, A., Ed.; Grupp Editoriale Faenza Editrice: Faenza, Italy, 1995; Volume 4, pp. 109–116.

31. Sternitzke, M.; Derby, B.; Brook, R.J. Alumina/silicon carbide nanocomposites by hybrid polymer/powder processing: Microstructures and mechanical properties. *J. Am. Ceram. Soc.* **1998**, *81*, 41–48.

32. Sawai, Y.; Yasutomi, Y. Effect of high-yield polycarbosilane addition on microstructure and mechanical properties of alumina. *J. Ceram. Soc. Jpn.* **1999**, *107*, 1146–1150.

33. Narisawa, M.; Okabe, Y.; Okamura, K.; Kurachi, Y. SiC-based fibers synthesized from hybrid polymer of polycarbosilane and polyvinylsilane. *Key Eng. Mat.* **1999**, *164*, 101–106.

34. Galusek, D.; Sedláček, J.; Riedel, R. Al2O3-SiC composites by warm pressing and sintering of an organosilicon polymer-coated alumina powder. *J. Eur. Ceram. Soc.* **2007**, *27*, 2385–2392.

35. Galusek, D.; Klement, R.; Sedláček, J.; Balog, M.; Fasel, C.; Zhang, J.; Crimp, M.A.; Riedel, R. Al2O3-SiC composites prepared by infiltration of pre-sintered alumina with a poly(allyl)carbosilane. *J. Eur. Ceram. Soc.* **2011**, *31*, 111–119.

36. Aguilar-Elguézabal, A.; Bocanegra-Bernal, M.H. Fracture behaviour of α-Al2O3 ceramics reinforced with a mixture of single-wall and multi-wall carbon nanotubes. *Compos. B* **2014**, *60*, 463–470.

37. Bi, S.; Hou, G.; Su, X.; Zhang, Y.; Guo, F. Mechanical properties and oxidation resistance of α-alumina/multi-walled carbon nanotube composite ceramics. *Mater. Sci. Eng. A* **2011**, *528*, 1596–1601.

38. Bakhsh, N.; Khalid, F.; Ahmad, H.; Saeed, A. Synthesis and characterization of pressureless sintered carbon nanotube reinforced alumina nanocomposites. *Mater. Sci. Eng. A* **2013**, *578*, 422–429.

39. Michálek, M.; Bodišová, K.; Michálková, M.; Sedláček, J.; Galusek, D. Alumina/MWCNTs composites by aqueous slip casting and pressureless sintering. *Ceram. Int.* **2013**, *39*, 6543–6550.

40. Hanzel, O.; Sedláček, J.; Šajgalík, P. New approach for distribution of carbon nanotubes in alumina matrix. *J. Eur. Ceram. Soc.* **2014**, *34*, 1845–1851.

41. Bi, S.; Su, X.J.; Hou, G.L.; Gu, G.Q.; Xiao, Z. Microstructural characterization of alumina-coated multi-walled carbon nanotubes synthesized by hydrothermal crystallization. *Phys. B* **2010**, *405*, 3312–3315.

42. Thaib, A.; Martin, G.A.; Pinheiro, P.; Schouler, M.C.; Gadelle, P. Formation of carbon nanotubes from the carbon monoxide disproportionation reaction over Co/Al2O3 and Co/SiO2 catalysts. *Catal. Lett.* **1999**, *63*, 135–141.

43. Mo, Y.H.; Kibria, A.K.M.F.; Nahm, K.S. The growth mechanism of carbon nanotubes from thermal cracking of acetylene over nickel catalyst supported on alumina. *Synth. Met.* **2001**, *122*, 443–447.

44. Chen, P.; Zhang, H.B.; Lin, G.D.; Hong, Q.; Tsai, K.R. Growth of carbon nanotubes by catalytic decomposition of CH4 or CO on a Ni–MgO catalyst. *Carbon* **1997**, *35*, 1495–1501.

45. Beitollahi, A.; Pilehvari, S.H.; Faghihi Sani, M.A.; Moradi, H.; Akbarnejad, M. *In situ* growth of carbon nanotubes in alumina-zirconia nanocomposite matrix prepared by solution combustion method. *Ceram. Int.* **2012**, *38*, 3273–3280.

46. Merchan-Merchan, W.; Saveliev, A.V.; Kennedy, L.; Jimenez, W.C. Combustion synthesis of carbon nanotubes and related nanostructures. *Prog. Energy Combust. Sci.* **2010**, *36*, 696–727.

47. Çelik, Y.; Suvacı, E.; Weibel, A.; Peigney, A.; Flahaut, E. Texture development in Fe-doped alumina ceramics via templated grain growth and their application to carbon nanotube growth. *J. Eur. Ceram. Soc.* **2013**, *33*, 1093–1100.

48. Stearns, L.C.; Zhao, J.; Harmer, M.P. Processing and microstructure development in Al_2O_3-SiC nanocomposites. *J. Eur. Ceram. Soc.* **1992**, *10*, 473–477.

49. Jeong, Y.-K.; Nakahira, A.; Niihara, K. Effects of additives on microstructure and properties of alumina-silicon carbide nanocomposites. *J. Am. Ceram. Soc.* **1999**, *82*, 3609–3612.

50. Borsa, C.E.; Ferreira, H.S.; Kiminami, R.H.G.A. Liquid phase sintering of Al_2O_3/SiC nanocomposites. *J. Eur. Ceram. Soc.* **1999**, *19*, 615–621.

51. Sedláček, J.; Galusek, D.; Riedel, R.; Hoffmann, M.J. Sinter-HIP of polymer-derived Al_2O_3-SiC composites with high SiC contents. *Mater. Lett.* **2011**, *65*, 2462–2465.

52. Ahmad, I.; Unwin, M.; Cao, H.; Chen, H.; Zhao, H.; Kennedy, A.; Zhu, Y.Q. Multi-walled carbon nanotubes reinforced Al_2O_3 nanocomposites: Mechanical properties and interfacial investigations. *Comp. Sci. Technol.* **2010**, *70*, 1199–1206.

53. Vasiliev, A.L.; Poyato, R.; Padture, N.P. Single-wall carbon nanotubes at ceramic grain boundaries. *Scripta Mater.* **2007**, *56*, 461–463.

54. Gao, L.; Jiang, L.; Sun, J. Carbon nanotube-ceramic composites. *J. Electroceram.* **2006**, *17*, 51–55.

55. Zhang, S.C.; Fahrenholtz, W.G.; Hilmas, G.E.; Yadlowsky, E.J. Pressureless sintering of carbon nanotube-Al_2O_3 composites. *J. Eur. Ceram. Soc.* **2010**, *30*, 1373–1380.

56. Rice, R.W. *Mechanical Properties of Ceramics and Composites: Grain and Particle Effects*; CRC Press: New York, NY, USA, 2000.

57. Yamamoto, G.; Shirasu, K.; Nozaka, Y.; Wang, W.; Hashida, T. Microstructure-property relationships in pressureless-sintered carbon nanotube/alumina composites. *Mater. Sci. Eng.* **2014**, *A617*, 179–186.

58. An, J.; You, D.; Lim, D. Tribological properties of hot-pressed alumina-CNT composites. *Wear* **2003**, *255*, 677–681.

59. Lim, D.-S.; An, J.-W.; Lee, H.J. Effect of carbon nanotube addition on the tribological behavior of carbon–carbon composites. *Wear* **2002**, *252*, 512–517.

60. Ahmad, K.; Pan, W. Hybrid nanocomposites: A new route towards tougher alumina ceramics. *Comp. Sci. Technol.* **2008**, *68*, 1321–1327.

61. Ahmad, K.; Pan, W. Dramatic effect of multiwalled carbon nanotubes on the electrical properties of alumina based ceramic nanocomposites. *Comp. Sci. Technol.* **2009**, *69*, 1016–1021.

62. Ahmad, K.; Pan, W.; Wan, C. Thermal conductivities of alumina-based multiwall carbon nanotube ceramic composites. *J. Mater. Sci.* **2014**, *49*, 6048–6055.

63. Carroll, L.; Sternitzke, M.; Derby, B. Silicon carbide particle size effects in alumina-based nanocomposites. *Acta Mater.* **1996**, *44*, 4543–4552.

64. Stearns, L.C.; Harmer, M.P. Particle-inhibited grain growth in Al_2O_3-SiC: I, experimental results. *J. Am. Ceram. Soc.* **1996**, *79*, 3013–3019.

65. Chou, I.A.; Chan, H.M.; Harmer, M.P. Machining-induced surface residual stress behavior in Al_2O_3-SiC. *J. Am. Ceram. Soc.* **1996**, *79*, 2403–2409.

66. Thompson, A.M.; Chan, H.M.; Harmer, M.P. Nanocomposites crack healing and stress relaxation in Al_2O_3 SiC "nanocomposites". *J. Am. Ceram. Soc.* **1995**, *78*, 567–571.

67. Xu, Y.; Zangvil, A.; Kerber, A. SiC nanoparticle-reinforced Al_2O_3 matrix composites: Role of intra- and intergranular particles. *J. Eur. Ceram. Soc.* **1997**, *17*, 921–928.

68. Pezzotti, G.; Sakai, M. Effect of a silicon carbide "nano-dispersion" on the mechanical properties of silicon nitride. *J. Am. Ceram. Soc.* **1994**, *77*, 3039–3041.

69. Jiao, S.; Jenkins, M.L. A quantitative analysis of crack-interface interactions in alumina-based nanocomposites. *Phil. Mag. A* **1998**, *78*, 507–522.

70. Tiegs, T.N.; Becher, P.F. Thermal shock behavior of an Alumina-SiC whisker composite. *J. Am. Ceram. Soc.* **1987**, *70*, C109–C111.

71. Sedláček, J.; Galusek, D.; Švančárek, P.; Riedel, R.; Atkinson, A.; Wang, X. Abrasive wear of Al$_2$O$_3$-SiC and Al$_2$O$_3$-(SiC)-C composites with micrometer- and submicrometer-sized alumina matrix grains. *J. Eur. Ceram. Soc.* **2008**, *28*, 2983–2993.

72. Walker, C.N.; Borsa, C.E.; Todd, R.I.; Davidge, R.W.; Brook, R.J. Fabrication, characterisation and properties of alumina matrix nanocomposites. *Br. Ceram. Proc.* **1994**, *53*, 249–264.

73. Anya, C.C. Wet erosive wear of alumina and its composites with SiC nano-particles. *Ceram. Int.* **1998**, *24*, 533–542.

74. Rodriguez, J.; Martin, A.; Pastor, J.Y.; Llorca, J.; Bartolome, J.F.; Moya, J.S. Sliding wear of alumina/silicon carbide nanocomposite. *J. Am. Ceram. Soc.* **1999**, *82*, 2252–2254.

75. Kara, H.; Roberts, S. Polishing behavior and surface quality of alumina and alumina/silicon carbide nanocomposites. *J. Am. Ceram. Soc.* **2000**, *83*, 1219–1225

76. Levin, I.; Kaplan, W.D.; Brandon, D.G.; Layyous, A.A. Effect of SiC submicrometer particle size and content on fracture toughness of alumina-SiC nanocomposites. *J. Am. Ceram. Soc.* **1995**, *78*, 254–256.

77. Winn, A.J.; Todd, R.I.; Microstructural requirements for alumina-SiC nanocomposites. *Br. Ceram. Trans.* **1999**, *5*, 219–224.

78. Luo, J.; Stevens, R. The role of residual stresses on the mechanical properties of Al$_2$O$_3$-5 vol% SiC nanocomposites. *J. Eur. Ceram. Soc.* **1997**, *17*, 1565–1572.

79. Todd, R.I.; Limpichaipanit, A. Microstructure-property relationships in wear resistant alumina/SiC "nanocomposites". *Adv. Sci. Technol.* **2006**, *45*, 555–563.

80. Deng, Z.Y.; Shi, J.L.; Zhang, Y.F.; Lai, T.R.; Guo, J.K. Creep and creep behaviour in silicon carbide particle reinforced alumina. *J. Am. Ceram. Soc.* **1999**, *82*, 944–952.

81. Reveron, H.; Zaafrani, O.; Fantozzi, G. Microstructure development, hardness, toughness and creep behavior of pressureless sintered alumina/SiC micro/nano-composites obtained by slip-casting. *J. Eur. Ceram. Soc.* **2010**, *30*, 1351–1357.

82. Sternitzke, M. Review: Structural ceramic nanocomposites, *J. Eur. Ceram. Soc.* **1997**, *17*, 1061–1082.

83. Ohji, T.; Hirano, T.; Nakahira, A.; Niihara, K. Particle/matrix interface and its role in creep inhibition in alumina/silicon carbide nanocomposites. *J. Am. Ceram. Soc.* **1996**, *79*, 33–45.

84. Ohji, T.; Nakahira, A.; Hirano, T.; Niikara, K. Tensile creep behaviour of alumina/silicon carbide nanocomposite. *J. Am. Ceram. Soc.* **1994**, *77*, 3259–3262.

85. Deng, Z.Y.; Zhang, Y.F.; Shi, J.L.; Guo, J.K.; Jiang, D.Y. Pinning effect of SiC particles on mechanical properties of Al$_2$O$_3$/SiC ceramic matrix composites. *J. Eur. Ceram. Soc.* **1998**, *18*, 501–508.

86. Descamps, P.; O'Sullivan, D.; Poorteman, M.; Descamps, J.C.; Leriche, A.; Cambier, F. Creep behavior of Al$_2$O$_3$-SiC nanocomposites. *J. Eur. Ceram. Soc.* **1999**, *19*, 2475–2485.

87. Thompson, A.M.; Chan, H.M.; Harmer, M.P. Tensile creep of almina-silicon carbide nanocomposites. *J. Am. Ceram. Soc.* **1997**, *80*, 2221–2228.

88. Parchoviansky, M.; Galusek, D.; Michálek, M.; Švančárek, P.; Kašiarová, M.; Dusza, J.; Hnatko, M. Effect of the volume fraction of SiC on the microstructure, and creep behavior of hot pressed Al_2O_3/SiC composites. *Ceram. Int.* **2014**, *40*, 1807–1814.

89. Hayashi, T.; Endo, M. Carbon nanotubes as structural material and their application in composites. *Compos. B* **2011**, *42*, 2151–2157.

90. Bocanegra-Bernal, M.H.; Echeberria, J.; Ollo, J.; Garcia-Reyes, A.; Domínguez-Rios, C.; Reyes-Rojas, A.; Aguilar-Elguezabal, A. A comparison of the effects of multi-wall and single-wall carbon nanotube additions on the properties of zirconia toughened alumina composites. *Carbon* **2011**, *49*, 1599–1607.

91. Li, G.H.; Hu, Z.X.; Zhang, L.D.; Zhang, Z.R. Elastic modulus of nano-alumina composite. *J. Mater. Sci. Lett.* **1998**, *17*, 1185–1186.

92. Yu, M.F.; Lourie, O.; Dyer, M.J.; Moloni, K.; Kelly, T.F.; Ruoff, R.S. Strength and breaking mechanism of multiwalled carbon nanotubes under tensile load. *Science* **2000**, *287*, 637–640.

93. Inam, F.; Peijs, T.; Reece, M.J. The production of advanced fine-grained alumina by carbon nanotube addition. *J. Eur. Ceram. Soc.* **2011**, *31*, 2853–2859.

94. Laurent, C.; Peigney, A.; Dumortier, O.; Rousset, A. Carbon nanotubes-Fe-alumina nanocomposites. Part II: Microstructure and mechanical properties of the hot-pressed composites. *J. Eur. Ceram. Soc.* **1998**, *18*, 2005–2013.

95. Michálek, M.; Sedláček, J.; Parchoviansky, M.; Michálková, M.; Galusek, D. Mechanical properties and electrical conductivity of alumina/MWCNT and alumina/zirconia/MWCNT composites. *Ceram. Int.* **2014**, *40*, 1289–1295.

96. Zhan, D.D.; Kuntz, J.D.; Wan, J.U.; Mukherjee, A.K. Single-wall carbon nanotubes as attractive toughening agents in alumina-based nanocomposites. *Nat. Mater.* **2003**, *2*, 38–42.

97. Fan, J.P.; Zhao, D.Q.; Wu, M.S.; Xu, Z.N.; Song, J. Preparation and microstructure of multi-wall carbon nanotubes-toughened Al_2O_3 composite. *J. Am. Ceram. Soc.* **2006**, *89*, 750–753.

98. Sun, J.; Gao, L.; Jin, X.H. Reinforcement of alumina matrix with multi-walled carbon nanotubes. *Ceram. Int.* **2005**, *31*, 893–896.

99. Fan, J.P.; Zhuang, D.M.; Zhao, D.Q.; Zhang, G.; Wu, M.S.; Wei, F.; Fan, Z.-J. Toughening and reinforcing alumina matrix composite with single-wall carbon nanotubes. *Appl. Phys. Lett.* **2006**, *89*, 121910:1–121910:3.

100. Duszova, A.; Dusza, J.; Tomasek, K.; Morgiel, J.; Blugan, G.; Kuebler, J. Zirconia/carbon nanofiber composite. *Scripta Mater.* **2008**, *58*, 520–523.

101. Cha, S.I.; Kim, K.T.; Lee, K.H.; Mo, C.B.; Hong, S.H. Strengthening and toughening of carbon nanotube reinforced alumina nanocomposite fabricated by molecular level mixing process. *Scripta. Mater.* **2005**, *53*, 793–797.

102. Sun, J.; Gao, L.; Li, W. Colloidal processing of carbon nanotube/alumina composites. *Chem. Mater.* **2002**, *14*, 5169–5172.

103. Siegel, R.W.; Chang, S.K.; Ash, B.J.; Stone, J.; Ajayan, P.M.; Doremus, R.W.; Schadler, L.S. Mechanical behavior of polymer and ceramic matrix nanocomposites. *Scripta Mater.* **2001**, *44*, 2061–2064.

104. Poyato, R.; Vasiliev, A.L.; Padture, N.P.; Tanaka, H.; Nishimura, T. Aqueous colloidal processing of single-wall carbon nanotubes and their composites with ceramics. *Nanotechnology* **2006**, *17*, 1770–1777.

105. Ahmad, I.; Cao, H.; Chen, H.; Zhao, H.; Kennedy, A.; Zhu, Y.Q. Carbon nanotube toughened aluminium oxide nanocomposite. *J. Eur. Ceram. Soc.* **2010**, *30*, 865–873.

106. Estili, M.; Kawasaki, A. Engineering strong intergraphene shear resistance in multi-walled carbon nanotubes and dramatic tensile improvements. *Adv. Mater.* **2010**, *22*, 607–610.

107. Wang, X.T.; Padture, N.P.; Tanaka, H. Contact-damage-resistant ceramic/single-wall carbon nanotubes and ceramic/graphite composites. *Nat. Mater.* **2004**, *3*, 539–544.

108. Samal, S.S.; Bal. S. Carbon nanotube reinforced ceramic matrix composite—A review. *J. Min. Mater. Char. Eng.* **2008**, *7*, 355–370.

109. Wong, E.W.; Sheehan, P.E.; Liebert, C.M. Nanobeam mechanics: Elasticity, strength, and toughness of nanorods and nanotubes. *Science* **1997**, *277*, 1971–1975.

110. Yamamoto, G.; Shirasu, K.; Nozaka, Y.; Sato, Y.; Takagi, T.; Hashida, T. Structure-property relationships in thermally-annealed multi-walled carbon nanotubes. *Carbon* **2014**, *66*, 219–226.

111. Corral, E.L.; Garary, J.; Barrera, E.V. Engineering nanostructure for single-walled carbon nanotubes reinforced silicon nitride nanocomposites. *J. Am. Ceram. Soc.* **2008**, *91*, 3129–3137.

112. Fan, J.; Zhao, D.; Song, J. Preparation and microstructure of multi-walled carbon nanotubes toughened Al_2O_3 composite. *J. Am. Ceram. Soc.* **2006**, *89*, 750–753.

113. Yamamoto, G.; Omori, M.; Hashida, T.; Kimura, H. A novel structure for carbon nanotube reinforced alumina composites with improved mechanical properties. *Nanotechnology* **2008**, *19*, 315708:1–315708:7.

114. Peigney, A.; Garcia, F.L.; Estournès, C.A.; Weibel, C.L. Toughening and hardening in double-walled carbon nanotube/nanostructured magnesia composites. *Carbon* **2010**, *48*, 1952–1960.

115. Xia, Z.; Riester, L.; Curtin, W.A.; Li, H.; Sheldon, B.W.; Liang, J.; Chang, B.; Xu, J.M. Direct observation of toughening mechanisms in carbon nanotube ceramic matrix composites. *Acta Mater.* **2004**, *52*, 931–944.

116. Xia, Z.; Curtin, W.A.; Sheldon, B.W. Fracture toughness of highly ordered carbon nanotube/alumina nanocomposites. *J. Eng. Mater. Technol. Trans. ASME* **2004**, *126*, 238–244.

117. Fu, Z.Y.; Huang, L.W.; Zhang, J.Y.; Todd, R. Ultra-fast densification of CNTs reinforced alumina based on combustion reaction and quick pressing. *Sci. China Technol. Sci.* **2012**, *55*, 484–489.

118. Hirota, K.; Takaura, Y.; Kato, M.; Miyamoto, Y. Fabrication of carbon nanofibers (CNF)-dispersed Al_2O_3 composites by pulsed electric current pressure sintering and their mechanical and electrical properties. *J. Mater. Sci.* **2007**, *42*, 4792–4800.

119. Blugan, G.; Michalkova, M.; Hnatko, M.; Šajgalík, P.; Minghetti, T.; Schelle, C.; Graule, T.; Kuebler, J. Processing and properties of alumina–carbon nano fibre ceramic composites using standard ceramic technology. *Ceram. Int.* **2011**, *37*, 3371–3379.

120. Borrell, A.; Torrecillas, R.; Rocha, V.G.; Fernández, A.; Bonache, V.; Salvador, M.D. Effect of CNFs content on the tribological behaviour of spark plasma sintering ceramic-CNFs composites. *Wear* **2012**, *274–275*, 94–99.

121. Borrell, A.; Alvarez, I.; Torrecillas, R.; Rocha, V.G.; Fernandez, A. Microstructural design form mechanical and electrical properties of spark plasma sintered Al$_2$O$_3$-SiC nanocomposites. *Mater. Sci. Eng.* **2012**, *A534*, 693–698.

122. Barea, R.; Belmonte, M.; Osendi, M.I.; Miranzo, P. Thermal conductivity of Al$_2$O$_3$/SiC platelet composites. *J. Eur. Ceram. Soc.* **2003**, *23*, 1773–1778.

123. Fabbri, L.; Scafé, E.; Dinelli, G. Thermal and elastic properties of alumina-silicon carbide whiskers composites. *J. Eur. Ceram. Soc.* **1994**, *14*, 441–446.

124. Reddy, P.; Castelino, K.; Majumdar, A. Diffuse mismatch model of thermal boundary conductance using exact phonon dispersions. *Appl. Phys. Lett.* **2005**, *87*, 315–322.

125. Hasselman, D.P.H.; Johnson, L.F. Effective thermal conductivity of composites with interfacial thermal barrier resistance. *J. Comp. Mater.* **1987**, *21*, 508–515.

126. Parchoviansky, M.; Galusek, D.; Švančárek, P.; Sedláček, J.; Šajgalík, P. Thermal behavior, electrical conductivity and microstructure of hot pressed Al$_2$O$_3$/SiC nanocomposites. *Ceram. Int.* **2014**, *14*, 14421–14429.

127. Rul, S.; Lefevre-Schlick, F.; Capria, E.; Laurent, C.; Peigney, A. Percolation of single-walled carbon nanotubes in ceramic matrix nanocomposites. *Acta Mater.* **2004**, *52*, 1061–1067.

128. Ahmad, K.; Pan, W.; Shi, S.L. Electrical conductivity and dielectric properties of multiwalled carbon nanotube and alumina composites. *Appl. Phys. Lett.* **2006**, *89*, 133122–133122.

129. Zhan, G.D.; Mukherjee, A.K. Carbon nanotube reinforced alumina-based ceramics with novel mechanical, electrical, and thermal properties. *Int. J. Appl. Ceram. Technol.* **2004**, *1*, 161–171.

130. Bergman, D.J. Exactly solvable microscopic geometries and rigorous bounds for the complex dielectric constant of a two-component composite material. *Phys. Rev. Lett.* **1980**, *44*, 1285–1287.

131. Kilbride, B.E.; Coleman, J.N.; Fraysse, J.; Fournet, P.; Cadek, M.; Drury, A.; Hutzler, S.; Roth, S.; Blau, W.J. Experimental observation of scaling laws for alternating current and direct current conductivity in polymer-carbon nanotube composite thin films. *J. Appl. Phys.* **2002**, *92*, 4024–4030.

132. Allaoui, A.; Bai, S.; Cheng, H.M.; Bai, J.B. Mechanical and electrical properties of a MWNT/epoxy composite. *Comp. Sci. Technol.* **2002**, *62*, 1993–1998.

133. Bauhofer, W.; Kovacs, J.Z. A review and analysis of electrical percolation in carbon nanotube polymer composites. *Comp. Sci. Technol.* **2009**, *69*, 1486–1498.

134. Zhou, X.W.; Zhu, Y.F.; Liang, J. Preparation and properties of powder styrene-butadiene rubber composites filled with carbon black and carbon nanotubes. *Mater. Res. Bull.* **2007**, *42*, 456–464.

135. Nan, C.-W. Physics of inhomogeneous inorganic materials. *Prog. Mater. Sci.* **1993**, *37*, 1–116.

136. Song, Y.; Noh, T.W.; Lee, S.-I.; Gaines, J.R. Experimental study of the three-dimensional AC conductivity and dielectric constant of a conductor-insulator composite near the percolation threshold. *Phys. Rev. B* **1986**, *33*, 904–908.

137. Kumari, L.; Zhang, T.; Du, G.H.; Li, W.Z.; Wang, Q.W.; Datye, A.; Wu, K.H. Thermal properties of CNT-Alumina nanocomposites. *Compos. Sci. Technol.* **2008**, *68*, 2178–2183.

138. Bakshi, S.R.; Balani, K.; Agarwal, A. Thermal conductivity of plasma-sprayed aluminum oxide-multiwalled carbon nanotube composites. *J. Am. Ceram. Soc.* **2008**, *91*, 942–947.

139. Zhang, H.L.; Li, J.F.; Zhang, B.P.; Yao, K.F.; Liu, W.S.; Wang, H. Electrical and thermal properties of carbon nanotube bulk materials: Experimental studies for the 328–958 K temperature range. *Phys. Rev. B* **2007**, *75*, 205407:1–205407:9.

140. Dresselhaus, M.S.; Eklund, P.C. Phonons in carbon nanotubes. *Adv. Phys.* **2000**, *49*, 705–814.

141. Biercuk, M.J.; Llaguno, M.C.; Radosavljevic, M.; Hyun, J.K.; Johnson, A.T.; Fischer, J.E. Carbon nanotube composites for thermal management. *Appl. Phys. Lett.* **2002**, *80*, 2767–2769.

142. Yang, D.J.; Zhang, Q.; Chen, G.; Ahn, J.; Wang, S.G.; Zhou, Q.; Li, J.Q. Thermal conductivity of multiwalled carbon nanotubes. *Phys. Rev. B* **2002**, *66*, 165440:1–165440:6.

143. Zhang, H.L.; Li, J.F.; Yao, K.F.; Chen, L.D. Spark plasma sintering and thermal conductivity of carbon nanotube bulk materials. *J. Appl. Phys.* **2005**, *97*, 114310:1–114310:5.

Synthesis, Characterization, and Mechanism of Formation of Janus-Like Nanoparticles of Tantalum Silicide-Silicon (TaSi₂/Si)

Andrey V. Nomoev [1,2,*], **Sergey P. Bardakhanov** [2,3,4], **Makoto Schreiber** [2], **Dashima Zh. Bazarova** [2],
Boris B. Baldanov [2] and **Nikolai A. Romanov** [2]

[1] Institute of Physical Materials Science, Siberian Branch of the Russian Academy of Sciences,
Sakhyanovoy str., 6, Ulan-Ude 670047, Russia

[2] Department of Physics and Engineering, Buryat State University, Smolina str., 24a,
Ulan-Ude 670000, Russia; E-Mails: bardnski@gmail.com (S.P.B.);
mschreib@mail.uoguelph.ca (M.S.); ars-d@mail.ru (D.Z.B.); boris.baldanov@mail.ru (B.B.B.);
nromanovv@mail.ru (N.A.R.)

[3] Institute of Applied and Theoretical Mechanics, Siberian Branch of the Russian Academy of
Sciences, Institutskaya str., 4/1, Novosibirsk 630090, Russia

[4] Department of Physics, Novosibirsk State University, Pirogova str., 2, Novosibirsk 630090, Russia

* Author to whom correspondence should be addressed; E-Mail: nomoevav@mail.ru

Academic Editor: Thomas Nann

Abstract: Metal-semiconductor Janus-like nanoparticles with the composition tantalum silicide-silicon (TaSi₂/Si) were synthesized for the first time by means of an evaporation method utilizing a high-power electron beam. The composition of the synthesized particles were characterized using high-resolution transmission electron microscopy (HRTEM), X-ray diffraction (XRD), selective area electron diffraction (SAED), and energy dispersive X-ray fluorescence (EDX) analysis. The system is compared to previously synthesized core-shell type particles in order to show possible differences responsible for the Janus-like structure forming instead of a core-shell architecture. It is proposed that the production of Janus-like as opposed to core-shell or monophase particles occurs due to the ability of Ta and Si to form compounds and the relative content of Ta and Si atoms in the produced vapour. Based on the results, a potential mechanism of formation for the TaSi₂/Si nanoparticles is discussed.

Keywords: Janus-like; nanoparticles; XRD analysis; mechanism of formation; TaSi₂/Si

1. Introduction

The unique properties of composite Janus-like (JL) nanoparticles have sparked the interest of researchers in recent years, resulting in a large number of publications about their synthesis (for example [1,2]). JL nanoparticles are particles with two halves in which the two halves of the particle are composed of two different materials. JL nanoparticles can have desirable properties which arise from the way in which the constituent materials are bound and spatially oriented with respect to each other. An example of such a property is the enhanced photocatalytic activity of JL particles composed of photocatalytic materials compared to the photocatalytic activity of the individual constituent materials. This enhancement is thought to stem from the heteroboundry interface between the two materials. This was shown for Janus Au-TiO_2 particles which possessed enhanced photocatalytic properties in comparison with TiO_2 or composite Au-TiO_2 nanoparticles [2]. JL nanoparticles can also possess a large dipole moment; caused by the different nature of the two constituent components [3,4]. A high dipole moment makes it possible to remotely control the position of these nanoparticles by means of electric and magnetic fields with a high spatial resolution [3,4].

Tantalum silicide ($TaSi_2$) has an attractive combination of properties, including a high melting point of 2200 °C [5], high thermal stability [6], low electrical contact resistance [6,7], a high modulus of elasticity [8], a high resistance to oxidation in air [6], and a good compatibility with silicon [6–8]. Due to its low electrical resistance and oxidation resistance, $TaSi_2$ has been utilized in switching devices as Schottky barriers, ohmic contacts, and connectors in integrated circuits [6,9]. In the present work, JL particles with a metallic $TaSi_2$ end and a semiconducting Si end were synthesized. This composition should lead to a spatial separation of electrons and the appearance of polarization charges, leading to the presence of a dipole moment. Thus, the spatial orientation of these JL nanoparticles should be controllable by the interaction of the dipole moment with an external electromagnetic field [10].

2. Results

This paper reports the first synthesis of composite JL $TaSi_2$/Si nanoparticles. X-ray diffraction (XRD) analysis (Figure 1) showed that the composite powders are composed of two phases: a $TaSi_2$ phase [11] and a Si phase [12]; the Si phase being more prominent. Additionally, an XRD of the material which remained in the graphite crucible after evaporation was measured; revealing that it was composed of Ta_5Si_3 [13]. Transmission electron microscopy (TEM) images of the obtained particles (Figure 2) show that sphere-like JL particles were obtained which were composed of two parts: a darker coloured half, likely $TaSi_2$, and a lighter coloured half, likely silicon; based on the densities of the two species. The obtained powder as a whole contained a mixture of these JL particles as well as monophase particles with the same colour as the lighter half of the JL particles; agreeing with the higher Si content in the powder as measured by XRD. Based on the high-resolution TEM (HRTEM) images, both the darker and lighter phases appeared to be crystalline. However, a thin outer layer (<10 nm) of the lighter phase was amorphous on all observed particles A few particles were observed to have two or three darker regions connected to a lighter phase instead of just one. A few of the darker phases were covered by a thin layer of the lighter phase. The average size of the JL particles was 150 nm ± 13 nm based on over 250 measurements and particle sizes were distributed in a lognormal distribution.

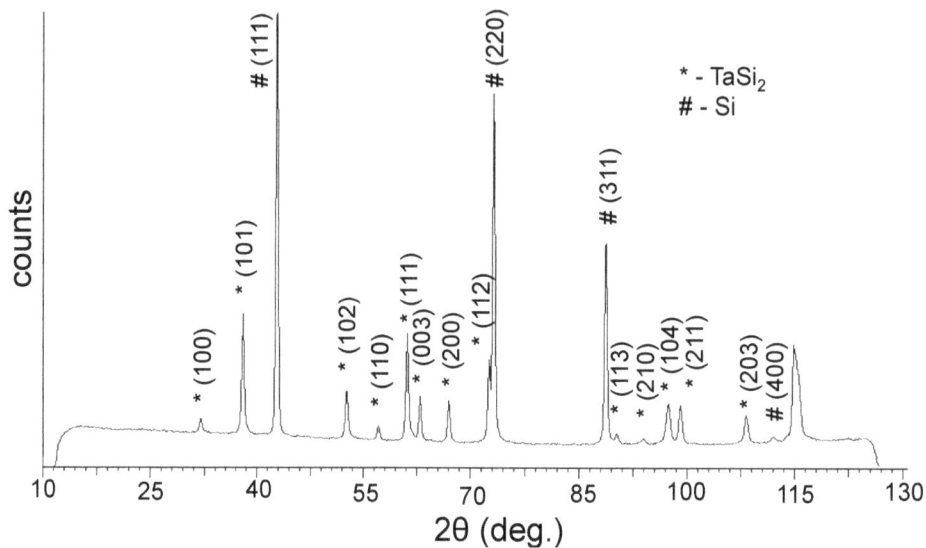

Figure 1. X-ray diffractogram (XRD) of the obtained powder.

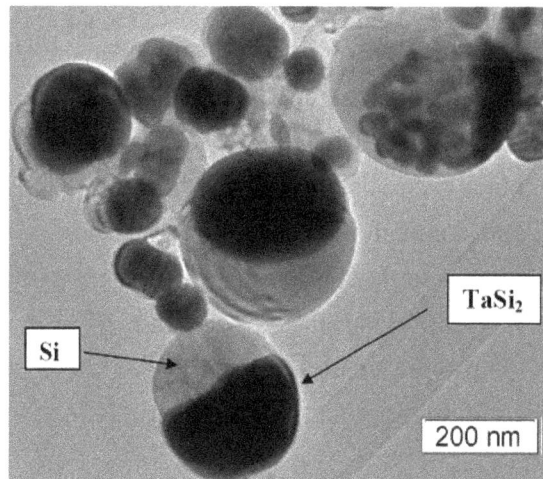

Figure 2. Transmission electron microscopy (TEM) image of the obtained TaSi₂/Si nanoparticles.

The hypothesis about the composition of the two halves of the JL particles was confirmed through energy dispersive X-ray fluorescence (EDX) (Figure 3) and selective area electron diffraction (SAED) (Figure 4) analysis on the two halves of the TaSi₂/Si nanoparticles. In the measured EDX spectra, a strong peak is present around 1.7 keV for both halves of the nanoparticles. As both silicon and tantalum have transitions around 1.7 keV (Kα transition of Si = 1.739 keV, M transition of Ta = 1.709 keV), the absence of Ta in the lighter coloured half of the particle could not be confirmed based on just that peak. However, a strong peak around 8 keV exists on the spectrum of the darker region of the nanoparticle. This corresponds to the Lα transition of Ta (8.145 keV). As this secondary Ta peak is very small on the spectrum of the lighter region of the particle, the EDX measurements support that the darker region is composed of TaSi₂ while the lighter region is composed of Si. SAED measurements on the dark regions of the nanoparticles measured d-spacings around 3.5 Å which correspond to the interlayer spacing of the (101) planes of TaSi₂ (3.51 ± 0.15 Å [10]). For the lighter regions, d-spacings of around 3.1 Å which correspond to the (111) planes of Si (3.1356 Å [12]) were also measured. Thus, on the basis of XRD, EDX, and SAED measurements, we can determine that the dark regions of the JL nanoparticles on the TEM images are composed of crystalline $TaSi_2$ and the light regions are composed of crystalline Si. SAED measurements also confirmed that the monophasic particles were composed of crystalline silicon.

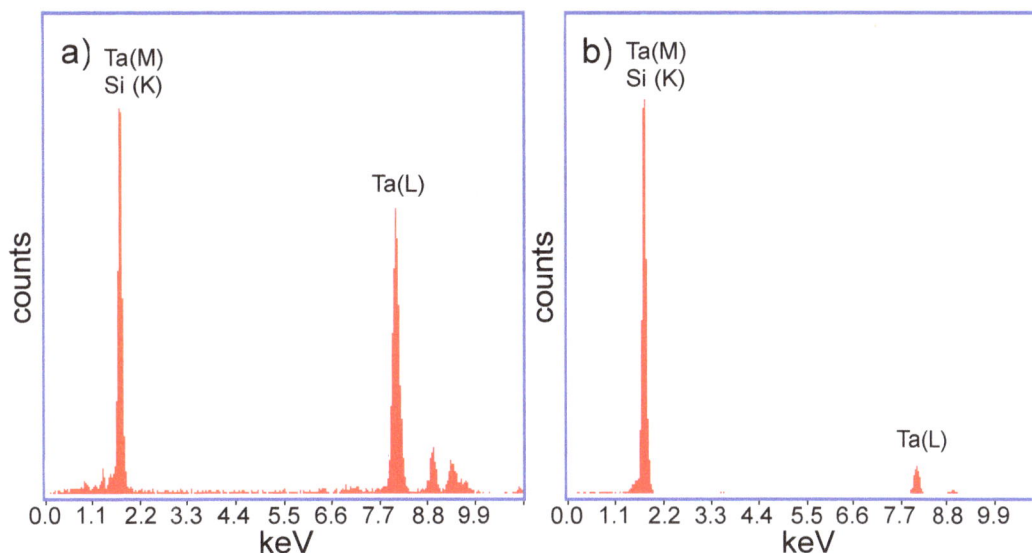

Figure 3. Energy dispersive X-ray fluorescence (EDX) spectra of (**a**) The darker ($TaSi_2$) region; and (**b**) The lighter (Si) region of the obtained $TaSi_2$/Si JL nanoparticles.

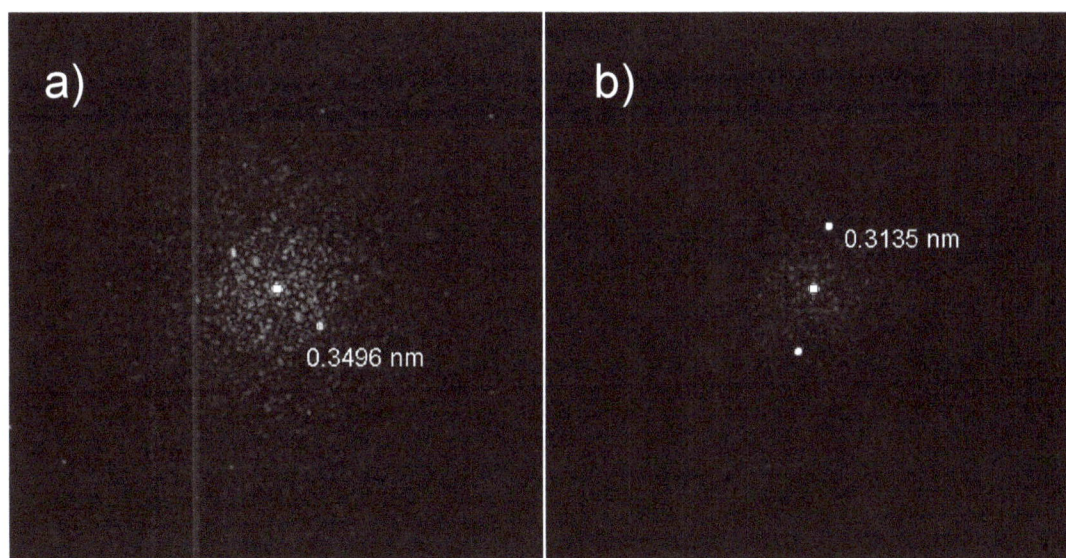

Figure 4. Representative selective area electron diffraction (SAED) diffraction patterns obtained from (**a**) The darker region ($TaSi_2$); and (**b**) Lighter region (Si) of the obtained JL nanoparticles.

In our initial experiments, pieces of Si and Ta were mixed (rather than layered as in the reported experiment) in the graphite crucible prior to evaporation. However, when the precursors were arranged in this fashion, it was found that the products were only Si nanopowders, with no Ta. This indicated that the silicon melted and evaporated before the Ta could melt. As the vapor pressure of Si ($P_v(Si) = 4.07 \times 10^6$ atm at $T = 1623$ K [14]) is higher than that of Ta ($P_v(Ta) = 6.66 \times 10^{-6}$ atm at $T = 3269$ K [15]) and has a much lower melting ($T_m(Si) = 1687$ K, $T_m(Ta) = 3293$ K [5]) and boiling

(T_b(Si) = 3538 K, T_b(Ta) = 5731 K) point, Si was placed at the bottom of the graphite crucible and Ta on top prior to evaporation in the reported experiment (Figure 5a).

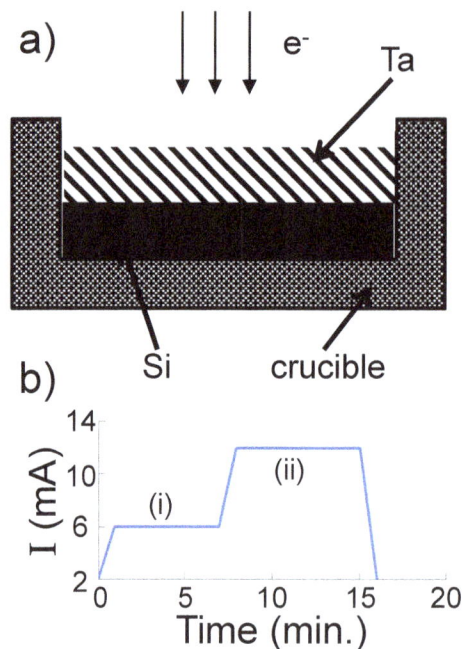

Figure 5. (**a**) Schematic of the Ta and Si arrangement in a graphite crucible before electron beam irradiation; (**b**) Dependence of the electron beam current on time. The materials in the crucible are melted in region (i) and evaporated in region (ii).

3. Discussion

We can speculate that the layered arrangement of materials used in the reported experiment compared to the mixed arrangement used in the initial experiments increased the heating rate of Ta and decreased the heating rate of Si. Thus, between about 1400–2000 °C, at the interface between the Ta and Si, a homogeneous Ta-Si liquid phase is thought to form as it is more thermodynamically favourable for Ta; based on the Ta-Si phase diagram [5]. By the end of the melting phase, the homogeneous Ta-Si liquid will have filled the crucible. During the evaporation phase, the evaporation rates of Ta and Si in the Ta-Si liquid would be expected to be higher and lower, respectively, to some degree compared to pure Ta (R_{exap}(Ta) = 6.80 × 10^{-5} g·cm^{-2}·s^{-1} at T = 3269 K [15]) and Si (R_{exap}(Si) = 14.1 × 10^{7} g·cm^{-2}·s^{-1} at T = 1623 K [14]). However, it is still expected that the evaporation rate of Si would still be significantly higher than that of Ta and thus the Si concentration in the vapour would be much higher than that of Ta. This is supported by the fact that the material which remained in the crucible after the experimental run was Ta$_5$Si$_3$ (a Ta-Si compound with less than 10 weight percent silicon) as well as that only monophasic silicon particles but no monophasic tantalum particles were observed in the powder.

A similar arrangement of substances, taking into account the difference in the vapor pressure of the evaporated species, has previously been used by us in the preparation of Cu@Si [16] and Ag@Si [17] core-shell particles by electron beam evaporation. During the evaporation of Ta and Si, similar conditions for the time variation in the electron beam strength, carrier gas flow rate, and geometry of installation were used as during the synthesis of Cu@Si and Ag@Si. Thus, it was surprising that instead of core-shell particles such as Cu@Si and Ag@Si, TaSi$_2$/Si JL particles formed for the Ta-Si system. Based on thermodynamic calculations [18], core-shell type nanoparticles tend to form either due to a large difference in the surface tension of the core and shell material (the core-shell configuration achieving the lowest surface energy for the system) or due to a large difference in the atomic sizes of the core and shell material. In the Ta-Si system, the surface tension of Ta ($\sigma(T)_{Ta} = \{2150 - 0.21(T - T_m(Ta))\}$ mN·m^{-1}, $T \geq T_m(Ta)$ [19]) is much higher than the surface tension of Si ($\sigma(T)_{Si} = \{820 - 0.3(T - T_m(Si))\}$ mN·m^{-1}, $T \geq T_m(Si)$ [20]) while their atomic radii ($r(Ta) = 146$ pm, $r(Si) = 111$ pm) are not significantly different. Thus, purely based on this information, core-shell particles would be expected to form.

However, there are a few differences in the Ta-Si system compared to the Ag-Si and Cu-Si systems. Unlike the Ag-Si and Cu-Si systems, which form no and very few (only within a narrow range of compositions) chemical compounds between the two components, respectively, the Ta-Si system forms a series of compounds (Ta, Ta$_3$Si, Ta$_2$Si, Ta$_5$Si$_3$, TaSi$_2$, and Si) as can be seen from the Ta-Si phase diagram [5]. Based on the melting temperatures of the Ta-Si phases (T_m(Ta$_3$Si) = 2340 °C, T_m(Ta$_2$Si) = 2440 °C, T_m(Ta$_5$Si$_3$) = 2550 °C, T_m(TaSi$_2$) = 2040 °C), it can be seen that it is much more thermodynamically favourable for Ta atoms to form into TaSi$_2$ particles rather than pure Ta or other Ta-Si phases. This has been shown previously by Ko *et al.* [21] who hot-pressed Ta and Si powders together. The resulting material was found to contain TaSi$_2$ and Si grains rather than Ta and Si grains. The expected high silicon ratio compared to tantalum in the evaporated vapour also supports the formation of TaSi$_2$ which is the last Ta containing phase to form before pure Si. In comparison, in the Cu-Si system, which can form Cu-Si compounds with lower melting points than Cu or Si, all the Cu-Si compounds require many atoms to form each compound (Cu$_{38}$Si$_7$, Cu$_9$Si$_2$, Cu$_{15}$Si$_4$, Cu$_{19}$Si$_6$) which makes them less likely to form from the gas phase and more likely that the Cu and Si vapour would nucleate into core-shell particles. Additionally, there is a stronger driving force for Ta to form TaSi$_2$ as the difference in the melting temperature of Ta and TaSi$_2$ is 980 °C, much larger than the difference in melting temperature between Cu and any of the Cu-Si compounds (around 600 °C).

Thus, it can be speculated that the high surface energy of Ta and the likely excess of Si in the gas makes it more energetically favourable for the Ta atoms to condense into TaSi$_2$ rather than pure Ta and Si particles or other Ta-Si compounds. Silicon also prefers to form pure Si particles rather than TaSi$_2$ (as it has a lower melting point) and so excess silicon condenses into pure Si particles. Thus, we propose that the TaSi$_2$/Si JL nanoparticles form by the following mechanism: Initial nucleation of the gas upon condensation proceeds according to the reaction:

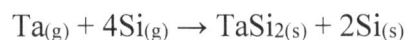

$$Ta_{(g)} + 4Si_{(g)} \rightarrow TaSi_{2(s)} + 2Si_{(s)}$$

where Ta forms TaSi$_2$ with Si and excess Si forms pure Si particles. The two sides of the nucleate are then built up as Ta and Si continue to condense onto the TaSi$_2$ side and Si condenses onto the Si side. Although to the authors' knowledge, there is no empirical data on the surface tension of TaSi$_2$, the simple surface tension model proposed by Ergy [22] for bimetallic systems shows that the surface tension of TaSi$_2$ will most likely lie between the surface tension values of Ta and Si; closer to Si. Further, the likely excess Si in the binary phase (darker shell half) will rapidly decreases the surface tension from the value of pure TaSi$_2$ due to surface segregation effects. Thus, the surface tension difference between the Si and TaSi$_2$ halves is likely to not be as great as between Si and Ta and has the possibility of being quite close; thus forming JL particles rather than core-shell type particles. Local fluctuations in the gas composition are expected to account for variations in the size ratios between the $TaSi_2$ and Si halves of the JL particles.

XRD analysis of the alloy remaining in the graphite crucible after Si and Ta were evaporated by electron beam irradiation gave a diffraction pattern corresponding to Ta_5Si_3. This may be due to the stoichiometry between the Ta and Si in the melt after the evaporation process was stopped and the melt solidified. The stoichiometry of Ta and Si in the melt at the end of the process is likely to be determined by the starting ratio of the two materials (1:1 by weight) and their relative evaporation rates as well as the evaporation time. Thus, if a different amount of each material were placed into the crucible prior to evaporation or different evaporation times were used, different Ta-Si phases or a solid solution of two phases may be left in the crucible at the end of the experiment. Further tests will be required to confirm this.

4. Experimental Section

Composite $TaSi_2/Si$ powders were obtained using an ELV direct-6 electron accelerator (Budker Institute of Nuclear Physics, Novosibirsk, Russia). Schematics of the device and its principles of operation are described in [16,23]. The incident electron energy was 1.4 MeV and the beam current was varied from 3 to 10 mA. In this accelerator, the target material can be irradiated by the electron beam at atmospheric pressure in a gas of the user's choice (argon in this case), rather than a vacuum. The maximum power density of the electron beam is 106 W/cm^2. In this work, the following procedure was used for the preparation of $TaSi_2/Si$ JL particles: A pure monocrystalline silicon ingot was melted by an electron beam into a graphite crucible. Upon termination of the electron beam, the silicon cooled down, solidifying and taking the form of the inside of the crucible. On top of the silicon, pieces of pure tantalum were placed (Figure 5a). The weight ratio of Ta to Si was 1:1. Ta and Si were then heated by the electron beam to their melting temperatures; resulting in a composite liquid. In accordance with the phase diagram of Ta and Si [5], at high temperatures, bulk Ta and Si are mixed in the liquid phase. Next, the beam current is increased to a value at which intense evaporation of the mixed composite liquid occurs. A plot of the electron beam current against time is shown in Figure 5b. The mixed vapors were transferred to a cold zone where they were condensed into nanoparticles by the argon carrier gas. The gas with the particles then passed through a special woven filter onto which the nanoparticles were deposited. The nanopowder sample was then collected from this filter. Previous nanopowder synthesis experiments using the high-powered electron beam evaporation process have produced powders at rates between hundreds of g/h [24] to kg/h [25–27] and thus similar production rates are expected for this synthesis. The obtained nanoparticles were characterized using XRD, TEM, HRTEM, SAED, and EDX analysis. These measurements were performed on a JEM-2010 TEM (JEOL, Tokyo, Japan; 200 kV accelerating voltage, 0.14 nm resolution) equipped with an LZ5 EDX spectrometer (Oxford Instruments, Wycombe, UK; 130 eV energy resolution, 1 nm spatial resolution) and a DIFRAY 401 (Scientific Instruments, Saint-Petersburg, Russia) XRD operated using a Cr X-ray source (λ = 2.2897 Å). To prepare the samples for microscopy measurements, the nanopowders were dispersed in ethanol by ultrasonication followed by precipitation of the sample onto a carbon film fixed on a copper grid.

5. Conclusions

Janus-like nanoparticles composed of a $TaSi_2$ half and a Si half were synthesized by electron beam evaporation of Ta and Si for the first time. The obtained powder contained a mixture of JL particles along with some silicon nanoparticles and silicon nanoparticles with more than one $TaSi_2$ phase connected to it. The composition of the JL particles was confirmed through TEM, diffraction, and EDX analysis techniques. JL particles are thought to form from the Ta-Si system instead of core-shell or monophasic particles due to the relative surface energy of Ta and Si, their ability to form into compounds, and the gas composition. Future work will investigate the properties of these $TaSi_2$/Si JL particles and their applications, in particular towards electrical applications.

Acknowledgments

This work was financially supported by the Russian Government under the project No. 16.1930.2014/K.

Author Contributions

The main idea for the experiment was thought up by A.V.N. Experimental work was performed by S.P.B, D.Zh.B, B.B.B, and N.A.R. Analysis of the experimental data and writing of the manuscript was carried out by A.V.N. and M.S.

Conflicts of Interest

The authors declare no conflict of interest.

References

1. Sotiriou, G.A.; Hirt, G.A.; Lozach, G.A.; Teleki, A.; Krumeich, F.; Pratsinis, S.E. Hybrid, silica-coated, Janus-like plasmonic-magnetic nanoparticles. *Chem. Mater.* **2011**, *23*, 1985–1992.

2. Fu, X.; Liu, J.; Yang, H.; Sun, J.; Li, X.; Zhang, X.; Jia, Y. BiVo4-graphene catalyst and its high photocatalytic performance under visible light irradiation. *Mater. Chem. Phys.* **2011**, *131*, 334–339.

3. Perro, A.; Reculusa, S.; Ravaine, S.; Bourgeat-Lami, E.; Duguet, E. Design and synthesis of Janus micro- and nanoparticles. *J. Mater. Chem.* **2005**, *15*, 3745–3760.

4. Hong, L.; Cacciuto, A.; Luijten, E.; Granick, S. Clusters of charged Janus spheres. *Nano Lett.* **2006**, *6*, 2510–2514.

5. Schlesinger, M.E. The Si-Ta (silicon-tantalum) system. *J. Phase Equilib.* **1994**, *15*, 90–95.

6. Ravindra, N.M.; Jin, L.; Ivanov, D.; Mehta, V.R.; Dieng, L.M.; Popov, G.; Gokce, O.H.; Grow, J.; Fiory, A.T. Electrical and compositional properties of $TaSi_2$ films. *J. Electron. Mater.* **2002**, *31*, 1074–1079.

7. Ino, K.; Taniguchi, Y.; Ohmi, T. Formation of ultra-shallow and low-reverse-bias-current tantalum-silicided junctions using a Si-encapsulated silicidation technique and low-temperature furnace annealing below 550 °C. *Jpn. J. Appl. Phys.* **1998**, *37*, 4277–4283.

8. Sidorenko, S.I. Formation of nanocrystalline structure of $TaSi_2$ films on silicon. *Powder Metall. Metal Ceram.* **2003**, *42*, 14–18.

9. Chen, H. Structure and Phase Transformation of Nanocrystalline and Amorphous Alloy Thin Films. Masters Thesis, University of Illinoice at Urbana-Champaign, Urbana and Champaign, IL, USA, 2007.

10. Gangwal, S.; Cayre, O.J.; Velev, O.D. Dielectrophoretic assembly of metallodielectric janus particles in AC electric fields. *Langmuir* **2008**, *24*, 13312–13320.

11. Joint Committee for Powder Diffraction Standards (JCPDS). Card No. 38–483; International Centre for Diffraction Data (ICDD): Newtown Square, PA, USA.

12. Joint Committee for Powder Diffraction Standards (JCPDS). Card No. 27–1402; International Centre for Diffraction Data (ICDD): Newtown Square, PA, USA.

13. Joint Committee for Powder Diffraction Standards (JCPDS). Card No. 6–594; International Centre for Diffraction Data (ICDD): Newtown Square, PA, USA.

14. Gulbransen, E.A.; Andrew, K.F.; Brassart, F.A. Oxidation of silicon at high temperatures and low pressure under flow conditions and the vapor pressure of silicon. *J. Electrochem. Soc.* **1966**, *113*, 834–837.

15. Langmuir, D.B.; Malter, L. The rate of evaporation of tantalum. *Phys. Rev.* **1939**, *55*, 748–749.

16. Nomoev, A.V.; Bardakhanov, S.P. Method of Producing Composite Nanopowders. *Patent RU 2412784 C2*, 3 February 2009.

17. Nomoev, A.V.; Bardakhanov, S.P. Synthesis and structure of Ag-Si nanoparticles obtained by the electron-beam evaporation/condensation method. *Tech. Phys. Lett.* **2012**, *38*, 375–378.

18. Ivanov, A.S.; Borisov, S.A. Surface segregation and concentration stresses in fine spherical particles. *Poverkhnost* **1982**, *10*, 140–145.

19. Paradis, P.-F.; Ishikawa, T.; Yoda, S. Surface tension and viscosity of liquid and undercooled tantalum measured by a containerless method. *J. Appl. Phys.* **2005**, *97*, 053506.

20. Shishkin, A.V.; Basin, A.S. Surface tension of liquid silicon. *Theor. Found. Chem. Eng.* **2004**, *38*, 660–668.

21. Ko, I.-Y.; Park, J.H.; Nam, K.-S.; Shon, I.-J. Pulsed current activated combustion synthesis and consolidation of nanostructured TaSi2. *J. Ceram. Res.* **2010**, *11*, 69–73.

22. Ergy, I. Surface tension of compound forming liquid binary alloys: A simple model. *J. Mater. Sci.* **2004**, *3*, 6365–6366.

23. Temuujin, J.; Bardakhanov, S.P.; Nomoev, A.V.; Zaikovskii, V.I.; Minjigmaa, A.; Dugersuren, G.; van Reissen, A. Preparation of copper and silicon/copper powders by a gas evaporation-condensation method. *Bull. Mater. Sci.* **2009**, *32*, 543–547.

24. Bardakhanov, S.P.; Gafner, Y.Y.; Gafner, S.L.; Korchagin, A.I.; Lysenko, V.I.; Nomoev, A.V. Preparation of nickel nanopowder through evaporation of the initial coarsely dispersed materials on an electron accelerator. *Phys. Solid State* **2011**, *53*, 854–859.

25. Lysenko, V.I.; Bardakhanov, S.P.; Korchagin, A.; Kuksanov, N.; Lavrukhin, A.; Salimov, R.; Fadeev, S.; Cherepkov, V.; Veis, M.; Nomoev, A. Possibilities of production of nanopowders with high power ELV electron accelerator. *Bull. Mater. Sci.* **2011**, *34*, 677–681.

26. Bardakhanov, S.P.; Korchagin, A.I.; Kuksanov, N.K.; Lavrukhin, A.V.; Salimov, R.A.; Fadeev, S.N.; Cherepkov, V.V. Nanopowder production based on technology of solid raw substances evaporation by electron beam accelerator. *Mater. Sci. Eng. B* **2006**, *132*, 204–208.

Synthesis, Characterization, and Mechanism of Formation of Janus-Like Nanoparticles of Tantalum...

151

27. Bardakhanov, S.P.; Korchagin, A.I.; Kuksanov, N.K.; Lavrukhin, A.V.; Salimov, R.A.; Fadeev, S.N.; Cherepkov, V.V. Nanopowders obtained by evaporating initial substances in an electron accelerator at atmospheric pressure. *Dokl. Phys.* **2006**, *51*, 353–356.

7

Recent Advances on Carbon Nanotubes and Graphene Reinforced Ceramics Nanocomposites

Iftikhar Ahmad [1], Bahareh Yazdani [2] and Yanqiu Zhu [2,*]

[1] Center of Excellence for Research in Engineering Materials, Advanced Manufacturing Institute, King Saud University, Riyadh 11421, Saudi Arabia; E-Mail: ifahmad@ksu.edu.sa
[2] College of Engineering, Mathematics and Physical Sciences, University of Exeter, Exeter EX4 4QF, UK; E-Mail: by219@exeter.ac.uk

* Author to whom correspondence should be addressed; E-Mail: y.zhu@exeter.ac.uk

Academic Editor: Emanuel Ionescu

Abstract: Ceramics suffer the curse of extreme brittleness and demand new design philosophies and novel concepts of manufacturing to overcome such intrinsic drawbacks, in order to take advantage of most of their excellent properties. This has been one of the foremost challenges for ceramic material experts. Tailoring the ceramics structures at nanometre level has been a leading research frontier; whilst upgrading via reinforcing ceramic matrices with nanomaterials including the latest carbon nanotubes (CNTs) and graphene has now become an eminent practice for advanced applications. Most recently, several new strategies have indeed improved the properties of the ceramics/CNT nanocomposites, such as by tuning with dopants, new dispersions routes and modified sintering methods. The utilisation of graphene in ceramic nanocomposites, either as a solo reinforcement or as a hybrid with CNTs, is the newest development. This article will summarise the recent advances, key difficulties and potential applications of the ceramics nanocomposites reinforced with CNTs and graphene.

Keywords: nanocomposites; mechanical properties, interface; graphene; carbon nanotubes (CNTs); ceramics

1. Introduction

Ceramics are potential contestants for diverse sophisticated engineering applications, and plenty of attentions have been focused to further improve their properties by adopting emerging technologies. As a result, much deeper understandings and significant amounts of improvements in their structures and properties have been achieved after decades of efforts. However, many challenge issues limit their wide applications, such as the degradation of high temperature mechanical properties of non-oxide silicon nitride (Si_3N_4) and silicon carbide (SiC), and low fracture toughness, poor creep, deprived thermal shock resistance of oxide ceramics like alumina (Al_2O_3) and zirconium oxide (ZrO_2) [1]. To date, ceramics have found some niche applications, from high speed cutting tools, dental implants, chemical and electrical insulators, to wear resistance parts and various coatings, due to their high hardness, chemical inertness and high electrical and thermal insulating properties [2]. Low fracture toughness restricts ceramics for applications in aircraft engine parts and in extreme environments for space engineering [3]. Presence of impurities, pores and cracks cause pure ceramics extremely brittle, and complex/expensive processing technology is needed to reduce such fatal drawbacks. For decades, the addition of a second reinforcing phase in ceramics has been an effective practice to improve their toughness, converting brittle ceramics to practical engineering materials. Recent advances in nanomaterials have offered the opportunity to tailor the ceramic structures at nanometre scale, for the development of new classes of stronger, tougher engineering ceramics with added functionalities. Chosen nanomaterials with distinct morphologies and properties have been used to reinforce monolithic ceramics [4–6]. In particular, the exceptional mechanical behaviour and outstanding multifunctional features of carbon nanotubes (CNTs) and graphene have made them the wonder materials, standing out from many other nanomaterials, among different research communities. There has been much documented research attempting to incorporate both types of CNT in brittle ceramics to convert them into tough, strong, electric and thermal conductive materials [7–15]. Graphene, known as a monolayer of carbon atoms arranged in a honeycomb lattice, has shown similar properties to carbon nanotubes with impressive thermal, mechanical, and electrical properties, and is a promising alternative of CNTs in various applications [16–18]. Compared with CNTs, graphene also have large specific surface areas and they do not form agglomerates in a matrix when handled appropriately, thus an ideal nano-filler for composite materials. In this regard, the low-cost, high quality and commercially more viable a-few-layer-thick graphene nanosheets, designated as graphene nanoplatelets (GNPs) are more promising for practical engineering applications, thus attracted considerable research interests for advanced ceramic matrices. Indeed, various crucial ceramics such as Al_2O_3, Si_3N_4 and ZrO_2 have been reinforced by the GNP fillers and obvious improvements in fracture toughness, thermal and electrical properties have been obtained [19–26]. However, research of ceramic-GNP nanocomposites is in its infancy, and more thorough and systematic studies are required.

Nevertheless, ceramics reinforced with CNTs, graphene and GNT (CNTs/graphene hybrid) have indeed showed significant improvements in the fracture toughness and other mechanical properties by following complex toughening mechanisms, although wide variations in the results still remain problematic. In fact, processing ceramic nanocomposites is complicated due to the introduction of a second reinforcement phase of nanometric scale. Conventional rules and benefits associated with microscopic reinforcement phases need to be modified carefully and validated fully before being applied directly. In this context, recent advances in the fabrication technology, mechanical properties and potential applications of typical ceramic nanocomposites reinforced with CNTs and graphene are presented in this paper. The main purpose of this review is to provide a comprehensive picture of the current state of research progresses and challenges concerning the graphene and CNTs-reinforced ceramic composites, to assist the ceramic community for further developments.

2. CNTs-Reinforced Ceramics Nanocomposites

2.1. Pre-Processing for Good Dispersion

A statistical summary of the varieties of processes opted to fabricate CNTs containing ceramic nanocomposites is graphically presented in Figure 1, and Table 1 gives further details of these processes. It is evident that 88% of the reported cases used the readily available and economically feasible multi-walled carbon nanotubes (MWCNTs) as the reinforcement, against single-walled carbon nanotubes (SWCNTs). Nearly 40% adopted a wet oxidation process (treatment with concentrated H_2SO_4 and HNO_3 in 3:1 ratio) to purify the CNTs in an effort to remove unwanted impurities including amorphous carbon nanofibres, carbon nanoparticles, amorphous carbon coating layers, and metallic catalyst residues; and about 33% of the reports attempted pristine CNTs; whilst the others tried oxidation through annealing [27–30]. The wet oxidation purification method for CNTs offers two folded advantages, realizing purification and simultaneous attachments of carboxyl functional groups onto the CNT surfaces which facilitates their mixing with and dispersions into matrices. It is a fact that CNT clusters prevent each individual CNT forming the ideal interconnection desired with the matrix, leading to ill-constructed interface and microstructures, detrimental to the final mechanical properties. Thus homogenous CNT dispersion within the matrix is extremely imperative. In addition, the actual quality of CNT dispersion is the foremost factor in tumbling the densities of CNT-reinforced ceramics, because homogenous dispersion is attainable only at low CNT concentrations (<2 wt%), and higher than that normally ended up with severe agglomerations [31].

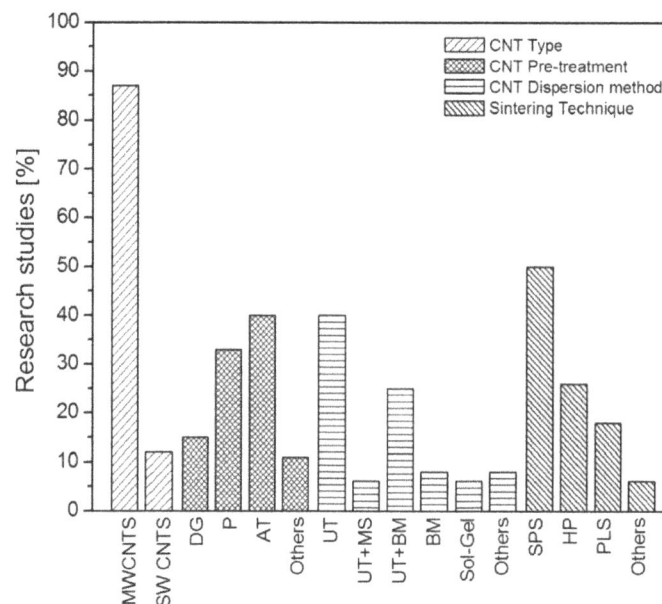

Figure 1. Statistical analysis of the carbon nanotubes (CNTs)-reinforced ceramic nanocomposites.

Table 1. Processing details of CNTs-reinforced ceramics nanocomposites.

Reference	Matrix	CNT types	Purification methods	Dispersion procedures	Sintering techniques
[10]	Si_3N_4	SW	P	UT of CNTs with surfactant (C16TAB) and Si_3N_4	SPS under vacuum
[12]	Al_2O_3	MW	Oxidation at 500 °C for 90 min	UT of CNTs in ethanol	SPS at 1500 °C for 10 min under 50 MPa
[32]	Al_2O_3	MW	AT (H_2SO_4 + HNO_3)	UT of CNTs into water and SDS then incubation for 2 weeks	HP at 1600 °C, 60 min, 40 MPa
[33]	Al_2O_3	MW	AT (H_2SO_4 + HNO_3) for 3 h	24 h BM of ball Al_2O_3 powder and 30 min UT of CNTs in water and then BM of CNTs/Al_2O_3 mixture	PLS at 1500–1600 °C, 120–240 min, Ar
[34]	Al_2O_3	MW	Pristine	UT of CNTs for 1 h in alcohol	CIP at150MPa and PLS at 1500 °C, and 1700 °C with 2 h
[35]	Al_2O_3	MW	AT (heating in 65% HNO_3 at 80 °C for 8 h)	BM and Surfactant (Darvan C–N)	PLS at 1500 °C for 2 h using Ar
[36]	Mulite	MW	P	CNTs dispersion into ethanol by MS and UT	HP at 1600 °C for 60 min under Ar atmosphere at 30 MPa
[37]	Si_3N_4	MW	P	24 h ball milling the CNTs and Si_3N_4 slurry	HP at 1750 °C for 60 min under 30 MPa
[38]	ZrB_2–SiC	MW	P	20 min UT of CNTs and matrix with subsequent 24 h ball milling	HP at 1900 °C for 60 min under 30 MPa
[39]	$BaTiO_3$	MW	P	-	HP, 1200 °C, 60 min
[40]	Al_2O_3	MW	-	DG (CVD at 750 °C for 15 min for direct CNTs growth on Al_2O_3 nano-particles)	SPS at 1150 °C for 10 min under 100 MPa
[41]	Al_2O_3	SW	Pristine	UT of CNTs in ethanol	SPS at 1520 °C under 80 MPa
[42]	Al_2O_3	MW	P	35 h UT in water	SPS at 1300 °C, 20 min, 90 MPa
[43]	Al_2O_3	MW	AT	UT of CNTs and Al_2O_3 in water followed by 2 h and BM of CNTs/Al_2O_3	PLS at 1600 °C, 15 min, Ar
[44]	Al_2O_3	MW	AT (HNO_3 for 30 min)	5 h BM of CNTs and 1 h UT of CNTs. 5 h BM of CNT/Al_2O_3 in ethanol	PLS at 1550 °C, Ar

Table 1. *Cont.*

Reference	Matrix	CNT types	Purification methods	Dispersion procedures	Sintering techniques
[45]	Al_2O_3	MW	AT (H_2SO_4 + HNO_3 in 3:1 for 7 h)	surfactant (SDS) using combination of UT and 24 h BM	HP at 1550 °C for 1 h under 30 MPa using Ar gas
[46]	Al_2O_3 + ZrO	MW	AT (heating in 65% HNO3 at 80 °C for 8 h))	2 min UT of CNTs with surfactant (SDS)and 24 BM then freezing with Nitrogen	HP at 1500 °C for 2 h under 30 MPa in Ar atmosphere
[47]	Al_2O_3	SW	AT (H_2SO_4 + HNO_3)	UT for 24 h	SPS at 1300 °C for 5 min under 75 MPa

Notes: SW: Single-walled CNTs; MW: Multi-walled CNTs; UT: Ultrasonication; BM: Ball milling; HP: Hot-pressing; SPS: Spark plasma sintering; PLS: Pressureless sintering; SDS: Sodium dodecyle sulphate; CIP: cold isostatic pressing; P: Pristine; MS: Magnetic stirring.

To combat this dispersion issue, as shown in Figure 1, the most dominant (40%) technique involving colloidal technology (ultra-sonication of CNTs for different durations into different solvents with or without surfactants). Until recently, attempts increasingly focused on a combined process (colloidal technique and ball milling) which have produced better results and was more reproducible than other techniques (ball milling, sol-gel, planetary centrifuge mixing, magnetic stirring, tape casting, *etc.*), as shown in Table 1. Moreover cationic, anionic and neutral surfactants have greatly contributed to the detangling of CNT clusters, of which SDS (sodium dodcyle sulphate) seems to be the most used one [11]. In addition, several reports described the growth of CNTs directly onto the surface of ceramics nanoparticles using a standard chemical vapour deposition (CVD) technique; however, this method failed to create high quality coverage, on top of the low yield issues [13].

Cultivation of CNTs in porous ceramics is intriguing process, and numerous efforts have been documented for highly ordered CNT growth within the pores of thin SiO_2 and Al_2O_3 membranes, which led to novel CNTs-reinforced porous ceramics with potential applications as field emitters, nanocapacitors, and scanning microscope probes [48–52]. To prepare CNTs carrying porous ceramic composites, Fan *et al.* [27] first embedded catalyst inside the SiO_2 pores of the ceramics pores, then allowed for the carbon source to diffuse into and deposit inside the pores; whilst Kyotani *et al.* [48] grew CNTs on an anodic aluminium oxide (AAO) porous membrane with and without catalyst. The CNT growth mechanism in a catalyst free CNTs-porous ceramic is still not completely understood. The AAO membrane template may itself catalysed the CNT growth by deposition of carbon atoms on the internal pore surface of the complex three dimensional structure, as proposed by Sui *et al.* [49]. Since catalyst facilities carbon source decomposition, thus further deposition of atomic carbon tends to result in more ordered or well-crystallised structures, leading to better quality nanocomposites than the non-catalysed process. Patterning and lithography technology enabled Bae *et al.* to deposit a thin Si layer on an AAO substrate for enhanced CNT growth [50]. Parham *et al.* have recently prepared a 3 wt% CNT-containing composite using Al_2O_3 and SiO_2 porous ceramics, and resulting composites exhibited a high efficiency for yeast cell filtration (98%), a 100% heavy metal ions removal from water and excellent particulate filtration performance from air [51,52].

Despite these achievements for CNT dispersion in various ceramic matrices, some known issues still remain. For example, SWCNTs are always produced in the form of bundles of tens of nanotubes, and their separation into individual tubes is still extremely difficult, concerning the nanocomposite fabrication. This area thus needs further investigations, because SWCNTs have promising applications in biomedical engineering, composite technology and nanodevices [30]. Dispersion of CNTs within ceramic matrices is generally assessed by the microscopic images of the fractured surfaces of nanocomposite samples taken from desired areas of interest, and the representativeness of this method sometimes is a concern, as it may not be a true reflection of the CNT dispersions for other locations [53–55]. Therefore, the standardization of CNT dispersion assessments in composites beyond ceramics is vital for quality control in manufacturing and industrial applications.

2.2. Densification Processes

Achieving near full density, without damaging the CNT structure and morphology, is a fundamental requirement and another important challenge in ceramic matrix nanocomposite technology, as most of the mechanical properties are strongly affected by the density. CNTs hinder the ceramic grains coalescence by existing at the grain boundaries, which tends to lead to poorly densified microstructures [56]. For this reason, pressure-assisted consolidification processes are generally be used to counter this problem. Figure 1 shows that about 76% nanocomposites were consolidated by pressure-assisted sintering processes, in which spark plasma sintering (SPS) and hot-pressing (HP) have a share of 50% and 26%, respectively. Hot-pressing provides simultaneous high pressures and high temperatures to powder systems, which in turn gives high densities, thus good mechanical properties to either pure ceramics and their composites. Coble *et al.* explained that the enhanced densities were associated with accelerate densification due to higher stresses caused by external pressure, and this phenomenon consolidated the grains to a desirable density; unfortunately, damage to the CNTs during the matrix grain growth could occur due to prolong sintering at extremely high temperatures which was a potential big shortcoming of HP [32,57]. The structural damage problem of CNTs during HP can be avoided by using SPS technique, in which near full densification is achievable at lower sintering temperatures with substantially short holding time. The microscopic images (Figure 2) showed that the CNTs were mainly located at the ceramic grain boundaries, well adhered with the matrix without apparent damage to the structure and morphology [58]. Pressureless sintering (PLS) offers convenient and cheaper consolidation alternations, but wide variation in earlier results have made this technique unattractive and debatable. For example, Zhan *et al.* [15] and Ahmad *et al.* [59] claimed widely different densities for similar samples, as high as 99% and as low as <90% for 1 wt% CNTs-reinforced Al_2O_3, respectively. However in recent reports, Sarkar *et al.* [34] densified Al_2O_3 containing 0.3 vol% of MWCNTs to >99% at 1700 °C using PLS sintering; and similarly Michalek *et al.* [35] and Ghobadi *et al.* [60] obtained 99.9% and >98% densities for Al_2O_3 reinforced with 0.1 wt% and 1 vol% MWCNTs, respectively. Regarding other ceramics, Tatami *et al.* [61] achieved Si_3N_4-MWCNTs composites by pressureless sintering during which the constituents including sintering additives were initially pressed uniaxially and subsequently sintered in furnace at 1700 °C under N_2 atmosphere. They recorded a drop in relative densities from 100% to 90% for 0 to 5 wt% CNT additions; whilst they obtained much higher densities >96% for 5 wt% MWCNT additions by using HP. Microwave sintering is another inspiring and "green" sintering technique, with the advantage of lower densification temperatures and shortened processing time. This technique has been successfully used to consolidate most mainstream industrial ceramics (e.g., Al_2O_3, ZrO_2, Si_3N_4), both pure and composite forms, and resulted in high densities, due to rapid microwaves heating characteristics [62,63]. Nevertheless, the mixed large and fine grained microstructure of the final ceramic consolidated by microwave sintering has made it bit divisive, but this technique exhibits great potentials for CNTs-reinforced ceramics' densification. The advantageous features such as short sintering time and low densification temperatures are not deleterious for CNT structures; furthermore the localize heating at the grain boundaries may be helpful in constructing strong interfaces between CNTs and the ceramic matrices; finally, the grain coarsening seems not a big issue in microwave sintered CNTs-reinforced ceramics, possibly owing to the grain refining tendency of CNTs [64,65]. Table 1 covers the HP, PLS and SPS these main methods.

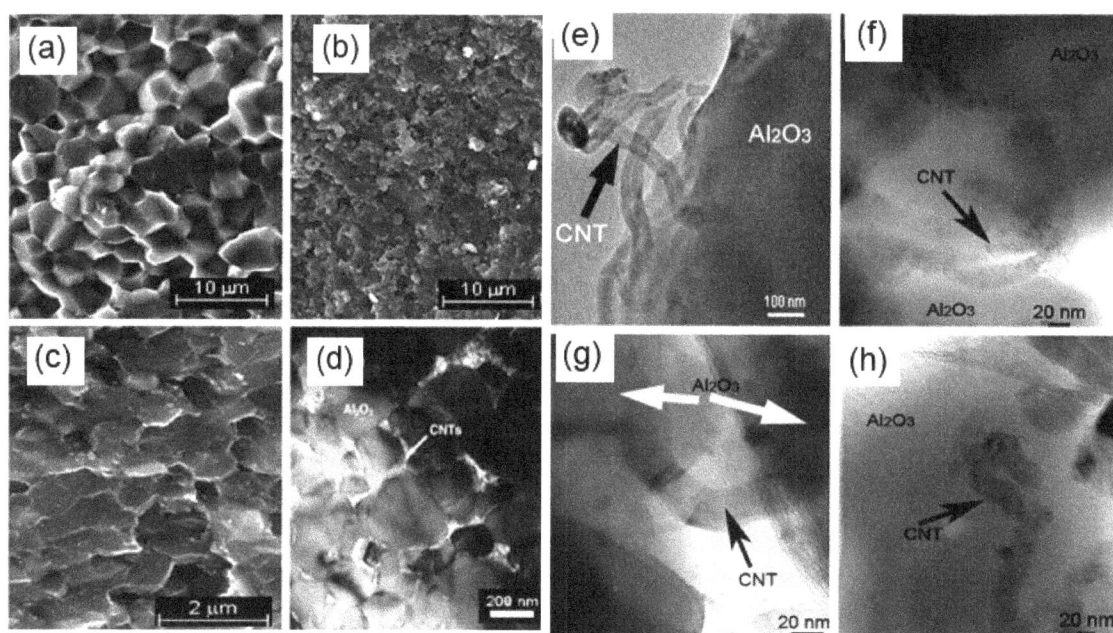

Figure 2. Structural features of (**a**) Monolithic Al_2O_3 showing large grains with inter-granular fracture; (**b**) CNTs/Al_2O_3 nanocomposites with fine grains; (**c**) Trans-granular fracture mode in CNTs/Al_2O_3 nanocomposites; and (**d**) Single-walled (SW)CNTs at grain boundary of Al_2O_3 matrix. TEM images exhibiting the CNT–ceramic interactions (**e**) Multi-walled (MW)CNTs (black arrow) showing their morphology in nanocomposite; (**f**) A single MWCNT existing at grain boundary; (**g**) in porosity and (**h**) Embedded within a single ceramic grain. Adapted from References [32] and [66] with permissions. Copyright 2010, Elsevier Ltd.

In seeking of highly dense composite structure, new techniques are always attempted, and we will summarise a few diverse and interesting methods here. For example, in order to protect the CNT structures by preventing reactions with SiC during high temperature integration, Thostenson *et al.* first prepared a preform of SiC nanoparticles and CNTs (duly dispersed in polymer matrix) then carbonized the perform followed by infiltrating molten Si into the preform under vacuum at 1400 °C to claim dense nanocomposites; whilst Wang *et al.* packed SiC nanoparticles and CNTs in a cylinder and arranged graphite heater inside the cylinder followed by heating up to 1700 °C to consolidate the CNTs-SiC mixture [36,67–69].

2.3. Microstructural Analysis

Sharp reduction from coarser grains in monolithic ceramics (Figure 2a) to finer grains in CNTs-reinforced ceramics (Figure 2b) is a principle feature of structural change, occurred due to the pinning of matrix grains by CNTs which restricted the grain growth during sintering [66]. Fracture mode alteration from inter-granular in monolithic ceramics (Figure 2a) to trans-granular in the CNTs-reinforced ceramics (Figure 2c) is another interesting change being revealed. The morphological analyses of fractured surfaces are helpful in depicting the mechanisms behind such changes. In the case of monolithic Al_2O_3, it shows clearly the edge and corner fractural features (Figure 2a), representing the typical inter-granular fracture mode; and conversely a blurry and glaze-like surface appears for CNTs-reinforced Al_2O_3 (Figure 2c), indicating the trans-granular mode of fracture [66]. These observations mean that CNTs, as the second phase, must be responsible for altering the fracture modes. Indeed, when CNTs were homogenously dispersed within the ceramic matrix, they arranged themselves at various locations such as along grain boundaries (Figure 2f), across grains boundaries (Figure 2g), inside single grains (Figure 2h), contributing to strengthening the composites at nanometre level by making bridges across grains and sharing the grains, as discussed in prior studies [66,70]. Presumably, all these interesting arrangements of CNTs in ceramic matrices promoted the trans-granular fracture, rather than inter-granular fracture as did in the pure ceramic. In a very recent report, Ahmad et al. obtained 5-fold finer grain size in MWCNTs/Al_2O_3 nanocomposites by 300 ppm Y_2O_3 doping than its undoped Al_2O_3 counterpart, and mixed inter/intra fracture mode in Y_2O_3 doped nanocomposites was observed [31,37]. However this fracture mode change phenomenon is another grey area that is not fully understood for CNTs-reinforced ceramics, which offers opportunities for prospective thinking and further research work.

2.4. Mechanical and Functional Properties

In view of the vast applications of the economically viable Al_2O_3 ceramics in industry, lots of studies have been done to improve their fracture toughness by CNT additions. However, inconsistent results (Table 2) put question marks on these triumphs and core issues in such discrepancies were found in the CNT dispersion methods, choice of sintering process and techniques adopted for characterisation [34,40,41,50,59,61,71,72]. For example, Table 2 shows that the higher fracture toughness values of MWCNTs-reinforced Al_2O_3 were obtained at lower CNT additions (<2 wt%), and declining trend can be seen at higher CNT levels in all cases, except from the values reported by Zhan et al. [15]. Furthermore, the composite fracture toughness reported by Zhan et al. [15] was the highest in Table 2. However, this value may be due to various factors: (1) the use of SPS techniques (positive); (2) reinforcement phase being SWCNTs (positive); and (3) the assessment of the fracture toughness by an unreliable direct crack method, DCM, (negative). Yamamoto et al. [12] used the SPS to sinter similar composite reinforced with MWCNTs, and used the single-edged notched beam (SENB) method to assess the fractures toughness, however the results were now as good as the results reported by Zhan et al. [15] and Ahmad et al. [31]. In case of the high values reported by Ahmad et al. [66], it is probably due to the better dispersion of CNTs within the matrix, as they adopted a unique method. Further, Huang *et al.* [39] showed tremendous improvements in fracture toughness (57%, 114% and 328%) values for BaTiO$_3$ ceramic after reinforced with (0.5, 1 and 3 wt%) MWCNTs; whereas a 15% improvement was recorded by Tian *et al.* [38] for 2 wt% MWCNTs-reinforced ZrB$_2$-SiC ceramics.

Table 2. Properties of CNTs-reinforced ceramics.

Reference	Matrix	CNT contents	Relative density (%)	Hardness (GPa)	Flexural strength (MPa)	Fracture toughness (MPa. m$^{1/2}$)
[10]	Si$_3$N$_4$	0	99.2	15.7	1046	4.8
		1 wt%MWCNTs	98.7	15.0	996	6.6
[12]	Al$_2$O$_3$	0	95.6	17.3	500	4.4
		0.5 wt% MWCNTs	99.2	16.8	685	5.9
		1 wt% MWCNTs	98.9	15.9	650	5.7
[15]	Al$_2$O$_3$	0	-	-	-	3.3
		3 wt% SWCNTs	-	-	-	7.9
[27]	Al$_2$O$_3$	0	97.7	-	326	3.08
		6 wt% MWCNTs	95.4	-	314	5.55
[32]	Al$_2$O$_3$	0	99.8	16	356	3.5
		2 wt% MWCNTs	99.5	18	402	6.8
		5 wt% MWCNTs	99.1	-	423	5.7
[34]	Al$_2$O$_3$	0	99.5	17.5	222	3.92
		0.15 vol% MWCNTs	98.4	21.4	242	5.27
[35]	Al$_2$O$_3$	0	-	16.9	-	5.5
		1 vol% MWCNTs	-	13.5	-	6.0
[36]	Mulite (3Al$_2$O$_3$ + 2SiO$_2$)	0	-	-	466	2.0
		2 wt% MWCNTs	-	-	512	3.3
[69]	SiC	0	939	-	303	3.3
		10 wt% MWCNTs	94.7	-	321	3.8
[38]	ZrB$_2$-SiC	0	-	15.8	582	4
		2 wt% MWCNTs	-	15.5	616	4.6
[39]	BaTiO$_3$	0	98.5			0.7
		98.50	98.5			0.7
		0.5 wt% MWCNTs	97.3	-	-	1.1
		1 wt% MWCNTs	99.2			1.5
		3 wt% MWCNTs	98.6			3.0
[73]	Al$_2$O$_3$	0	-	-	395	4.41
		20 vol% MWCNTs	-	-	403	4.62
[74]	Al$_2$O$_3$	0	-	-	-	3
		1 wt% MWCNTs	-	-	-	5
[75]	Al$_2$O$_3$	0	-	15.71	-	3.24
		5 wt% MWCNTs	-	0.72	-	4.14
[76]	Al$_2$O$_3$	0	-	18.2	-	4.5
		2.5 wt% MWCNTs	-	15.75	-	11.4
[77]	Al$_2$O$_3$	0	99.9	22.9	-	3.54
		10 vol% MWCNTs	97.4	11	-	2.76

Recently, Sarkar *et al.* [34] calculated fracture toughness values of the Al_2O_3–MWCNT nanocomposites by employing DCM method using Niihara and Liang models, and reported better fracture toughness values than those obtained using SENB technique; whereas Ahmad *et al.* [66], reported higher fracture toughness values attained from SENB method than those obtained from DCM method using Chantikul model. These conflicting reports suggest that engineering components cannot be validated for structural load-bearing applications using DCM method; however, this convenient method is widely employed for fracture toughness comparisons [31]. Similar inconclusive and controversial fracture toughness values regarding CNTs-reinforced Si_3N_4 were also reported by Corral *et al.* [10] and Pasupuleti *et al.* [37] Both consolidate Si_3N_4 with 1 wt% CNTs and obtained 30% reduction (by SENB method) and 40% increment (by ISB method) in fracture toughness, respectively.

Regarding other mechanical properties such as hardness and elastic modulus, Yamamoto *et al.* [12] investigated a range of MWCNT additions in Al_2O_3 ceramics and concluded a drop in hardness and rise in flexural strengths at low MWCNT additions and further reduction in both properties by adding more MWCNTs, and consistent results were reported by many others for the same material system, as shown in Table 2. Pasupuleti *et al.* [37] showed a small decrease in hardness (4%) and flexural strength (5%) for 1 wt% MWCNTs-reinforced Si_3N_4, however, Corral *et al.* [10] reported a much severer 45% reduction in hardness for same reinforcement contents in Si_3N_4.

The dual role of CNTs, indirectly enhancing the mechanical properties and directly acting as lubricant, converts ceramic composites into an attractive wear resistance material, and various reports demonstrated the steady reduction of friction coefficient with CNT additions [78]. High thermal and electrical properties of the CNTs have been predicted and several attempted to incorporate CNTs into insulated ceramics in order convert them into highly electrical and thermally conductive materials [73]. Ceramics exhibited higher electric conductivity (EC) when reinforced with SWCNTs (10^6 S/m) than with MWCNTs (10^3–10^5 S/m) [40,79]. Sarkar *et al.* reported that the EC of MWCNTs-reinforced composites was dependent on the formation of electrically conductive networks by dispersing the CNTs homogenously in the matrix, and on the grain sizes of the final nanocomposites, as larger grain size with less grain boundaries showed better results [80]. So far Estili *et al.* has obtained the highest EC of 4816 S/m for Al_2O_3-20 vol% MWCNTs, which is 43% higher than that reported by Zhan *et al.* [73,79]. In addition, Kumari *et al.* obtained an exceptional value of 3336 S/m by reinforcing Al_2O_3 with 19 wt% MWCNTs nanocomposites, however, at the cost of poor mechanical properties [41]. In contrast to MWCNTs, the SWCNTs reinforcement into the Al_2O_3 matrix offered better conductivity of 3345 S/m without compromising mechanical properties [71]. Zaman *et al.* studied the effects of surface functionalization of the SWCNTs on EC and reported that the hydroxyl group functionalized SWCNT offered ~10 times higher EC in 1 wt% SWCNTs-reinforced Al_2O_3 nanocomposites than those functionalized by carboxylic acid group [72]. Moreover, Bi *et al.* reported a drop in the electrical percolation by increasing the aspect ratios of MWCNTs [81]. Although the thermal conductivity (TC) of SWCNTs and MWCNTs are ranges from 3000 to 6000 W/m·K, however, their nanocomposite with ceramics barely demonstrated good thermal performance. Compared to unreinforced Al_2O_3, Zhan *et al.* reported lower (7.3 W/m·K) TC in nanocomposites reinforced with 15 vol% SWCNTs than their monolithic counterpart (27.3 W/m·K) [71]. Both Kumari *et al.* and Bakshi *et al.* reported higher TC values (63.52 and 6 W/m·K) in nanocomposites with (8 and 4 wt%) MWCNTs than those of pure Al_2O_3 (19.96 and 5.37 W/m·K) samples, respectively [82]. This area of research is therefore more complicated and interesting.

2.5. CNTs/Ceramic Interface and the Toughening Mechanism

Reinforcement (fibres or whiskers) pullout is the main toughening mechanism in conventional ceramics, which is further associated with weak interfacial connection of reinforcement with matrix. Same classical approach was proposed for toughness mechanism in CNTs-reinforced ceramics in several initial reports [83–85]. However, in later research Padture *et al.* and many others observed that micro-structural features of CNTs-ceramics were immensely dissimilar from conventional composites, and these observations strongly suggested that existing microscale mechanism may not be fully applicable to CNTs–ceramic systems [18,23,61]. Microstructure of conventional ceramic composites consists of inflexible and straight reinforcement, and the interface is optimally designed in such a way that it debonds on applied load [86]. Imagine, when a reinforcement encounters a crack then it bridges the crack in its wake, pullout does frictional work and these together effectively make crack propagation more difficult, in addition to this large reinforcement dimensions lead to longer crack-wake bridging zones and consequently resulted in higher toughness [85,87]. Meanwhile, these large reinforcements prompt larger flaws and turn strength to lower values. In contrast, CNTs are highly flexible, hollow nanometre sized fibres, therefore the toughening mechanism may be entirely different from conventional ones. Frictional pullout of fibres occurred in classical composites may not be the only toughening mechanism in CNTs-reinforced ceramics. Accordingly, new concepts and philosophies of uncoiling and elastic stretching of CNTs during the crack propagation were proposed as main toughness mechanisms by Padture *et al.* [7]. During crack propagation, initial uncoiling of CNTs occurs in the crack wake, and when the crack further propagates the uncoiled CNT stretches elastically serving as stretched CNT bridges instead of conventional frictional pull-out bridges, thus impedes the crack propagation, as shown in Figure 2g [66]. These concepts are convincingly identified the role of CNTs as an individual entity, also applied to the cluster form. Surface damages to SWCNTs during purification and subsequent sintering process are well-known, and in this picture the role of CNT's elastic stretching in toughness is slightly litigious. In this regard, mathematical modelling will be a helpful tool in explaining the role of CNTs in strengthening ceramics and predicting their behaviour in services.

Recent developments in the electron microscope technology are changing the research approaches, and attentions are now tending to focus on tailoring the interface structure at atomic level, to construct defect-free structures with interesting functionalities. FIB-SEM (focused ion beam scanning electron microscopy) has made the scientists' life not as hard as ages ago. A TEM (transmission electron microscope) sample of hard materials, e.g., ceramics can be prepared in hours, which was a laborious task to arrange for days and even weeks earlier. For CNTs-Al_2O_3 nanocomposites, due to the interesting reaction of Al_2O_3 with alkaline, a simple powder etching process can always be used to collect CNTs with a thin layer of Al_2O_3 residue, for interface study under TEM. Similar results are obtained when compared with FIB-SEM [66].

Back to CNTs-reinforced ceramics, where the interface controls the CNT debonding, pullout and crack-bridging at micron and nanometre level, these different mechanisms act as energy dissipative processes during mechanical loading. Physically, an interface is a complex transitional region layered between the reinforcement and parent matrix thus, the control of interface chemistry and tailoring smart microstructures are essential steps for producing exceptionally tough and strong nanocomposites. Dedicated efforts have been done to explore the CNTs-ceramic transition region and each addressed in interesting way [12,66]. Indeed, rough surface of CNTs produces the required frictional forces which resist in detaching CNTs from the ceramic matrix. Yamamoto *et al.* proposed that acid treatment does not significantly damage the overall structures of MWCNTs however, localized etches of the cylindrical body at different locations create nanoscale defects (nano-pits) along the tube axis, as shown in Figure 3b [12]. These nano-pits having depths of ~15 nm are anchored by the matrix grains (Figure 3c), forming locks and resistance in MWCNTs' sliding over the matrix, thus leading to good connection of composite constituents at the interface [12]. Further, a close cross-sectional look of the MWCNT shown in Figure 3d of the high resolution TEM image reveals its uneven surface, hollow core and graphitic layers. These layers are not concentrical on a long distance and many compartments exist, which is a typical feature of MWCNTs synthesized by CVD. Ahmad *et al.* [78] postulated that high surface roughness of the CNTs could result in two potential advantages like chemically highly reactive and physically difficult to slide out of the matrix, compared with a smooth surface. The former could help to improve the interfacial bonding with the matrix and the latter can pose much larger friction forces to stop the CNT pullout [78].

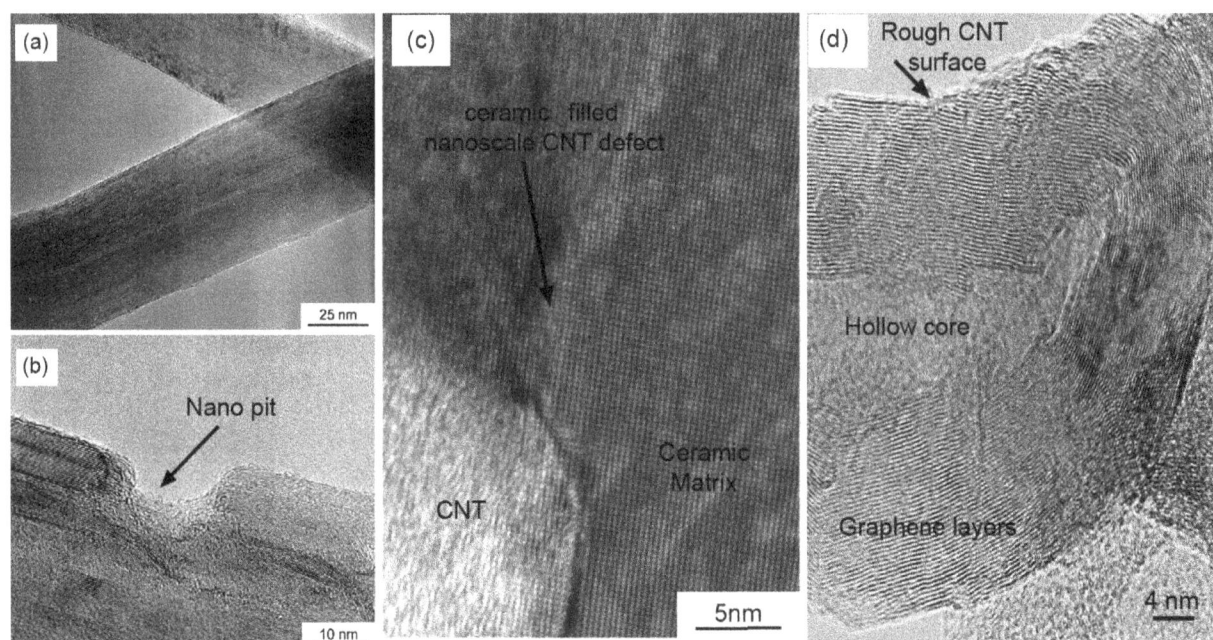

Figure 3. (**a**) TEM image of the pristine MWCNTs; (**b**) High-magnification TEM image of the acid-treated MWCNT surface, arrow indicates nano-pit; (**c**) Nano-pit on the acid-treated MWCNTs is filled up with Al_2O_3 crystal; and (**d**) Rough surface of MWCNT produced by chemical vapour deposition (CVD) method. Adapted from References [12] and [32] with permissions. Copyright 2009, Elsevier Ltd. and 2008 IOP Publishing Ltd.

The CNT's surface unevenness and its anchoring with the ceramics matrix are a good physical explanation of enhanced frictional forces at the interface. However, the chemical interactions of CNTs with the ceramics remained unattended for several years. Estili *et al.* [88] rigorously studied the interfacial areas of CNTs/Al_2O_3 nanocomposites using high resolution-TEM, but unable to identify any interfacial phases or intermediate compounds at the CNTs/Al_2O_3 interface. A recent attempt addressed this topic and explained the chemical activity took place at the CNTs/Al_2O_3 interface during HP process, and reported the formation of an extremely thin (1–2 nm) intermediate phase of Al_2OC, which is possibly produced due to the carbothermal reduction of Al_2O_3 by CNTs [78]. Figure 4a–b show clear evidence of a CNT sticking with Al_2O_3 at the interface.

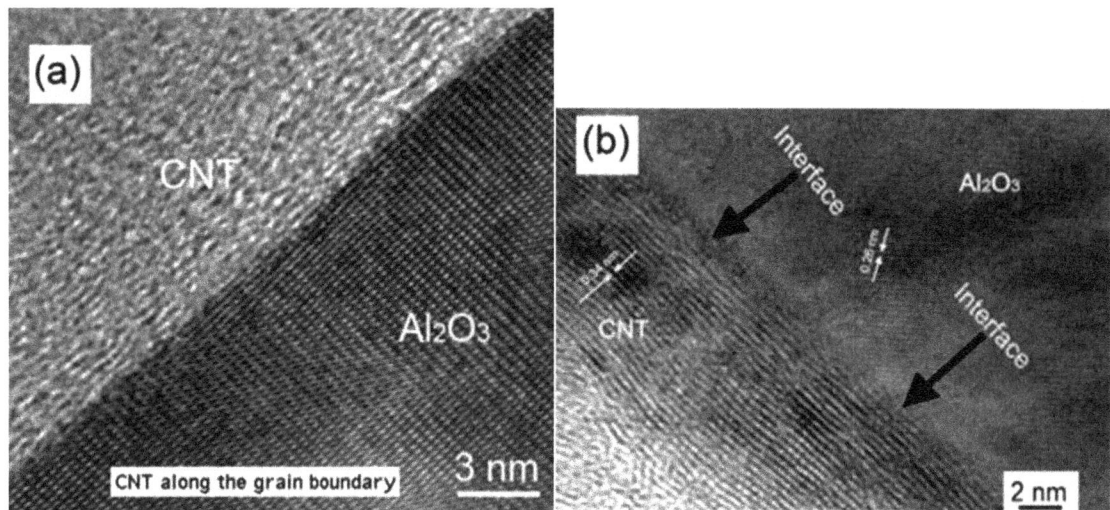

Figure 4. (**a,b**) High-resolution TEM images showing CNT/ceramic interfaces. Adapted from References [15] and [32] with permissions. Copyright 2005 Advanced Study Center Co. Ltd. and 2010 Elsevier Ltd.

Owning to the multi-layer graphene structure of MWCNTs, the possibility of such chemical and physical reactions with the accommodation of nano-pits and eating of few outer layers for Al_2OC or Al_4C_3 formation can be justified. However, this may not be true for SWCNTs which contain only a single graphene layer while forming the tubular structure, even plenty of studies claimed tremendous improvements in ceramics properties [22,26]. This raises one big question as to being only one layer how it reacts with the matrix to form a good interface following the toughening mechanisms proposed above. Therefore, this mystery remains unresolved. The understanding of the nanostructure characteristics and the interfacial relationship between SWCNTs and the ceramic matrices is far from satisfactory, which opens new windows of potential research in this advanced area of nanotechnology [18,26,40,61].

3. Graphene Reinforced Ceramic Nanocomposites

3.1. Raw Materials

As a cousin of CNTs, the 2D graphene bears many similarities to CNTs in terms of nanocomposites application. For bulk engineering nanocomposite applications which require large volume amount, a few layered graphene platelets or flakes, including the reduced graphene oxide (GO), are far more viable and economical than single layered graphene. Therefore, the term graphene in this context refers to graphene nanoplatelets (GNPs).

In composite applications, both the mechanical exfoliation and reduction from GO have been used successfully [89–94]. In the mechanical cleavage method, commercial graphite powder (Aldrich) has been milled intensively in high efficient attritor mill in the presence of ethanol for 10 h, then the produced GNPs were mixed with ceramic powder [89]. The second method uses the Hummers' process to produce GO, then using this water soluble GO to mix with ceramic powders [90]. In general, both mechanical milling and hummers method suffers from various sizes and thicknesses for the former due to lack of control on the milling energy and from surface structural damage for the latter originated from the oxidation [91], which will have negative effects on the final properties of the ceramic composites. Therefore, better quality control of the GNPs is of fundamentally importance for high quality nanocomposites development.

3.2. GNS Dispersions Processes

As discussed above for CNTs, mixing step is an equally challenging step in preparing graphene-reinforced ceramics composites. To avoid any damage and reduce agglomeration of GNPs will help to achieve high mechanical and physical properties. In essence, the dispersion of GNPs in fact is easier than CNTs, as the difficulties accompanied in CNT's dispersion such as high aspect ratio and van der Waals interactions which cause CNT bundling are absence for GNPs. In addition, high specific area and 2D geometry of GNPs offer better disperseability in ceramic matrices. As a younger cousin to CNTs, the gained knowledge for CNTs can generally be used as a reference for GNPs-reinforced ceramic composites. Thus in this context, focus will be mainly on the different features with comparison.

Wet powder mixing are successful to disperse CNTs in ceramic matrixes [92–95], whilst for GNPs the choice of solvents are much wider than processing CNTs. Isopropyl alcohol NMP, DMF have all be used to mix with various ceramic matrices such as Al_2O_3, Si_3N_4, and ZrO_2 powders. This drawback of this technique is energy consuming, and might cause damage to the GNP reinforcements. Colloid processing is a modified wet mixing process, and the key is to produce stabilized suspensions from GNTs and ceramic particles by changing their surface chemistry which facilitates homogeneous dispersion of GNPs. Anionic or cationic surfactants are generally used to alter the surface charge of GNPs, to positive or negative respectively, followed by adding them to a ceramic suspension with the same/opposite charges to form an homogenous ceramic-GNP dispersion. This hetero-coagulation process is a very effective route for well-dispersed ceramic composites [12,91]. Starting with GO, Wang et al. [20] used such electrostatic attractions between GO and Al_2O_3 particles to obtain homogenous dispersions of GO in Al_2O_3 powder first, followed by subsequent reduction of GO, who achieved a 53% and 13 orders of magnitude improvement in fracture toughness and conductivity. Walker et al. [25] used CTAB in both the GNPs and Si_3N_4 suspensions for mixing, and resulted in a 235% improvement in fracture toughness with only 1.5 vol% GNP addition.

3.3. Sintering Techniques

The densification of GNPs-reinforced ceramic nanocomposites also includes pressureless sintering, HP, SPS and HIP (hot-isostic pressing). The low temperature requirement and fast sintering rate advantages of SPS made it widely used for ceramic nanocomposites filled with carbon nano-fillers [20,24,43,95–98]. However, there are a few groups reported very good GNPs-reinforced ceramic nanocomposites based on HP densification [97–100]. For example, the GNPs-Si_3N_4 nanocomposites reported by Rutkowski et al. [99] showed improved thermal properties. After comparing the HP and SPS processes for GNPs-Al_2O_3 nanocomposites, Inam et al. [98] found out that the structural integrity of graphene from HP process is better than SPSed samples, with higher crystallinity, thermal stability and electrical conductivity, and was attributed to the thermally induced graphitization caused by longer sintering condition in a HP.

3.4. Structural Features, Mechanical Properties and Toughening Mechanisms

Toughening ceramic is one of the main research objectives for GNP nanocomposites, whist other benefits such as flexural strength and hardness can also be obtained. Using only 1.5 vol% the flexible 2D GNPs as a reinforcement for Si_3N_4, Walker et al. reported a 235% improvement in the toughness, and found GNPs anchoring with or wrapping around Si_3N_4 grains [25], thus blocking the crack propagation through the GNPs. This is the first time that such toughening mechanism was observed, and is a major different from the 1D CNTs. The same effective anchoring toughening was also confirmed by Liu et al. [24] in their GNPs/Al_2O_3 system, documented a 30.75% increase in flexural strength and a 27.20% increase in fracture toughness. These securely anchored GNPs around Al_2O_3 grains can form large area of interfaces with the matrix, increasing the interfacial friction, therefore the energy required for pulling out GNPs from the matrix will be greater than pulling out CNTs. They also successfully extended their process to a GNPs-reinforcing the ZrO_2-Al_2O_3 system using SPS [24], in comparison with CNTs. The authors believed that due to similar mechanical properties to CNTs, and better dispersability GNPs are an effective alternative for CNTs in ceramic composites. In Si_3N_4 matrix, Tapaszto et al. [100] showed that GNPs indeed outperformed CNTs. However, it should be noted that, due to the larger contact area between GNPs and the matrix grains, the interface quality plays a more important role in toughening the ceramics than CNTs and other reinforcement phases.

The different roles of CNTs and GNPs separately in ceramic matrices were well-documented, however within the same matrix could be more complex. A combination of the various advantages of different reinforcement phases, the very nature of the composite concept, could lead to superior properties, however this has rarely been investigated so far in ceramics. Very recently, Yazdani et al. [26] reported both 63% and 17% improvements in fracture toughness and flexural strength by using such a hybrid (MWCNTs + GNPs = GNTs) reinforcement phase in Al_2O_3 matrix. In their report, the role of GNPs and CNTs has been investigated separately. As evident in Figure 5, a large GNP rolled around Al_2O_3 grain due to its flexibility and produced large area of interface with Al_2O_3 matrix; therefore, it increased the required pull out energy during fracture and strengthened the grain boundaries so the fracture occurred through the Al_2O_3 grains rather than along the grain boundaries, whist MWCNTs contributed more to the bridging effect due to their higher aspect ratio. It is believed that MWCNTs can be stretched much longer than GNPs before collapsing during crack propagation. These roles are complementary with each other at appropriate concentrations, allowing for absorbing more energy during crack propagations.

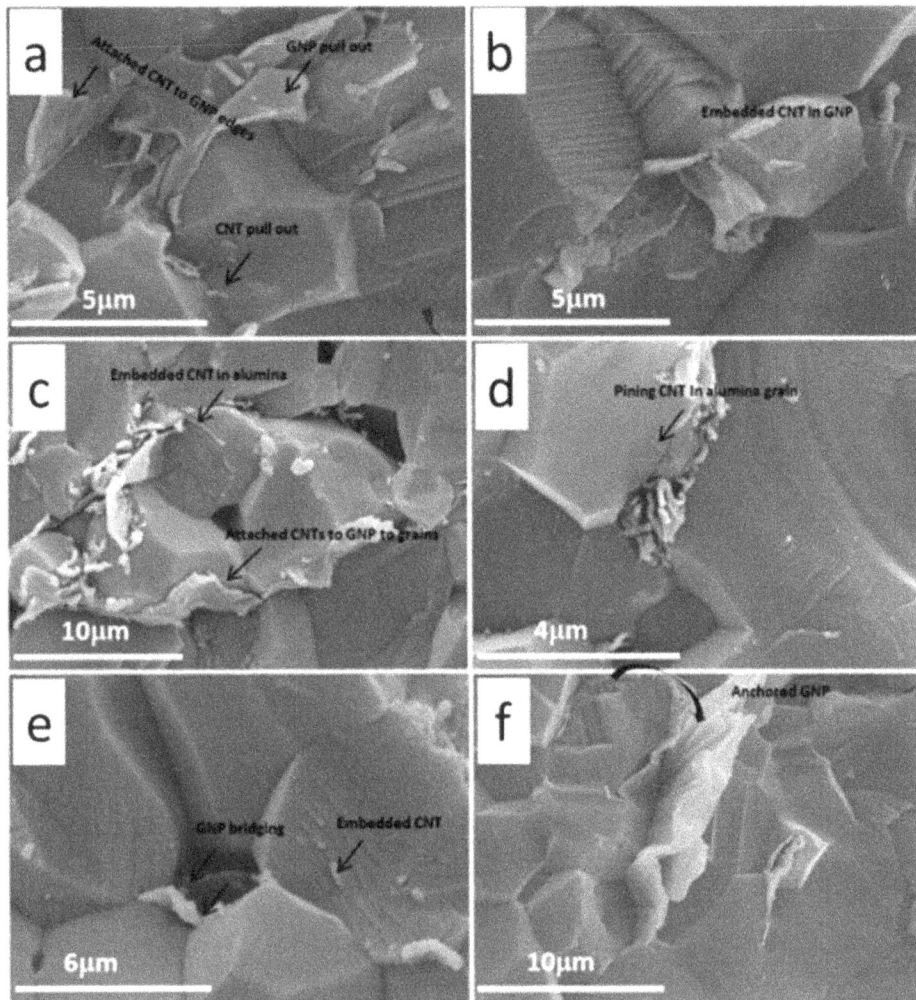

Figure 5. SEM images from fractured surface of GNT-Al$_2$O$_3$ nanocomposites with various GNP/CNT ratio, (**a** and **b**); Al$_2$O$_3$-(0.5 wt% GNP + 1 wt% CNT), (**c–e**); Al$_2$O$_3$-(0.5 wt% GNP + 0.5 wt% CNT), (**f**); Al$_2$O$_3$-0.5 wt% GNP. Adapted from Reference [26] with permission. Copyright 2014 Elsevier Ltd.

4. Potential Applications

Owing to the improved fracture toughness and ancillary benefits of electrical and thermal properties, ceramics reinforced with CNTs and GNPs are promising for numerous prospective applications in the field of photonics, biomedical, automotive and aerospace engineering. Firstly, associated with the enhanced mechanical performance of Al$_2$O$_3$, the significantly improved wear resistance property of these composites could be suitable for a number of wear and sliding applications in automobile industry like cylinder lines, valve seat and piston rings [101]. Secondly, the SiC, Si$_3$N$_4$ and BaTiO$_3$ systems filled with CNTs made them suitable for structural applications, such as bearings, seals, armour, liners, nozzles and cutting tools. Thirdly, the thermally and chemically stable ceramic composites could revise their high thermal conductivity and be suitable for high temperature components such as in jet engine and brake disks for aircrafts [102]. Further, CNTs/GNPs can also convert ceramics into functional materials for aerospace and automobile industries, such as knock sensors, seat pressure sensors, temperature sensors, oil sensors, impact sensors and road surface sensors, whilst the outstanding electrical properties of CNTs/GNPs make Al_2O_3 ceramic attractive for specific applications like heating elements, electrical igniters, electromagnetic/antistatic shielding of electronic components, electrode for fuel cells, crucibles for vacuum induction furnaces and electrical feed through [44,74,86,103–105]. Table 3 summarises the potential industries may have benefits from ceramic nanocomposites reinforced with CNTs and graphene. As the research is progressing in this important area, novel CNTs-reinforced ceramics with stunning properties are expected and may substituted several automobile and aerospace components in future furthermore, owning functionalities these have potential for third generation nanodevices.

Table 3. Potential application of key ceramics nanocomposites reinforced with CNTs and graphene.

References	Ceramic matrix	Reinforcing agent	Key properties	Parts/Components	Potential industries
[101]	Al_2O_3	CNTs/graphene	Wear resistance, high toughness, electrical properties, thermal properties	Cutting tools, corrosion/erosion resistance pipes, electrical contacts, armour plates	Automobile, petrochemical industry, electric component manufacturing, defence industry
[106]	Si_3N_4	CNTs/graphene	Excellent mechanical, chemical, and thermal properties	Gas turbines, aircraft engine components and bearings	Power generation, aerospace, automobile sector
[107]	$BaTiO_3$	CNTs/graphene	Ferroelectrics, piezoelectric and colossal magnetoresistor properties	Electric generator, computer hard disks, sensors	Renewable energy, power generation, electronic, computer manufacturing, data storage, aerospace industry
[108–110]	ZrO_2	CNTs/graphene	High mechanical properties, excellent fracture toughness, elevated temperature stability, high breakdown electrical field and large energy bandgap	Solid oxide fuel cells, oxygen sensors and ceramic membranes	Renewable energy, chemical industry, water desalination sectors
[111–113]	TiN and FeN	CNTs/graphene	Excellent electrical properties	Capacitors, electronic conductor in electronic devices	Electrochemical industry, power and electronic sector, aerospace and automobile industries
[114]	Mulite	CNTs/graphene	High in electric and optical properties	Sensor	Electronic industry, aerospace sector and automobile industry

5. Conclusions

Advanced in the ceramics reinforced with carbon nanostructures (CNTs and graphene) have been thoroughly reviewed. Successes in the purification and dispersions of MWCNTs are somehow satisfactory, however SWCNTs need further research and standards for CNT dispersion are vital for addressing the quality and reliability with confidence. Microwave sintering has potential for producing dense nanocomposites and may eliminate the CNT damage problem associated with the hot-pressing, and by adopting to standard testing methods fracture toughness discrepancies could be reduced. CNTs-reinforced ceramics follow the combined advanced toughening mechanisms of CNT's stretching/uncoiling and the classical fibre pullout theory, as an energy dissipating process. Rough surface and nanopits of MWCNTs explain the strong interface connections with ceramic matrix and the confirmation of the formation of intermediate Al_2OC or Al_4C_3 phases at the interface further strengthens these explanations. Conclusively, problems of reinforcing MWCNTs into ceramics have been solved to some extend; however, the addition of SWCNTs still carries questions. Despite challenges and controversial issues, CNTs have successfully enhanced the toughness and other properties of brittle ceramics and converted them into useful materials for next generation applications.

It is clear that graphene can play an important role as filler in ceramics according to publications. In addition to the exceptional mechanical properties of GNPs which are similar to CNTs, researches have shown that GNPs can be more easily dispersed in ceramic matrix than CNTs which is the key challenge in preparing ceramic composites. Additionally its 2D and flexible microstructure introduced a new toughening mechanism in the ceramic matrix (anchoring around the grain) that could significantly absorb energy against crack propagation and delay the fracture. However, work on graphene ceramic composites is in its early stages and there are still considerable works that need to be done in order to optimise their processing, microstructure and interfacial properties to obtain better multifunctional properties from graphene-ceramic composites.

Acknowledgments

Iftikhar Ahmad gratefully acknowledges the technical and financial support of Research Center of College of engineer, Deanship of Scientific Research, King Saud University, Riyadh, Kingdom of Saudi Arabia.

Author Contributions

Iftikhar Ahmad contributed to the carbon nanotube part, Bahareh Yazdani contributed to the graphene part, and Yanqiu Zhu involved in all stages of the article preparation.

Conflicts of Interest

The authors declare no conflict of interest.

References

1. Niihara, K. New design concept of structural ceramics-ceramics nanocomposites. *J. Ceram. Soc. Jpn.* **1991**, *99*, 974–982.

2. Osayande, L.; Okoli, O.I. Fracture toughness enhancement for Al_2O_3 system: A review. *Int. J. Appl. Ceram. Technol.* **2008**, *5*, 313–323.

3. Ohnabe, H.; Masaki, S.; Sasa, T. Potential application of ceramics matrix composites to aero-engine components. *Compos. A* **1999**, *30*, 489–496.

4. Llorca, J.; Elices, M.J.; Celemin, A. Toughness and microstructural degradation at high temperature in SiC fiber-reinforced ceramics. *Acta Mater.* **1998**, *46*, 2441–2453.

5. Yongqing, F.; Gu, Y.W.; Hejun, D. SiC whisker toughened Al_2O_3-(Ti,W)C ceramic matrix composites. *Scr. Mater.* **2001**, *44*, 111.D–116.D.

6. Garcıa, E.; Schicker, S.; Bruhn, J.; Janssen, R.; Claussen, N. Processing and mechanical properties of pressureless-sintered niobium—Al_2O_3-matrix composites. *J. Am. Ceram. Soc.* **1998**, *81*, 429–432.

7. Padture, N.P. Multifunctional composites of ceramics and single-walled carbon nanotubes. *Adv. Mater.* **2009**, *21*, 1767–1770.

8. Sheldon, B.W.; Curtin, W.A. Nanoceramics composites tough to test. *Nat. Mater.* **2004**, *3*, 505–506.

9. Peigney, A. Tougher ceramics with nanotubes. *Nat. Mater.* **2003**, *2*, 15–16.

10. Corral, E.; Bell, N.; Stuecker, J.; Perry, N.; Garay, J.; Barrera, E.V. Engineered nanostructures for multifunctional single-walled carbon nanotube reinforced silicon nitride nanocomposites. *J. Am. Ceram. Soc.* **2008**, *91*, 3129–3137.

11. Fan, J.; Zhao, D.; Song, J. Preparation and microstructure of multi-walled carbon nanotubes toughened Al_2O_3 composite. *J. Am. Ceram. Soc.* **2006**, *89*, 750–753.

12. Yamamoto, G.; Omori, M.; Hashida, T.; Kimura, H. A novel structure for carbon nanotube reinforced Al_2O_3 composites with improved mechanical properties. *Nanotechnology* **2008**, *19*, 315708.

13. Xia, Z.H.; Lou, J.; Curtin, W.A. A multiscale experiment on the tribological behavior of aligned carbon nanotube/ceramic composites. *Scr. Mater.* **2008**, *58*, 223.

14. Wie, T.; Fan, Z.; Wie, F. A new structure for multi-walled carbon nanotubes reinforced Al_2O_3 nanocomposite with high strength and toughness. *Mater. Lett.* **2008**, *62*, 641–644.

15. Zhan, G.D.; Mukherjee, A. Processing and characterization of nanoceramic composites with interesting structural and functional properties. *Rev. Adv. Mater. Sci.* **2005**, *10*, 185–196.

16. Costa, J.; Flacker, A.; Nakashima, M.; Fruett, F.; Zampieri, M.; Longo E. Integration of microfabricated capacitive bridge and thermistor on the alumina substrates. ECS *Trans.* **2012**, *49*, 451–458.

17. Martin, C.A.; Lee, G.F.; Fedderly, J.J. Composite Armor System Including a Ceramic-Embedded Heterogeneously Layered Polymeric Matrix. *U.S. Patent 8387510 B1*, 2013.

18. Baron, B.; Kumar, C.; le Gonidec, G.; Hampshire, S. Comparison of different alumina powders for the aqueous processing and pressureless sintering of Al_2O_3-SiC nanocomposites. *J. Eur. Ceram. Soc.* **2002**, *22*, 1543–1552.

19. Ipek, M.; Zeytin, S.; Bindal, C. An evaluation of Al₂O₃-ZrO₂ composites produced by coprecipitation method. *J. Alloys Compd.* **2011**, *509*, 486–489.

20. Wang, K.; Wang, Y.F.; Fan, Z.J.; Yan, J.; Wei, T. Preparation of composites by spark plasma sintering. *Mater. Res. Bull.* **2011**, *46*, 315–318.

21. Ramirez, C.; Garzón, L.; Miranzo, P.; Osendi, M.; Ocal, C. Electrical conductivity maps in graphene nanoplatelet/silicon nitride composites using conducting scanning force microscopy. *Carbon* **2011**, *49*, 3873–3880.

22. Balazsi, C. Silicon nitride composites with different nanocarbon additives. *J. Korean Ceram. Soc.* **2012**, *49*, 352–362.

23. Kvetkova, L.; Duszova, A.; Hvizdos, P.; Dusza, J.; Kun, P.; Balazsi, C. Fracture toughness and toughening mechanisms in graphene platelet reinforced Si₃N₄ composites. *Scr. Mater.* **2012**, *66*, 793–796.

24. Liu, J.; Yan, H.; Reece, M.J.; Jiang, K. Toughening of zirconia/alumina composites by the addition of graphene platelets. *J. Eur. Ceram. Soc.* **2012**, *32*, 4185–4193.

25. Walker, L.S.; Marotto, V.R.; Rafiee, M.A.; Koratkar, N.; Corral, E.L. Toughening in graphene ceramic composites. *ACS Nano* **2011**, *5*, 3182–3190.

26. Yazdani, B.; Xia, Y.; Ahmad, I.; Zhu, Y. Graphene and carbon nanotube (GNT)-reinforced alumina nanocomposites. *J. Eur. Ceram. Soc.* **2015**, *35*, 179–186.

27. Fan, Y.; Wang, L.; Li, J.; Sun, S.; Chen, F.; Chen, L.; Jiang, W. Preparation and electrical properties of graphene nanosheet/Al₂O₃ composites. *Carbon* **2010**, *48*, 743–1749.

28. Ebbesen, T.W.; Ajyan, P.M. Large scale synthesis of carbon nanotubes *Nature* **1992**, *358*, 220–222.

29. Hiura, H.; Ebbesen, T.W.; Tanigaki, K. Opening and purification of carbon nanotubes in high yields. *Adv. Mater.* **1995**, *7*, 275–276.

30. Tohji, K.; Takashaki, H.; Nishina, Y. Purification procedure for single-walled nanotubes. *J. Phys. Chem.* **1997**, *B101*, 1974–1978.

31. Ahmad, I.; Islam, M.; Almajid, A.; Yazdani, B.; Zhu, Y.Q. Investigation of yttria-doped Al₂O₃ nanocomposites reinforced by multi-walled carbon nanotubes. *Ceram. Int.* **2014**, *40*, 9327–9335.

32. Ahmad, I.; Kennedy, A.; Zhu, Y.Q. Carbon nanotubes reinforced Al₂O₃ nanocomposites: Mechanical properties and interfacial investigations. *J. Comput. Sci. Technol.* **2010**, *70*, 1199–1206.

33. Zhang, S.C.; William, G.; Hilmas, G.E.; Edward, J.Y. Pressureless sintering of carbon nanotube-Al₂O₃ composites. *J. Eur. Ceram. Soc.* **2010**, *30*, 1373–1380.

34. Sarkar, S.; Das, P.K. Microstructure and physic-mechanical properties of pressure-less sintered multi-walled carbon nanotubes/Al₂O₃ nanocomposites. *Ceram. Int.* **2012**, *38*, 423–432.

35. Michalek, M.; Lkova, M.; Sedla, J.; Galusek, D. Al₂O₃/MWCNTs composites by aqueous slip casting and pressureless sintering. *Ceram. Int.* **2013**, *L39*, 6543–6550.

36. Wang, J.; Kou, H.; Liu, X.; Pan, Y.; Guo, J. Reinforcement of mullite matrix with multi-walled carbon nanotubes. *Ceram. Int.* **2007**, *33*, 719–722.

37. Pasupuleti, S.; Peddetti, R.; Halloran, J.P. Toughening behavior in a carbon nanotube reinforced silicon nitride composite. *Mater. Sci. Eng.* **2008**, *A491*, 224–229.

38. Tian, W.; Kan, Y.; Wang, P. Effect of carbon nanotubes on the properties of ZrB₂-SiC ceramics. *Mater. Sci. Eng.* **2008**, *487*, 568–573.

39. Huang, Q.; Gao, L.; Sun, J. Effect of adding carbon nanotubes on microstructure, phase transformation and mechanical properties of BaTiO₃ ceramics. *J. Am. Ceram. Soc.* **2005**, *88*, 3515–3518.

40. Kumari, L.; Zhang, T.; Du, G.H.; Li, W.Z.; Wang, Q.W.; Datye, A.; Wu, K.H. Synthesis, microstructure and electrical conductivity of carbon nanotube–alumina nanocomposites. *Ceram. Int.* **2009**, *35*, 1775–1781.

41. Echeberria, J.; Rodríguez, N.; Bocanegra-Bernal, M.H. Hard and tough carbon nanotube-reinforced zirconia-toughened Al₂O₃ composites prepared by spark plasma sintering. *Carbon* **2012**, *50*, 706–717.

42. Kim, S.W.; Chung, W.S.; Sohn, K.S.; Son, C.Y.; Lee, S. Improvement of flexure strength and fracture toughness in alumina matrix composites reinforced with carbon nanotubes. *Mater. Sci. Eng.* **2009**, *A517*, 293–299.

43. Bakhsh, N.; Khalid, F.A.; Hakeem, A.S. Effect of sintering temperature on densification and mechanical properties of pressureless sintered CNT–Al₂O₃ nanocomposites. *Mater. Sci. Eng.* **2014**, *60*, 012059.

44. Li, T. Improving the antistatic ability of polypropylene fibers by inner antistatic agent filled with carbon nanotubes. *Comput. Sci. Tech.* **2004**, *64*, 2089–2096.

45. Hanzel, O.; Sedlácek, J.; Sajgalík, P. New approach for distribution of carbon nanotubes in Al₂O₃ matrix. *J. Eur. Ceram. Soc.* **2014**, *34*, 1845–1851.

46. Michalek, M.; lkova, M.; Sedla, J.; Galusek, D. Mechanical properties and electrical conductivity of Al₂O₃/MWCNT and Al₂O₃/zirconia/MWCNT composites. *Ceram. Int.* **2014**, *40*, 1289–1295.

47. Poyato, R.; Gallardo-López, A.; Gutiérrez-Mora, F.; Morales-Rodríguez, A.; Muñoz, A.; Domínguez-Rodríguezb, A. Effect of high SWNT content on the room temperature mechanicalproperties of fully dense 3YTZP/SWNT composites. *J. Eur. Ceram. Soc.* **2014**, *34*, 1571–1579.

48. Kyotani, T.; Tsai, L.F.; Tomita, A. Preparation of ultrafine carbon tubes in nanochannels of an anodic aluminum oxide film. *Chem. Mater.* **1996**, *8*, 2109–2113.

49. Sui, Y.C.; Acosta, D.R.; Cui, B.Z. Structure, thermal stability, and deformation of multibranched carbon nanotubes synthesized by CVD in the AAO template. *J. Phys. Chem.* **2001**, *B105*, 1523–1527.

50. Bae, E.J.; Choi, W.B.; Park, G.S.; Song, S. Selective growth of carbon nanotubes on pre-patterned porous anodic aluminum oxide. *Adv. Mater.* **2002**, *14*, 277–279.

51. Parhama, H.; Bates, S.; Xia, Y.; Zhu, Y. A highly efficient and versatile carbon nanotube/ceramic composite filter. *Carbon* **2013**, *54*, 215–223.

52. Parhama, H.; Kennedy, A.; Zhu, Y. Preparation of porous Al₂O₃—Carbon nanotube composites via direct growth of carbon nanotubes. *J. Comput. Sci. Technol.* **2011**, *71*, 1739–1745.

53. Sun, J.; Gao, L. Development of a dispersion process for carbon nanotubes in ceramic matrix by hetero-coagulation. *Carbon* **2003**, *41*, 1063–1068.

54. Chan, B.; Seung, I. Fabrication of CNT-reinforced Al₂O₃ matrix nanocomposites by sol-gel. *Mater. Sci. Eng.* **2005**, *395*, 124–128.

55. Sun, J.; Gao, L. Reinforcement of Al₂O₃ matrix with multi-walled CNT. *Ceram. Int.* **2004**, 893–896.

56. Gao, L.; Jiang, L.; Sun, J. Carbon nanotube-ceramic composites. *J. Electroceram.* **2006**, *17*, 51–55.

57. Coble, R.L. Diffusion Models for hot pressing with surface energy and pressure effects as driving forces. *J. Appl. Phys.* **1970**, *41*, 4798–4808.

58. Legorreta, G.; Estournes, C.; Peigney, A.; Weibel, A.; Flahaut, E.; Laurent, C. Spark-plasma-sintering of double-walled carbon nanotube–magnesia nanocomposites. *Scr. Mater.* **2009**, *60*, 741–744.

59. Ahmad, I.; Dar, M.A. Structure and properties of Y_2O_3-doped Al_2O_3-MWCNT nanocomposites prepared by PL-sintering and hot-pressing. *J. Mater. Eng. Perform.* **2014**, *23*, 2110–2119

60. Ghobadi, H.; Ali, N.; Ebadzade, T.; Sadeghian, Z.; Barzegar-Bafrooei, H. Improving CNT distribution and mechanical properties of MWCNT reinforced Al_2O_3 matrix. *Mater. Sci. Eng.* **2014**, *A617*, 110–114.

61. Tatami, J.; Katashima, T.; Komeya, K.; Meguro, T.; Wakihara, T. Electrically conductive CNT-dispersed silicon nitride ceramics. *J. Am. Ceram. Soc.* **2005**, *88*, 2889–2893.

62. Katz, J.D.; Blake, R.D. Microwave sintering of multiple alumina and composite components. *J. Am. Ceram. Soc.* **1991**, *70*, 1304.

63. Sheppard, L.M. Firing technology heats up for the 90s. *J. Am. Ceram. Soc.* **1988**, *67*, 1656.

64. De, A.; Ahmad, I.; Whitney, E.D.; Clark, D.E. Microwaves theory and applications. *Mater. Process.* **1991**, *21*, 329–339.

65. Fujitsu, S.; Ikegami, M.; Hyashi, T. Sintering of partially stabilized zirconia by microwave heating using $ZnO–MnO_2–Al_2O_3$ plates in a domestic microwave oven. *J. Am. Ceram. Soc.* **2000**, *83*, 2085–2087.

66. Ahmad, I.; Cao, H.; Chen, H.; Zhao, H.; Kennedy, A.; Zhu, Y. Carbon nanotube toughened aluminium oxide nanocomposites. *J. Eur. Ceram. Soc.* **2009**, *30*, 865–873.

67. Valecillos, M.C.; Hirota, M.; Brite, M.E.; Hirao, K. Sintering of alumina by 2.45 GHz microwave heating. *J. Eur. Ceram. Soc.* **2004**, *24*, 387–391.

68. Thostenson, P.G.; Karandikar, T.W. Fabrication and characterization of reaction bonded silicon carbide/carbon nanotube composites. *J. Appl. Phys.* **2005**, *38*, 3962–3965.

69. Wang, Y.; Voronin, G.A.; Zerda, T.W.; Winiarski, A. SiC–CNT nanocomposites: High pressure reaction synthesis and characterization. *J. Phys. Condens. Matter.* **2006**, *18*, 275–282.

70. Kristen, H.B.; Gary, L.M.; Dinesh, K.A. Microwave sintering of alumina at 2.45 GHz. *J. Am. Ceram. Soc.* **2003**, *86*, 1307–1312.

71. Lopez, A. Hardness and flexural strength of single-walled carbon nanotubes/Al_2O_3 composites. *J. Mater. Sci.* **2014**, *49*, 7116–7123.

72. Zaman, A.C.; Kaya, F.; Kaya, C. OH and COOH functionalized single walled carbon nanotubes-reinforced alumina ceramic nanocomposites. *Ceram. Int.* **2012**, *38*, 1287–1293.

73. Estili, M.; Kawasaki, A.; Sakka, Y. Highly concentrated 3D macrostructure of individual carbon nanotubes in a ceramic environment. *Adv. Mater.* **2012**, *24*, 4322–4326.

74. Martinlli, J.R.; Sene, F.F. Electrical resistivity of ceramic-metal composite materials: Application in crucibles for induction furnaces. *Ceram. Int.* **2000**, *26*, 325–335.

75. Puchy, V.; Hvizdos, P.; Dusza, J.; Kovac, F.; Inam, F.; Reece, M.J. Wear resistance of Al_2O_3–CNT ceramic nanocomposites at room and high temperature. *Ceram. Int.* **2013**, *39*, 5821–5826.

76. Lee, K.; Mo, C.B.; Park, S.B.; Hong, S.H. Mechanical and electrical properties of multiwalled CNT–alumina nanocomposites prepared by a sequential two-step processing of ultrasonic spray pyrolysis and spark plasma sintering. *J. Am. Cream. Soc.* **2011**, *94*, 3774–3779.

77. Thomson, K.E.; Jiang, D.; Yao, W.; Ritchie, R.O.; Mukherjee, A.K. Characterization and mechanical testing of alumina-based nanocomposites reinforced with niobium and/or carbon nanotubes fabricated by spark plasma sintering. *Acta Mater.* **2012**, *60*, 622–632.

78. Ahmad, I.; Kennedy, A.; Zhu, Y. Wear resistance properties of multi-walled carbon nanotubes reinforced Al₂O₃ nanocomposite. *Wear* **2010**, *269*, 71–78.

79. Zhan, G.D.; Mukherjee, A.K. Carbon nanotube reinforced alumina-based ceramics with novel mechanical, electrical, and thermal properties. *Int. J. Appl. Ceram. Technol.* **2004**, *1*, 161–171.

80. Sarkar, S.; Das, P.K. Effect of sintering temperature and nanotube concentration on microstructure and properties of carbon nanotube/alumina nanocomposites. *Ceram. Int.* **2014**, *40*, 7449–7458.

81. Bi, S.; Su, X.; Hou, G.; Liu, C.; Song, W.L.; Cao, M.S. Electrical conductivity and microwave absorption of shortened multi-walled carbon nanotube/alumina ceramic composites. *Ceram. Int.* **2013**, *39*, 5979–5983.

82. Bakshi, S.R.; Balani, K.; Agarwal, A. Thermal conductivity of plasma-sprayed aluminum oxide—Multiwalled carbon nanotube composites. *J. Am. Cream. Soc.* **2008**, *91*, 942–947.

83. Laurent, C.; Peigney, A.; Rousset, A. Carbon nanotubes-Fe-Al₂O₃ nanocomposites. Part II: Microstructure and mechanical properties of the hot-pressed composites. *J. Eur. Ceram. Soc.* **1998**, *18*, 2005–2013.

84. Siegel, R.W.; Chang, S.K.; Ajayan, P.M. Mechanical behaviour of polymer and ceramic matrix nanocomposite. *Scr. Mater.* **2001**, *44*, 2061–2064.

85. Hoagland, R.G. A treatment of inelastic deformation around a crack tip due to micro cracking. *J. Am. Ceram. Soc.* **1980**, *63*, 404–410.

86. Kawamura, H.; Yamamoto, S. *Improvement of Diesel Engine Startability by Ceramic Glow Plug Start System*; Society of Automotive Engineers: New York, NY, USA; 1983.

87. Zheng, G.; Sano, H.; Cheng, H.M. A TEM study of the microstructure of carbon fiber/polycarbosilane-derived SiC composites. *Carbon* **1999**, *37*, 2057–2062.

88. Estili, M. The homogeneous dispersion of surfactantless, slightly disordered, crystalline, multiwalled carbon nanotubes in α-alumina ceramics for structural reinforcement. *Acta Mater.* **2008**, *56*, 4070–4079.

89. Knieke, C.; Berger, A.; Voigt, M.; Taylor, R.N.K.; Röhrl, J.; Peukert, W. Scalable production of graphene sheets by mechanical delamination. *Carbon* **2010**, *48*, 3196–3204.

90. Hummers, J.; William, S.; Offeman, R.E. Preparation of graphitic oxide. *J. Am. Chem. Soc.* **1958**, *80*, 1339–1339.

91. Inam, F.; Yan, H.; Reece, M.J.; Peijs, T. Dimethylformamide: an effective dispersant for making ceramic-carbon nanotube composites. *Nanotechnology* **2008**, *19*, 195710.

92. Balázsi, C.; Wéber, F.; Arató, P.; Fényi, B.; Hegman, N.; Kónya, Z.; Kiricsi, I.; Vértesy, Z.; Biró, L.P. Development of CNT-silicon nitrides with improved mechanical and electrical properties. *Adv. Sci. Technol.* **2006**, *45*, 1723–1728.

93. Milsom, B.; Viola, G.; Gao, Z.; Inam, F.; Peijs, T.; Reece, M.J. The effect of carbon nanotubes on the sintering behaviour of zirconia. *J. Eur. Ceram. Soc.* **2012**, *32*, 4149–4156.

94. Guo, S.; Sivakumar, R.; Kitazawa, H.; Kagawa, Y. Electrical properties of silica-based nanocomposites with multiwall carbon nanotubes. *J. Am. Ceram. Soc.* **2007**, *90*, 1667–1670.

95. Liu, J.; Yan, H.; Jiang, K. Mechanical properties of graphene platelet-reinforced alumina ceramic composites. *Ceram. Int.* **2013**, *39*, 6215–6221.

96. Ramirez, C.; Osendi, M.I. Characterization of graphene nanoplatelets-Si$_3$N$_4$ composites by Raman spectroscopy. *J. Eur. Ceram. Soc.* **2013**, *33*, 471–477.

97. Kvetková, L.; Duszová, A.; Kašiarová, M.; Orčáková, F.; Dusza, J.; Balázsi, C. Influence of processing on fracture toughness of Si$_3$N$_4$ + graphene platelet composites. *J. Eur. Ceram. Soc.* **2013**, *33*, 2299–2304.

98. Inam, F.; Vo, T.; Bhat, B.R. Structural stability studies of graphene in sintered ceramic nanocomposites. *Ceram. Int.* **2014**, *40*, 16227–16233.

99. Rutkowski, P.; Stobierski, L.; Górny, G. Thermal stability and conductivity of hotpressed Si$_3$N$_4$–graphene composites. *J. Therm. Anal. Calorim.* **2014**, *116*, 321–328.

100. Tapaszto, O.; Kun, P.; Weber, F.; Gergely, G.; Balazsi, K.; Pfeifer, J.; Arato, P.; Kidari, A.; Hampshire, S.; Balazsi, C. Silicon nitride based nanocomposites produced by two different sintering methods. *Ceram. Int.* **2011**, *37*, 3457–3461.

101. Evans, A.G. Perspective on the development of high-toughness ceramics. *J. Am. Ceram. Soc.* **1990**, *73*, 187–206.

102. Ritchie, R.O. The quest for stronger tougher materials. *Science* **2008**, *320*, 448–452.

103. Curtin, W.A.; Sheldon, B.W. CNT-reinforced ceramics and metals. *Mater. Today* **2004**, *7*, 44–49.

104. Kramer, P.; White, K. Effect of sintering parameters and composition on the resistivity of a cermet used as an electrical feed through. *Ceram. Eng. Sci. Proc.* **1982**, *3*, 512–518.

105. Tajima, Y. Development of high performance silicon nitride ceramics and their applications. *Mater. Res. Soc. Symp. Proc.* **1993**, *287*, 98–201.

106. Komeya, H.K. Development of ceramic antifriction bearing. *JSAE Rev.* **1986**, *7*, 72–79.

107. Huang, Q.; Gao, L. Manufacture and electrical properties of multiwalled carbon nanotube/BaTiO$_3$ nanocomposite ceramics. *J. Mater. Chem.* **2004**, *14*, 2536–2541.

108. Lee, S.Y.; Kim, H.; McIntyre, P.C.; Saraswat, K.C.; Byun, J.S. Atomic layer deposition of ZrO$_2$ on W for metal-insulator-metal capacitor application. *Appl. Phys. Lett.* **2003**, *823*, 2874–2876.

109. Dusza, J.; Tomasek, K.; Blugan, G.; Kuebler, J. Microstructure and properties of carbon nanotube/zirconia composite. *J. Eur. Ceram. Soc.* **2008**, *28*, 1023–1027.

110. Patsalas, P.; Logothetidis, S. Optical, electronic, and transport properties of nanocrystalline titanium nitride thin films. *J. Appl. Phys.* **2001**, *90*, 4725–4734.

111. Carmalt, C.J.; Whaley, S.R.; Lall, P.S.; Cowley, A.H.; Jones, R.A. Titanium (IV) azido and imido complexes as potential precursors to titanium nitride. *J. Chem. Soc. Dalton Trans.* **1998**, *1998*, 553–558.

112. Janes, R.A.; Aldissi, M.; Kaner, R.B. Controlling surface area of titanium nitride using metathesis reactions. *Chem. Mater.* **2003**, *15*, 4431–4435.

113. Kim, S.; Kumta, P.N. Hydrazide sol–gel synthesis of nanostructured titanium nitride: Precursor chemistry and phase evolution. *J. Mater. Chem.* **2003**, *13*, 2028–2035.

114. Cao, M.S.; Wang, R.G.; Fang, X.Y.; Cui, Z.X.; Chang, T.J.; Yang, H.J. Preparing γ′-Fe4N ultrafine powder by twice-nitriding method. *Powder Technol.* **2001**, *115*, 96–98.

8

Rare Earth Ion-Doped Upconversion Nanocrystals: Synthesis and Surface Modification

Hongjin Chang [1], **Juan Xie** [2], **Baozhou Zhao** [1], **Botong Liu** [1], **Shuilin Xu** [1], **Na Ren** [1], **Xiaoji Xie** [1], **Ling Huang** [1,*] **and Wei Huang** [1,2,*]

[1] Key Laboratory of Flexible Electronics (KLOFE) and Institute of Advanced Materials (IAM), Jiangsu National Synergistic Innovation Center for Advanced Materials (SICAM), Nanjing Tech University (NanjingTech), Nanjing 211816, China; E-Mails: 460653533@njtech.edu.cn (H.C.); 823184209@njtech.edu.cn (B.Z.); liubotong201304@gmail.com (B.L.); xu_shuilin@njtech.edu.cn (S.X.); iamnren@njtech.edu.cn (N.R.); iamxjxie@njtech.edu.cn (X.X.)

[2] Key Laboratory for Organic Electronics and Information Displays and Institute of Advanced Materials (IAM), National Synergistic Innovation Center for Advanced Materials (SICAM), Nanjing University of Posts and Telecommunications, Nanjing 210023, China; E-Mail: 15050528303@163.com

* Authors to whom correspondence should be addressed; E-Mails: iamlhuang@njtech.edu.cn (L.H.); iamwhuang@njtech.edu.cn & iamwhuang@njupt.edu.cn (W.H.).

Academic Editor: Thomas Nann

Abstract: The unique luminescent properties exhibited by rare earth ion-doped upconversion nanocrystals (UCNPs), such as long lifetime, narrow emission line, high color purity, and high resistance to photobleaching, have made them widely used in many areas, including but not limited to high-resolution displays, new-generation information technology, optical communication, bioimaging, and therapy. However, the inherent upconversion luminescent properties of UCNPs are influenced by various parameters, including the size, shape, crystal structure, and chemical composition of the UCNPs, and even the chosen synthesis process and the surfactant molecules used. This review will provide a complete summary on the synthesis methods and the surface modification strategies of UCNPs reported so far. Firstly, we summarize the synthesis methodologies developed in the past decades, such as thermal decomposition, thermal coprecipitation, hydro/solvothermal, sol-gel, combustion, and microwave synthesis. In the second part, five main streams of surface modification strategies for converting hydrophobic UCNPs into hydrophilic ones are elaborated. Finally, we consider the likely directions of the future development and challenges of the synthesis and surface modification, such as the large-scale production and actual applications, stability, and so on, of the UCNPs.

Keywords: rare earth; nanocrystal; upconversion; synthesis; surface modification

---

1. Introduction

On 29 December 1959, Richard Feynman predicted at the annual American Physical Society meeting that: "*if we use a method to control the arrangement of things at a small scale, so that we can get a lot of features beyond imagination, also can see the performance of the material to produce rich change.*" The tiny material is what we call nanomaterial today, which is usually referred to as that in the three dimensional space, where at least one dimension (1D) is at the nanoscale range (1–100 nm) or materials are composed of them as a basic unit. As the size of the material decreases to the nanometer scale, the material will show a lot of new features that the bulk counterpart does not possess, such as the small size effect, surface effect of nanomaterials, quantum confinement effect, and macroscopic quantum tunneling effect [1–3]. In the past decades, nanomaterials have gained wide attention from all over the world, and have now been widely applied in many fields.

Rare earth (RE) elements possess unique electronic configuration where the 4f electrons are effectively shielded by the closely lied 5s and 5p subshells. Typically, the electron transitions of RE ions are mainly derived from the inner *4f-4f* or *4f-5d* transitions, and thus the spectroscopic properties of RE ions are barely perturbed by the local chemical microenvironment, which imparts RE compounds including the complexes' and nanomaterials' unique spectroscopic characteristics, such as rich energy levels, long luminescence decay time, narrow emission line, and high color purity in contrast to those of quantum dots and organic dyes [4–6]. Due to the preserved electron transitions, the emission wavelengths of each lanthanide ion largely depend on its own electronic configuration, and the combination (at certain ratio) of different lanthanide ions is also widely adopted to realize various luminescent materials with adjustable luminescence for different purposes.

If these RE ions were doped into proper nanocrystals, a series of new luminescent characteristics related with the original features of according RE ions may be observed. In addition, upconversion nanocrystals (UCNPs) have shown high chemical stability, biological compatibility, long luminescence lifetime, and tunable emission wavelength [7–9], making them tremendously exploited in biolabeling [10], bio-detection [11,12], bioimaging [13,14], FRET-based sensing [15], drug delivery [16], and volumetric 3D display [17].

Despite the substantially shielded transitions of RE ions, the luminescent properties of UCNPs are also affected by their size, shape, crystal structure, and chemical composition of the materials. For example, reduced particle size will cause increased surface area, which would introduce more defects on the nanoparticle surface and consequently hamper the luminescent efficiencies, though the small nanoparticles are more advantageous to the biological applications. To facile the applications of UCNPs, it is of primary necessity to develop according viable and robust methodologies to synthesize target nanocrystals with desired size, shape, crystal structure, chemical composition, and most importantly, the proper surface functional groups for anticipated applications.

In recent years, various attempts have been reported for synthesizing UCNPs in a controlled manner, including thermal decomposition, thermal coprecipitation, hydro/solvothermal, combustion, microwave, and so on. In the meantime, the nanocrystals, after surface modification, such as SiO_2 encapsulation, polymer encapsulation, ligand oxidation, and ligand exchange, can be easily coupled with DNA, protein, and other functional molecules, and facilitate expected applications. However, there are still some challenges in the synthesis of desired nanocrystals, and the surface modification usually involves extra experimental steps and lowers the luminescence efficiency at certain extent depending on the chosen method. Thus, a proper synthesis method and a suitable strategy for designed surface modification are highly desired. Herein, we attempt to provide a comprehensive overview of the state-of-the-art synthetic methods and the surface modification strategies for UCNPs reported in the past decades.

2. Synthetic Approaches

2.1. Thermal Decomposition

The thermal decomposition process, comprising of dissolution of organic and/or inorganic precursors in organic solvent with high-boiling point, is a traditional method for preparation of inorganic nanocrystals. The typical experimental procedure of thermal decomposition method is composed of: (1) a given amount of $RE(CF_3COO)_3$ precursors is added into a mixture of oleic acid (OA), 1-octadecene (OD), and sometimes oleylamine (OM) at room temperature; (2) the solution is heated to 165 °C for 30 min with vigorous magnetic stirring to remove water and oxygen under argon protection; (3) the solution is heated to high temperature (usually >300 °C) for a certain period of time under argon protection and the nanocrytals are then collected from the reaction mixture after cooling down to room temperature. Yan [18,19] first prepared high-monodisperse LaF_3 triangular nanoplates (Figure 1a) and hexagonal SmF_3 nanoparticles (Figures 1b,c) via the thermal decomposition process using $La(CF_3COO)_3$ and $Sm(CF_3COO)_3$ as precursors. Using this method, Capobianco [20] and Nann [21] synthesized $NaYF_4$ nanocrystals with narrow size-distribution. Later, this approach was extended as a common process to synthesize high-quality UCNPs including but not limited to $NaLaF_4$ [22], $NaGdF_4$ [23], $LiYF_4$ [24], KY_3F_{10} [25], $BaYF_5$ [26], REOF [19,27,28], and REOCl [29]. The reaction temperature, time, and the molar ratio of OA, OD, and sometimes OM in the reaction mixture have been demonstrated to exert different effects on the final nanocrystals. It should be noted that sometimes OM shall be introduced as a necessary component to adjust the reaction environment so that the final product with different morphologies and dimensions or a brand new product can be obtained [27–29].

The most prominent advantages of this method are that the products are of high quality, with pure crystal phase, and strong upconversion emission. However, this method also suffers from disadvantages including: (1) presynthesis of $RE(CF_3COO)_3$ precursors is typically required; (2) the decomposition of trifluoroacetates simultaneously produces toxic fluorinated and oxyfluorinated carbon species and thus careful handling of the reactions in fully ventilated chemical hood is required; (3) the anaerobic and water-free reaction environment further increases the operation difficulty. In the themolysis method, one of the key factors for achieving size-tunable and monodispersed UCNPs requires a proper selection of the coordinating ligands. Yan and co-worker reported that oleylamine ligand is a delicate buffer for F^- ions, the lighter the rare earth, the more OM it requires, owing to the fact that the basicity of the RE oxide gradually decreases along with the increasing of the atomic number of the RE series [19].

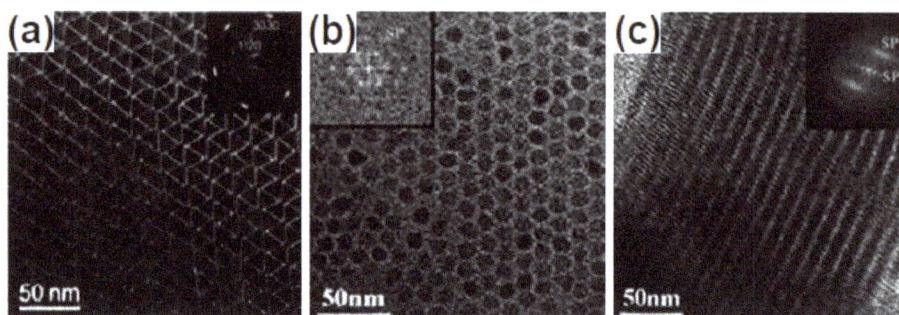

Figure 1. (**a**) TEM image of edge-to-edge super lattices of LaF_3 triangular nanoplates. Reproduced from [18]. Copyright 2005, American Chemical Society. (**b,c**) TEM images of edge-to-edge and face-to-face super lattices of SmF_3 hexagonal nanoplates, respectively. Reproduced from [19]. Copyright 2006, John Wiley and Sons.

2.2. Thermal Coprecipitation

Due to the limitations of the thermal decomposition method, thermal coprecipitation approach is developed and has now been used as one of the most convenient methods for UCNPs synthesis. The experimental procedure is generally composed of: (1) RE salts were mixed with a solution of OA, OD, and sometimes OM, at certain ratio, which was heated to 165 °C for 30 min and then cooled down to room temperature; (2) a methanol solution of NH_4F and AOH (A = Li, Na, K) was added to the mixture and stirred for 30 min; (3) after removal of the methanol and residual water by evaporation, the reaction mixture was heated to high temperature (usually >300 °C) under argon protection, which produces desired nanocrystals. The benefits of the coprecipitation method include operational simplicity, the lack of toxic by-products, and the wide application across various materials.

With continuous efforts, many kinds of UCNPs have been directly synthesized by this method. For example, in 2002, van Veggel and co-workers [30] have synthesized the down conversion LaF_3 nanocrystals doped with RE^{3+} (RE = Eu, Er, Nd, and Ho) ions. This approach was later expanded by Yi and Chow [31], who prepared upconversion LaF_3 nanophosphors with smaller particle size (5 nm). In addition to LaF_3, $LuPO_4$:Yb/Tm, $YbPO_4$:Er, $NaYF_4$:Yb/Er(Tm), $NaGdF_4$:Yb/Er, and $Y_3Al_{15}O_{12}$(YAG):Yb/Tm nanophosphors were synthesized using this method [5,32–34]. Recently, our group has synthesized the first Scandium-based fluoride nanocrystals Na_xScF_{3+x}:Ln using the coprecipitation method [35] (Figure 2). Interestingly, the crystal structure evolution from pure monoclinic Na_3ScF_6 to pure hexagonal $NaScF_4$ phase was observed by tuning the ratio of OA and OD (Figure 2a). In addition, the hexagonal $NaScF_4$:Yb/Er crystals emit strong red upconversion (665 nm) under 980 nm laser excitation, different from those of the traditional hexagonal $NaYF_4$:Yb/Er nanocrystals, which usually emit strong green (547 nm) upconversion luminescence (Figure 2b).

Despite the general usefulness and obvious improvement of the thermal coprecipitation method for UCNPs synthesis, it suffers from the long-time and continuous operation of the experimental process, which usually takes more than 5 h, including the removal of methanol solvent, water generated during the synthesis, and the controlled crystal growth at a certain high-temperature. Moreover, large scale synthesis of UCNPs using this method is still a great challenge.

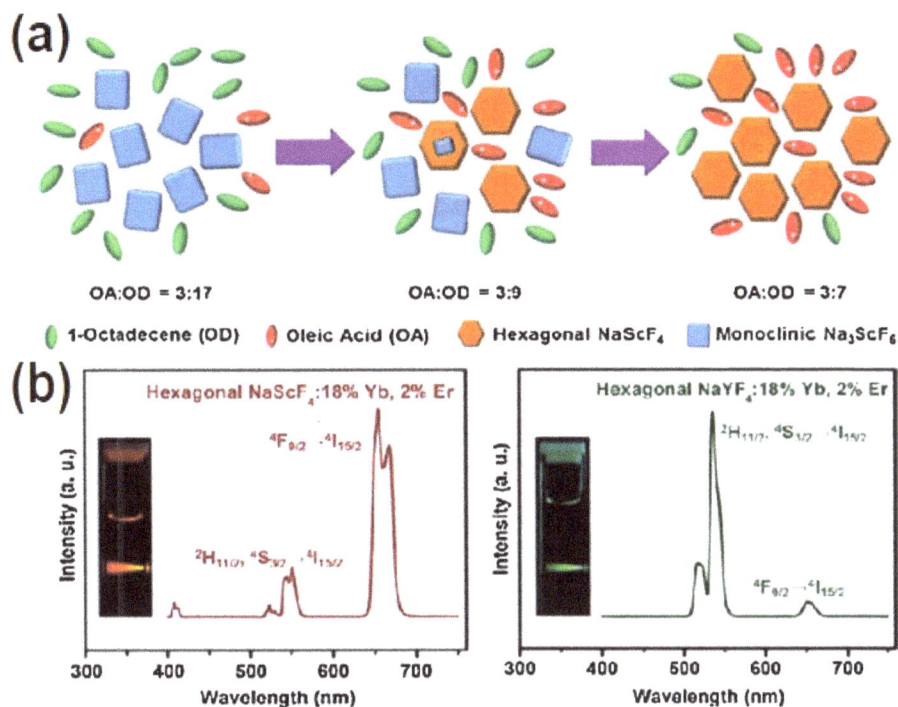

Figure 2. (**a**) Schematic illustration of the crystal structure evolution at varying polarities of the reaction medium. (**b**) Upconversion luminescence spectra of hexagonal-phase NaScF$_4$:Yb/Er and NaYF$_4$:Yb/Er nanocrystals. Reproduced from [35]. Copyright 2012, American Chemical Society.

2.3. Hydro/Solvothermal Synthesis

Hydro/solvothermal is a process of chemical reaction between negative ions and positive ions that usually precipitate from the solvent under high temperature and high pressure, generating nanoscale materials in the solvent after proper processing. This method has now become a widely-employed synthetic approach for UCNPs since it is easy to operate, does not require stringent operation of the experimental process, and moreover, the reaction temperature of hydro/solvothermal method is usually lower than those used for thermal decomposition and thermal co-precipitation synthesis for UCNPs.

The commonly used surfactants for UCNPs preparation include ethylenediamine tetraacetic acid (EDTA) [36,37], cetyltrimethylammonuim bromide (CTAB) [38], trisodium citrate (Na$_3$Cit) [39], linoleate acid [40], oleic acid [41–46]. As early as 2002, Li and coworkers [47–49] used RE nitrate and KOH to synthesize Ln(OH)$_3$ nanowires, nanotubes, and nanoparticles (Figure 3a–c). Long and uniform nanobelts were obtained when NaOH/KOH and RE acetates were used (Figures 3d,e). Later on, EDTA was used as complexing agent and CTAB as structure-directing agent to synthesize β-NaYF$_4$:Yb/Er nanotubes, nanorods, and nanospheres [34] (Figure 4a–d). Different from Li's work, Wang and co-workers [50] synthesized upconversion NaYF$_4$ nanocrystals using oleic acid-mediated hydrothermal synthesis (Figure 4e–h).

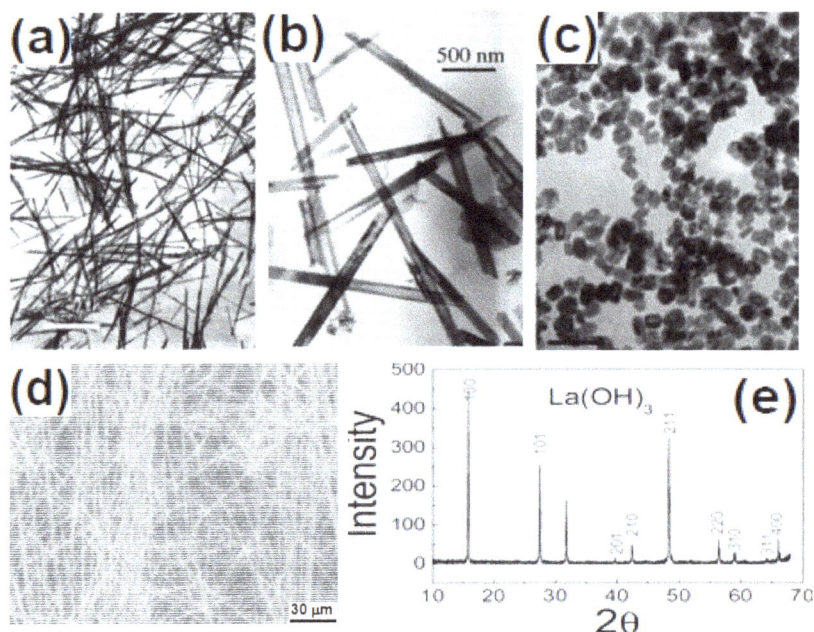

Figure 3. TEM images of (**a**) La(OH)$_3$ nanowires. Reproduced from [47]. Copyright 2002, John Wiley and Sons; (**b**) Y$_2$O$_3$ nanotubes. Reproduced from [48]. Copyright 2003, John Wiley and Sons; (**c**) LaF$_3$ nanoparticles. Reproduced from [49]. Copyright 2003, John Wiley and Sons; (**d**) La(OH)$_3$ nanobelt; (**e**) a typical XRD pattern of the as-synthesized La(OH)$_3$. Reproduced from [50]. Copyright 2007, John Wiley and Sons.

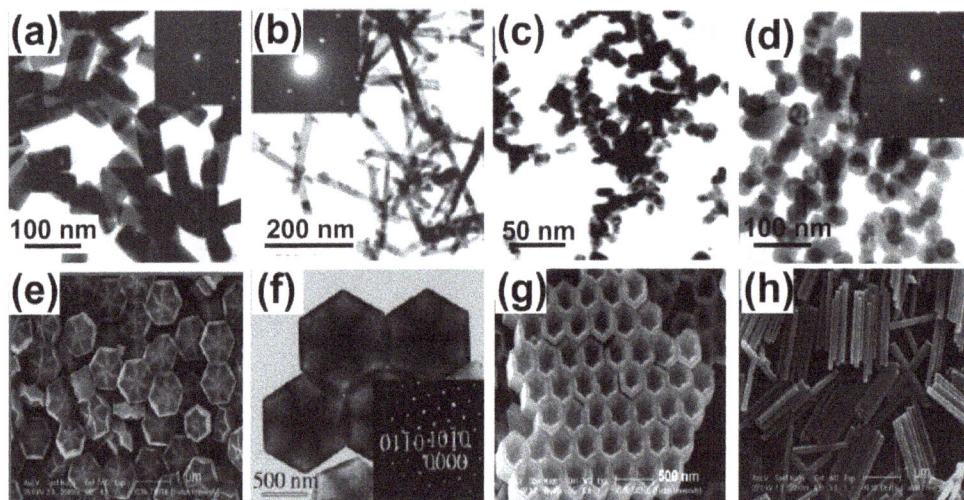

Figure 4. TEM and SEM images of NaYF$_4$:Yb/Er nanocrystals prepared under different hydro/solvothermal conditions. (**a,b**) TEM images of NaYF$_4$:Yb/Er nanocrystals synthesized in acetic acid and ethanol in the presence of CTAB, respectively; (**c,d**) TEM images of NaYF$_4$:Yb/Er nanocrystals using EDTA in acetic acid and ethanol, respectively. Reproduced from [34]. Copyright 2005, John Wiley and Sons; SEM images of (**e,f**) flower-patterned hexagonal disks; (**g**) hexagonal nanotubes; and (**h**) nanorods of β-NaYF$_4$. Reproduced from [51]. Copyright 2007, John Wiley and Sons.

Recently, via a hybrid thermal decomposition/solvothermal method, LaF$_3$:Yb/Er/Tm/Ho nanoplates with multicolor upconversion luminescence were synthesized [51,52]. The feasibility of the hydro/solvothermal methods was demonstrated by several groups. For example, Zhang has synthesized $BaYF_4$ [53], $CePO_4$ [54], YPO_4 [55], Gd_2O_3 [56], $GdVO_4$ [57], and Lin's group reported the uniform microstructured YPO_4 [58], YVO_4 [59], LnF_3 [60], and $NaYF_4$ [45] via the sodium citrate assisted hydrothermal route. In recent years, Liu and co-workers reported the synthesis of $KMnF_3$ nanocrystals with only single-band UC emission [61]. Zhao and Hao's group [62,63] represented a strategy for the rationale manipulation of green and red upconversion emission, and the pure red emission of $NaYF_4$:Yb/Er nanocrystals has been achieved by Mn^{2+} doping.

However, the disadvantages of this method are substantially difficult to overcome. Firstly of all, there are too many parameters including the reaction temperature, surfactant type and concentration, reactant concentration, solvent and the composition of it, and reaction time, to consider in order to filter out the most optimized experimental conditions for one specific reaction. Furthermore, the obtained nanocrystals usually have large size distribution, and sometimes the by-product residues stay on the surface of the nanocrystals, which are difficult to remove. Therefore, it remains challenging to develop a general and facile hydro/solvothermal method for the synthesis of high-quality UCNPs.

2.4. Sol-Gel Method

Sol-gel process is a typical wet-chemical technique for synthesizing UCNPs, which can be generally divided into three types: (1) sol-gel route based on the hydrolysis and condensation of molecular precursors; (2) gelation route based on condensation of the aqueous solutions containing metal-chelates; and (3) polymerizable complex route [64]. In sol-gel process, a RE nitrate salt or metal alkoxide is generally used as the starting reactants. The reaction is started by mixing the reactants in liquid phase through hydrolysis and condensation reactions, followed by annealing at high temperature for a certain period of time.

In 2002, Prasad et al. [65] developed a sol emulsion-gel method that produces Er^{3+} doped ZrO_2 nanophosphors. In order to reduce the segregation of particles and ensure compositional homogeneity, Lin and co-workers fabricated an inorganic YVO_4:Eu thin film phosphor by combining the pechini-type sol-gel process with inkjet printing. The mixed solution of metal salt precursors, citric acid, and poly(ethylene glycol) was directly used as ink to deposit patterns on ITO-coated glass substrate. After calcination at 600 °C in air, the YVO_4:Eu patterns at the micrometer-scale were formed on the substrate [66]. In another report, Song and co-works successfully fabricated many kinds of inverse opal photonic crystals (PCs) by the sol-gel method with a PMMA latex sphere template, including YVO_4:Dy [67], TiO_2:Sm [68], YBO_3:Eu [69], and $LaPO_4$:Ce,Tb [70]. In addition, the sol-gel process was also developed for synthesizing various UCNPs using metal oxides as host materials such as TiO_2:Er, $BaTiO_3$:Er (Figure 5), $Lu_3Ga_5O_{12}$:Er, and YVO_4:Yb/Er [71–73].

For the sol-gel method, the annealing procedure (temperature and time) is the key step in the preparation process, which can seriously determine the quality of the samples. It should be noted that, although the sol-gel method can be used for large-scale production and the products usually offer high luminescence intensity due to the high crystallinity formed at high annealing temperature, the sol-gel derived nanocrystals generally have broad particle size distribution, irregular morphology, and are insoluble in water, which compose the shortcomings of this method.

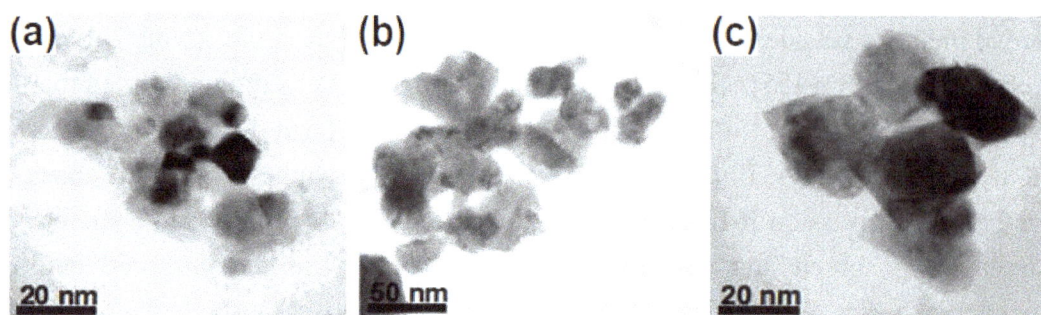

Figure 5. TEM images of the 1.0 mol% Er^{3+}-doped $BaTiO_3$ nanoparticles obtained after heating to three different temperatures: (**a**) 700 °C; (**b**) 850 °C; and (**c**) 1000 °C. Reproduced from [71]. Copyright 2003, American Chemical Society.

2.5. Microemulsion

To synthesize nanomaterials via the microemulsion method, it usually requires surfactant, co-surfactant, organic solvent, water, and the initial reagents. In a typical microemulsion solution, amphiphilic surfactants form a monolayer at the oil-water interface, with the hydrophobic tails of the surfactant molecules dissolved in the oil phase and the hydrophilic head groups in the aqueous phase. Preparation of RE fluoride nanomaterials usually needs two separate microemulsion systems and the two microemulsions containing RE ions and fluorine ions, respectively, get mixed to initiate the reaction. Over the past few years, Lemyre and coworkers [74] have reported that upconversion YF_3 nanoparticles can be prepared in a reverse water-in-cyclohexane microemulsion system stabilized by polyoxyethylene isooctylphenyl ether (IGEPAL CO-520 or NP-5was supplied by Aldrich) (Figure 6a). Qin and coworkers [75,76] prepared YF_3 upconversion nanophosphors, using CTAB and 1-pentanol instead of NP-5 (Figure 6b).

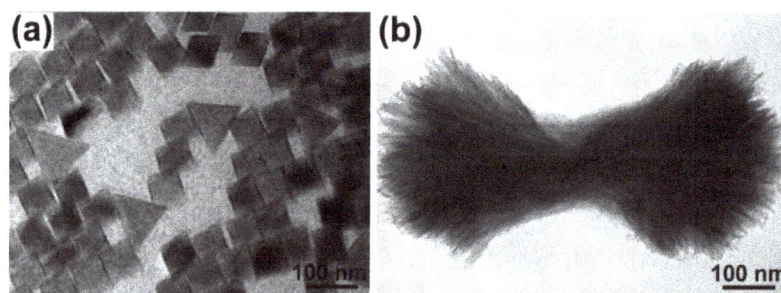

Figure 6. TEM images of (**a**) YF_3 nanophosphors prepared in the NP-5 stabilized microemulsion system. Reproduced from [74]. Copyright 2005, American Chemical Society. (**b**) YF_3 nanobundles synthesized in the CTAB and NP-5 stabilized microemulsion system. Reproduced from [75]. Copyright 2008, American Chemical Society.

The use of microemulsion has many advantages, such as the low-cost for equipment, easy operation, the small size of the UCNPs, and the controlled morphology of products by adjusting the dosage of the surfactant, solvent, as well as the aging time. However, this technique has many challenges including the small amount of products generated, difficulty of sample separation and narrow scope synthesis. More importantly, it is difficult to achieve massive production to meet industry requirements.

2.6. Combustion Synthesis

Compared to sol-gel and hydro/solvothermal synthesis, combustion method for UCNPs synthesis can be finished in a short period of time. The process of combustion synthesis is an oxidation-reduction reaction in essence. Metal nitrates are usually selected as oxidizer and the source for metal ions while the organic compounds as reducing agent and fuel. There are two requirements for organic fuel selection: (1) the reaction occurred between the fuel and nitrate must be relatively mild, producing nontoxic gases; (2) it is better to select the fuels that can complex with the metal ions, enhance the solubility of metal ions, and prevent the separation by crystallization of the metal salts in the precursor solution.

During the combustion synthesis process, combustion wave spreads the reaction materials in a self-sufficient situation without requiring extra heat in the process of the whole reaction. This time- and energy-saving method was used to synthesize oxide and oxysulfide nanomaterials. For example, Capobianco, Luo, and Zhang's groups have respectively synthesized a variety of oxide and oxysulfide nanophosphors (Y_2O_3, $Gd_3Ca_5O_{12}$, La_2O_2S, and Gd_2O_3) [77–80] via this method.

2.7. Flaming Synthesis

Flaming synthesis is another powerful method for producing RE oxide nanomaterials, which can be divided into four stages: (1) precursor reaction; (2) nucleation; (3) growth and polymerization; and (4) ion deposition. Notably, the flaming synthesis differs from the typical combustion synthesis that all reactions take place in the gas-phase and form fine powders. The core advantages of the technique are time-saving and low-cost. In 2007, Ju [81] reported the synthesis of Y_2O_3:Yb/Er(Tm, Ho) nanophosphors using this method. However, since the oxidation reaction happens along with the flaming process, this method is mainly limited to oxide nanomaterial synthesis, and it is almost impossible to synthesize other types of UCNPs, such as fluoride, phosphate, vanadate, and so on. It should also be noted that the flaming synthesis offers an opportunity for large-scale synthesis of RE doped oxides.

2.8. Electrospinning

During the electrospinning process, the precursor solution is spun through four stages (cocoons, stretching, refinement, and curing) under the effect of high voltage electrostatic field. Simple operation, good repeatability, and wide application scope are the advantages of this approach. For example, Song [82] have reported $NaYF_4$/PVP composite nanofibers, with diameter in the range of 300–800 nm, prepared through electrospinning. Nevertheless, this method is still in its infantile stage. New recipes for the synthesis of uniform and small-sized (<20 nm) nanoparticles with controllable morphology are still required.

2.9. Microwave Synthesis

Microwave synthesis contains solid microwave and liquid microwave method. The former mixes up RE oxides with ammonium bifluoride and ammonium fluoride, and the nanoparticles were directly synthesized via microwave. The latter dissolves the RE salts and the fluoride source into solvent. Then, these raw materials react with each other when the solvent is heated by microwave. The method was extended to produce PrF_3 hemispheres with diameter of about 31 nm [83].

The advantage of microwave synthesis is generally composed of: (1) it can selectively heat the samples to high temperature while the rest of the microwave device remains at room temperature; (2) microwave can heat the reaction system uniformly, cause less side reaction and the product is relatively simple; (3) fast heat and low energy consumption; (4) improve the structure and properties of synthetic material by adding proper surfactant.

Besides the above synthetic strategies elaborated, combinatorial approaches have also been employed to produce micro- or nano-scale RE fluoride crystals, which after proper lanthanide doping, can upconvert near-infrared light to visible frequencies, enabling the applications of such materials to biological imaging, telecommunication, and solar energy conversion [84]. For example, Liu [85] have synthesized $NaYF_4$:Yb,Er upconversion nanocrystals with characteristic upconversion luminescence spectra in a continuous capillary. Later, the research group improved the condition of the reaction prevented the growth of β-$NaYF_4$ [86]. On the other side, Zhu *et al.* synthesized LaF_3 [87] and $LaPO_4$ [88] nanoparticles doped with Ce^{3+} and Tb^{3+} using microcapillary flow reactors heated with microwaves. However, they also suffer from the wide quality variation of the UCNPs synthesized from different vessels and the large-scale production.

3. Surface Modification of UCNPs

Because of the influence of impurities and lattice defects on the synthesized UCNPs, the quantum yield of UCNPs is lower than the corresponding bulk materials. In addition, the UCNPs are mostly insoluble in water since they are prepared from the organic environment and usually surrounded by hydrophobic surfactant molecules. Thus, it is important to develop appropriate strategies to make them hydrophilic and in the meantime maintain their upconversion efficiency to satisfy various purposes. For example, the ideal luminescent nanocrystals used for biocompatible purposes should meet several requirements, including: (1) high luminescence efficiency and low background noise; (2) good solubility and stability in biological environment; (3) good biological compatibility; and (4) proper size (below 100 nm).

However, a notorious weak point of UCNPs is the inherent low upconversion efficiency. Currently, the urgent task is to improve the upconversion efficiency, and there have been several groups working on this topic via the core-shell strategy.

The introduction of an inert crystalline shell of an undoped material around each doped nanocrystal provides an effective option to improve the luminescence efficiency of UCNPs. The shell usually has the same composition as the core host crystal, which can effectively reduce the surface quenching effect. In such structures, all dopant ions are confined in the interior core of the nanocrystals, effectively suppressing the non-radiative energy transfer from RE ions to the surface quenching sites, which results in improved upconversion luminescence efficiency. A significant demonstration was made by Yi and Chow [89] who reported a luminescence enhancement of nearly 30 times on 8 nm $NaYF_4$:Yb/Tm nanocrystals coated with a 1.5 nm thick $NaYF_4$ shell. Later on, Chen and co-works exploited a strategy to achieve dual-mode luminescence from identical Eu^{3+} ions in monodisperse hexagonal-phase $NaGdF_4$ nanocrystals that consist of the $NaGdF_4$:Yb/Tm core and the $NaGdF_4$:Eu shell. Typical red downconversion luminescence of Eu^{3+} has been detected via the sensitization of Gd^{3+} ions. By using Yb^{3+} and Tm^{3+} embedded in the cores as double sensitizers, intense upconversion luminescence of Eu^{3+} in the shells can be achieved in $NaGdF_4$:Yb/Tm@$NaGdF_4$:Eu core-shell nanocrystals upon excitation at 976 nm. The upconversion intensity of Eu^{3+} in core-shell nanocrystals is found about one order of magnitude higher than the counterparts of the triply-doped core only, due to the inhibition of the deleterious cross-relaxations between Tm^{3+} and Eu^{3+} ions in core-shell nanocrystals that are reasonably separated in space [90].

Recently, Capobianco [91] proposed a strategy to significantly enhance the intensity of the upconversion by employing a novel $NaGdF_4$:Yb/Er@$NaGdF_4$:Yb active-core@active-shell architecture (Figure 7). The active-shell serves two purposes: (i) minimize the quenching centers and channels; and (ii) transfer absorbed near infrared energy to the luminescence centers (emitters).

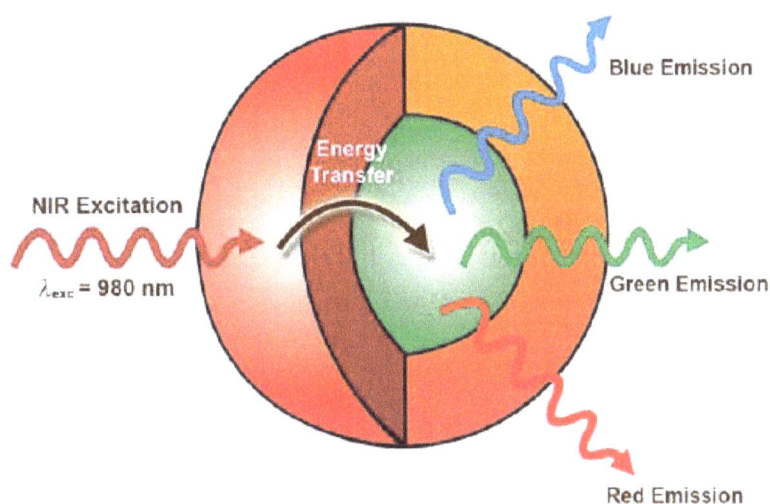

Figure 7. Schematic illustration of the active core-active shell nanoparticle architecture showing the absorption of NIR near infrared light by the Yb^{3+}-rich shell (red) and subsequent energy transfer to the Er^{3+},Yb^{3+}-doped core (green), which leads to upconverted blue, green, and red emissions. Reproduced from [91]. Copyright 2009, John Wiley and Sons.

Utilizing the core-shell strategy, Zhang and co-workers [92] have broken through the well accepted upper limit of the concentration quenching threshold, e.g., from ~2 mol% to 5 mol% for Er^{3+}, by a designed multi-layer core-shell structure, which contains four parts: the illuminating core ($NaYF_4$:Yb/Er), the first separating shell ($NaYF_4$:Yb), the second illuminating shell ($NaYF_4$:Yb/Er) and the final inactive shell ($NaYF_4$:Yb). The separating layer effectively inhibits the energy transfer process between the Er^{3+} ions in the inside and outside layers and largely reduces the possibility of excitation energy trapping by defects, resulting in an effective upconversion luminescence enhancement [92]. Other core-shell examples include CeF_3:Tb@LaF_3 [93], $NaYF_4$:Yb/Er@$NaYF_4$ [89,94], $NaGdF_4$:Yb/Er@$NaGdF_4$ [95], and KYF_4@KYF_4:Yb/Er [96]. Liu [97] introduced a rationale core-shell strategy that provides precise control over the concentration of dopant in the core and shell layers of nanoparticles. A small amount of Nd^{3+} ions is doped into the core, while a high concentration of Nd^{3+} ions (~20 mol%) is selectively doped in the shell layers for effective harvesting of light at 800 nm. Chen and co-workers developed a facile strategy based on successive layer-by-layer (LBL) injection of shell precursors for the synthesis of LiLuF$_4$:Ln^{3+} core-shell UCNPs [98]. It is reported that the size of nanocrystals can be tuned by adjusting the amount of NH$_4$F (the more the amount of NH$_4$F solution used, the bigger the size of crystals) and the shell thickness of core-shell nanocrystals can also be controlled by the volume of the core [99]. Besides, Yan has proved in the structure of NaYF$_4$:Yb,Er@CaF$_2$ that the thickness of the CaF$_2$ shell were controlled by adjusting the [Ca]/[RE] ratio [100]. Although this approach provides tunable emission intensity in nanocrystals, the luminescence quantum yields of the nanocrystals are limited due to the weak ligand fields and high energy oscillations. Further improvement of the luminescence efficiency of UCNPs can be expected through the controlled growth of a rationally designed crystalline inner shell. Generally speaking, such core-shell structure improves the optical properties of the nanocrystals, but usually does not change the chemical functional group binding on the surface. So, proper surface modification is needed in order to make practical applications of UCNPs, especially for bio-related purposes.

3.1. SiO$_2$ Encapsulation

SiO$_2$ encapsulation involves the growth of an amorphous silica shell on the UCNPs core. Due to the rich –OH groups on the SiO$_2$ coated UCNPs, it becomes possible to further functionalize the UCNPs with expected chemical groups such as –NH$_2$, –COOH, polymers, and more importantly a wide variety of biomolecules. Nann and Capobianco [101] obtained surface-functionalized YF$_3$ nanocrystals using this technique. One year later, Zhang and coworkers [102] succeeded in growing a silica shell with adjustable thickness in the range of 2–10 nm on the surface of PVP stabilized cubic NaYF$_4$:Yb/Er/Tm nanocrystals. Later on, they used the microemulsion method to coat NaYF$_4$ nanocrystals with SiO$_2$, which resulted in monodispersed SiO$_2$-coated UCNPs (Figure 8) [103].

Alternatively, Song and co-workers [104] showed the conversion of hydrophobic NaYF$_4$:Yb/Er upconversion nanophosphors into hydrophilic ones by amphiphilic silane modification with ultrathin thickness (~1 nm) at room temperature. In this strategy, the coating layers can also provide the possibility for loading Eu(TTA)$_3$(TPPO)$_2$ complex with downconversion luminescence where they realized the dual mode temperature sensing and dual mode cell imaging within the physiological environment [104]. There are other successful examples to modify the surface of the UCNPs [105–108] via the SiO$_2$ encapsulation method, such as LaF$_3$:Ln@SiO$_2$, NaYF$_4$:Yb/Er/Fe$_3$O$_4$@SiO$_2$, NaYF$_4$:Er,Tm,Ho@SiO$_2$, NdF$_3$@SiO$_2$, and NaYF$_4$@SiO$_2$@ quantum dots [103].

The SiO$_2$ coating provides a versatile platform for multi-functionalization of UCNPs. Nevertheless, the method is time consuming and difficult for large scale synthesis. On the other hand, the coated SiO$_2$ layer may also affect the luminescence intensity of the UCNPs through light scattering, which prevents it from becoming the ideal surface modification recipe because of the already low upconversion efficiency of the UNCPs. Thus, there is still much work to do to improve the general applicability of this technique.

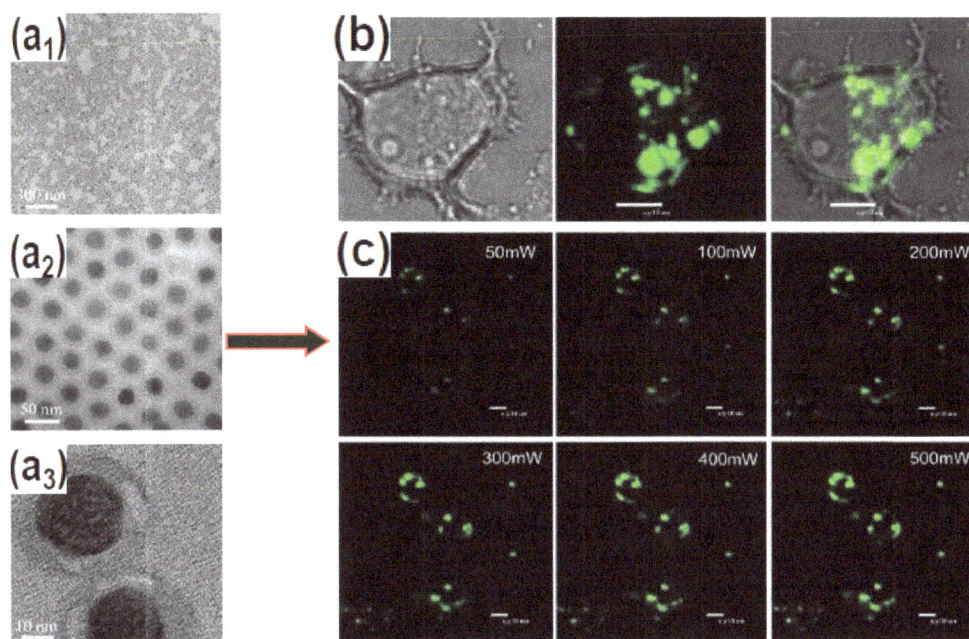

Figure 8. Silica-coated NaYF$_4$:Yb/Er nanocrystals and their application for cell imaging. (a1–a3) TEM images of silica-coated NaYF$_4$:Yb/Er UCNPs upconversion nanoparticles at different magnifications; (**b**) Confocal fluorescence image of MCF-7 cells using silica-coated NaYF$_4$:Yb/Er nanospheres (Left: bright-field, middle: upconversion image under 980 nm excitation, and Right: superimposed images of MCF-7 cells incubated with the nanoparticles for 24 h); (**c**) Confocal fluorescence images of MCF-7 cells with the nanospheres, excited by a 980 nm laser with different power intensities. Reproduced from [103]. Copyright 2008, John Wiley and Sons.

3.2. Polymer Encapsulation

Polymer encapsulation typically involves the absorption of an additional amphiphilic polymer onto the nanophosphor surface through the hydrophobic-hydrophobic attraction between the original surfactant molecules and the hydrocarbon chains of the polymer. Using this strategy, Chow and co-workers successfully coated a polyacrylicacid (PAA) layer on NaYF$_4$:Yb/Er@NaYF$_4$ core-shell nanoparticles [89]. The hydrophobic core-shell nanoparticles were rendered hydrophilic by amphiphilic PAA coating.

Through amphiphilic coating, the multi-functionalization of upconversion nanoparticles could also be realized. For instance, Feng and co-workers fabricated upconversion detection nanocomposites, which were formed by coating the amphiphilic polymer (C18PMH-PEG) on the NaYF$_4$:Yb/Er,Tm nanophosphors based on the hydrophobic–hydrophobic interaction. The polymer modified nanoparticles were then assembled for the selective luminescence detection of mercury ions in water. Using the ratiometric upconversion luminescence emission as a detection signal, the detection limit of Hg^{2+} for this nanoprobe in aqueous solution is 8.2 ppb, which is much lower than that (329 ppb) determined by the UV/Vis technology [109]. Liu modified UCNPs with a polyethylene glycol (PEG) grafted amphiphilic polymer, via a hydrophobic interaction-based supramolecular chemistry strategy, for targeted intracellular drug delivery and UCL imaging. This work reveals the great potential of

UCNPs for multifunctional drug delivery and biomedical imaging applications (Figure 9) [110]. In contrast to the conventional methods, K. Prud'homme group [111] reported the successful preparation of colloidal UCNPs stable in buffers and serum media (Leibovitz L-15 media with added fetal bovine serum) using FNP and PEG surface coatings. These polymer-modified UCNPs provide promising new materials for applications in bioimaging and photodynamic therapy.

The polymer-coating method is easier when compared to SiO_2 encapsulation. The *in-situ* coating of polymers on the surface of UCNPs sheds light on the development of multi-functional upconversion nanoparticles. However, the coating through hydrophobic interaction is not stable, and better strategies for a robust polymer coating are needed to satisfy various application requirements.

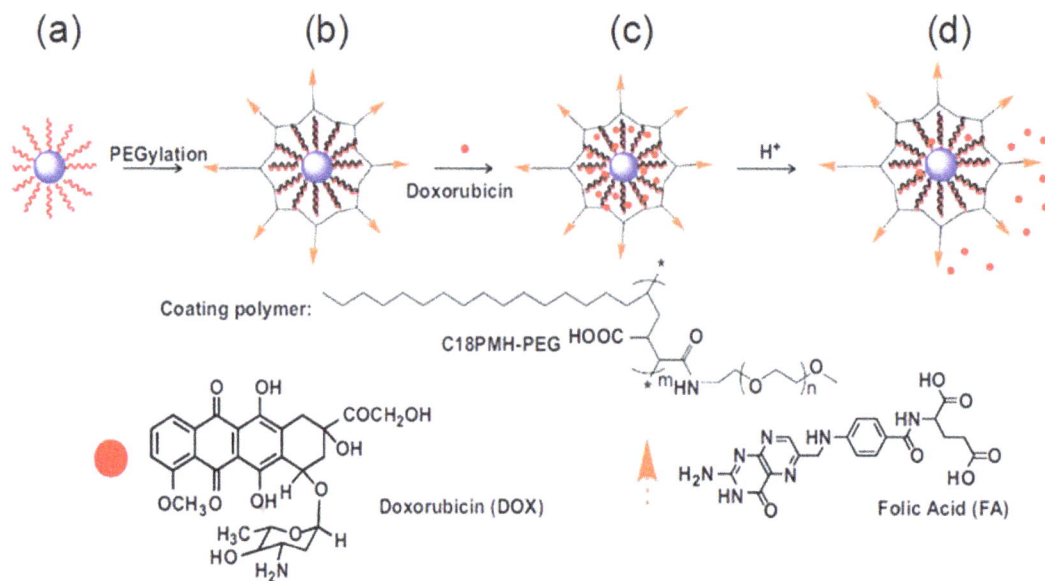

Figure 9. Schematic illustration of the UCNP-based drug delivery system. (**a**) As-synthesized oleic acid capped UCNPs; (**b**) C18PMH-PEG-FA functionalized UCNPs; (**c**) DOX loading on UCNPs. DOX molecules are physically adsorbed into the oleic acid layer on the nanoparticle surface by hydrophobic interactions; (**d**) Release of DOX from UCNPs triggered by decreasing pH. Reproduced from [110]. Copyright 2011, Elsevier Ltd.

3.3. Ligand Oxidation

The ligand oxidation technique involves oxidation of the native ligands that contain unsaturated carbon-carbon bonds to generate a pendant hydrophilic functional group. It is reported that a large number of carboxylic acid groups can be generated on the UCNPs surface after oxidation, which not only renders the nanocrystals good solubility in water, but also provides tailorability to biological molecules via direct chemical coupling (Figure 10). It is also reported that the oxidation process has no obvious negative effect on the shape, crystal structure, chemical composition, and luminescence properties of the upconversion nanomaterials [112]. However, this method is limited only to those ligands that contain unsaturated carbon-carbon bonds. The oxidation process may also cause ligand removal due to the harsh experimental conditions. It would be ideal if the modulation of ligand could be achieved under mild experimental conditions, which cause no or less damage to the surfaces of the UCNPs. Biomedical research faces a major challenge in that the lengthy period of oxidation may lead to the excessive formation of the brown MnO_2 side product, which is not easy to separate and further weakens the upconversion fluorescence [23]. So, the surface functionalization of UCNPs to render these nanoparticles dispersible in aqueous media is very important. Capobianco and co-workers show that the Ln-OA (Ln = lanthanide; OA = oleate) surface of the Ln-UCNPs is replaced by Ln-OH whose state of charge can be tuned by pH and the efficiency of upconversion luminescence can be enhanced by replacing OH with OD. Furthermore, they have studied the effect of different acids, *i.e.*, HCl, H_3PO_4, and HF on the surface properties of the oleate-free Ln-UCNPs [113].

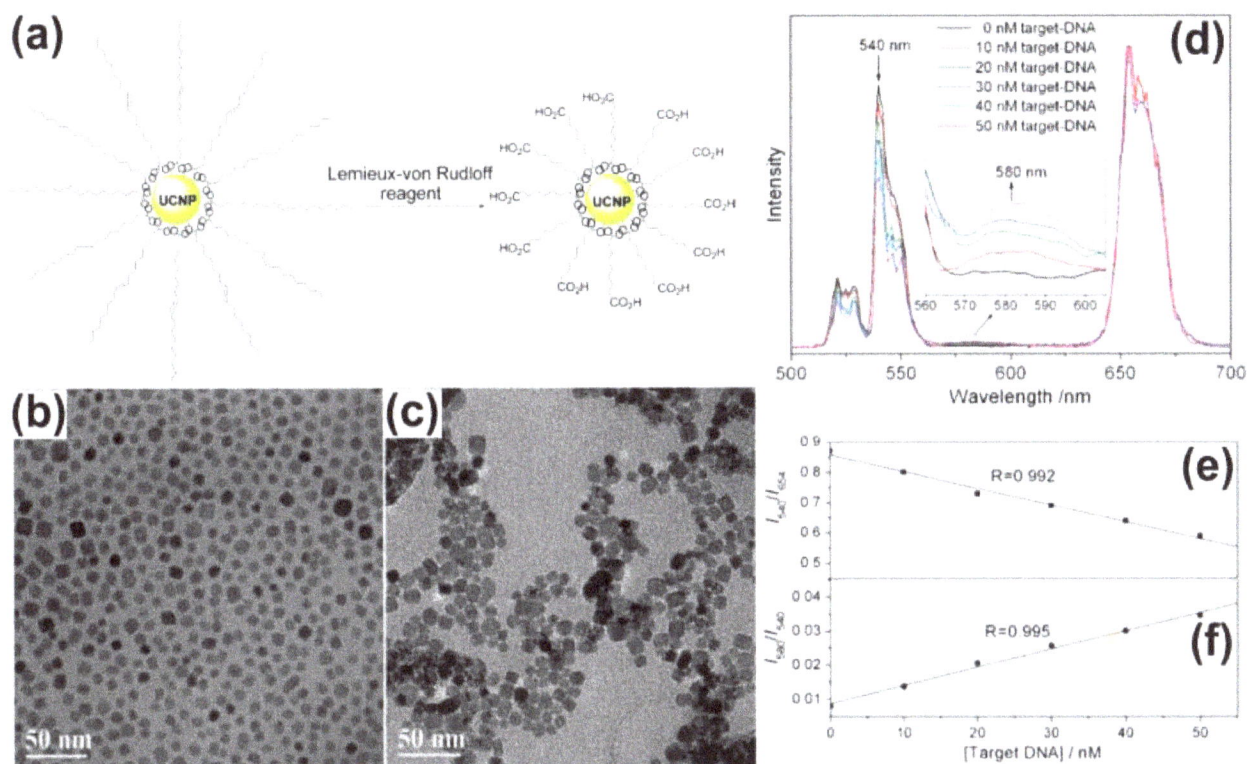

Figure 10. (a) Schematic illustration of the ligand oxidization process; (b,c) TEM images of the NaYF4:Yb/Er nanoparticles before and after ligand oxidization, respectively; (d) Luminescence spectra of a mixture of streptavidin-functionalized NaYF4:Yb/Er nanoparticles, capture-DNA, and reporter-DNA in the presence of different concentrations of target-DNA under continuous-wave excitation at 980 nm. The linear relationships between target-DNA concentration and the intensity ratios of (e) I540/I654 and (f) I580/I540. Reproduced from [112]. Copyright 2008, American Chemical Society.

3.4. Ligand Exchange

Ligand exchange involves the displacement of original hydrophobic ligands that have weak coordination with the RE ions on the surface of UCNPs, by ligands that have stronger coordination capability (with the RE ions) and hydrophilic functional groups. As an early example, Chow and co-workers [114] have demonstrated the preparation of water-soluble NaYF4:Yb/Er nanoparticles with oleylamine ligands via the ligand exchange method (Figure 11). A large variety of ligands have been reported, including poly(acrylicacid) (PAA) [115], poly(ethyleneglycol) (PEG)-phosphate [116], mercaptopropionic acid (MPA) [117], citrate [118]. Li's research group [119] demonstrated a new generation of ^{18}F-labeled lanthanide nanoparticles of NaYF$_4$ co-doped with Gd^{3+}/Yb^{3+}/Er^{3+} as multimodality nanoprobes for PET, MR and UCL imaging. The presence of Yb^{3+} and Er^{3+} co-doped in the NaYF$_4$ nanoparticles gives rise to intense UCL emission in the visible region for luminescent imaging, and 60% Gd^{3+} doping provides the paramagnetic relaxation for MRI. Successful labeling of the lanthanide nanoparticles with ^{18}F gave a product suitable for PET imaging [119]. It is noted that after the ligand exchange, most of these commonly used ligand molecules on the UCNP surface carry additional functional groups to facilitate further biofunctionalization and bioconjugation. However, the exchange efficiency of the technique is difficult to evaluate since the surfactant molecules cannot be completely replaced by the oleylamine molecules, and the exchange process is also affected by the pH value of the solution, the concentrations of both the surfactant and the oleylamine molecules, and even the ionic strength.

Besides the above surface modification strategies, Prabhas V. Moghe and coworkers [120] have developed a new approach to rendering NaYF$_4$:Yb,Er nanoparticles stable in coacervated HSA nanoshells functionalized with cyclic arinine-glycine-aspartic acid (cRGD) tripeptide. They observed that the composite particles were highly biocompatible *in vitro*, capable of selectively targeting cancerous cell lines exhibiting higher expression of cancer-specific integrin markers, and amenable to fluorescence imaging with high fidelity.

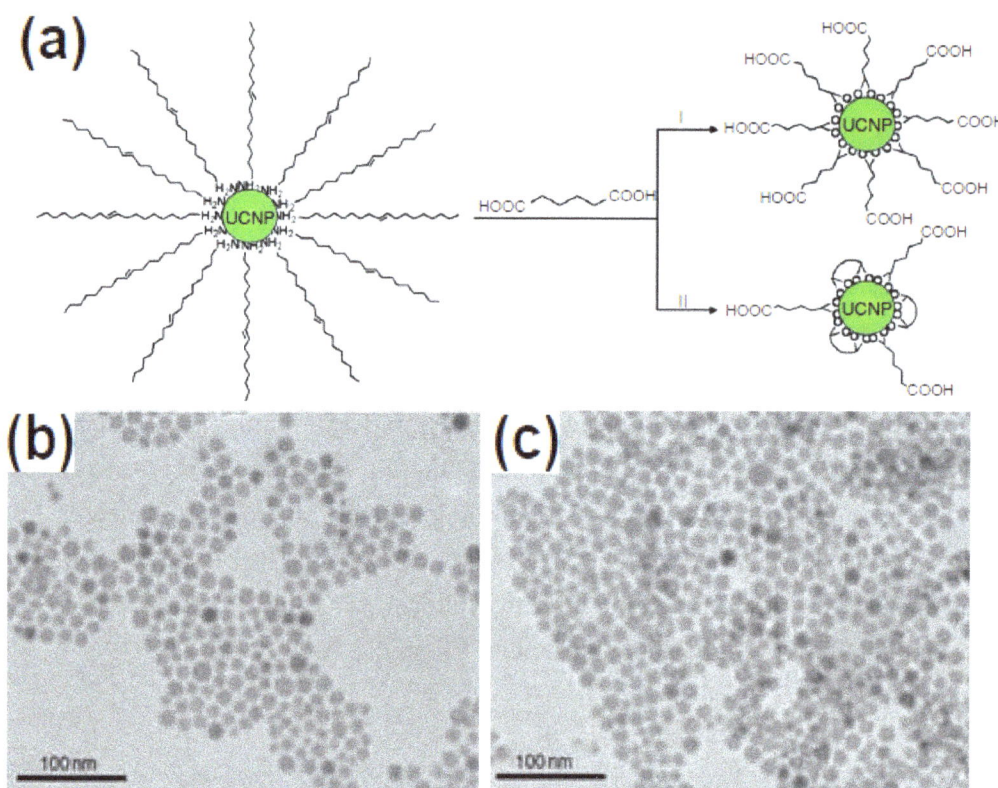

Figure 11. (a) Schematic illustration showing the ligand-exchange reactions on OM-stabilized upconversion nanophosphors. (b,c) TEM images of NaYF$_4$:Yb/Er nanophosphors prior to and after ligand exchange reactions, respectively. Reproduced from [114]. Copyright 2006, John Wiley and Sons.

4. Conclusions

In summary, this review has summarized the recent development of UCNPs with emphasis on the synthetic methods and the surface modification strategies. With careful control of the reaction conditions, we can now obtain high quality, monodispersed UCNPs with various chemical components. However, each individual synthetic strategy has their unique advantages associated with substantial shortcomings. Our subsequent goal is to develop a general synthetic approach which could meet the requirements of large scale synthesis and multifunctionalities for UCNPs.

In parallel, the controlled surface modification method now can produce UCNPs with high colloidal stability, biocompatibility, and tailorable chemical functionalities. The above discussion has also demonstrated that the rapid development of surface modification method facilitates the applications of UCNPs in detection, bioimaging, therapy, and solar cells. Despite the rapid development, these exiting methods still can only be used in a narrow scope. Strategies with universal comparability for various UCNPs may be the next goal of this field. Alternatively, methods that provide easy surface modulation capability for certain types of nanocrystals could be another option for multifunctionalization.

Acknowledges

We thank the financial support from the National Natural Science Foundation of China (No.: 21301090) and the Natural Science Foundation of Jiangsu Province (No.: BK20130923).

Author Contributions

Hongjin Chang, Ling Huang, and Wei Huang proposed the structure of the manuscript and wrote the paper. Juan Xie, Baozhou Zhao, Botong Liu, Shuilin Xu, and Na Ren helped to search for the reference and copyright collection. Xiaoji Xie performed the manuscript editing and language coloring.

Conflicts of Interest

The authors declare no conflict of interest.

Reference

1. Law, M.; Goldberger, J.; Yang, P.D. Semiconductor nanowires and nanotubes. *Annu. Rev. Mater. Sci.* **2004**, *34*, 83–122.

2. Yoffe, A.D. Low-dimensional systems: Quantum size effects and electronic properties of semiconductor microcrystallites (zero-dimensional systems) and some quasi-two-dimensional systems. *Adv. Phys.* **2002**, *42*, 173–266.

3. Su, V.C.; Chen, P.H.; Lin, R.M.; Lee, M.L.; You, Y.H.; Ho, C.I.; Chen, Y.C.; Chen, W.F.; Kuan, C.H. Suppressed quantum-confined stark effect in InGaN-based LEDs with nano-sized patterned sapphire substrates. *Opt. Exp.* **2013**, *21*, 30066–30071.

4. Wang, F.; Liu, X.G. Recent advances in the chemistry of lanthanide-doped upconversion nanocrystals. *Chem. Soc. Rev.* **2009**, *38*, 976–989.

5. Yi, G.S.; Lu, H.C.; Zhao, S.Y.; Yue, G.; Yang, W.J.; Chen, D.P. Synthesis, characterization, and biological application of size-controlled nanocrystalline NaYF₄:Yb,Er infrared-to-visible up-conversion phosphors. *Nano Lett.* **2004**, *4*, 2191–2196.

6. Carlos, L.D.; Ferreira, R.A.; Ribeiro, S.J. Lanthanide-containing light-emitting organic-inorganic hybrids: A bet on the future. *Adv. Mater.* **2009**, *21*, 509–534.

7. Zhou, J.; Liu, Z.; Li, F.Y. Upconversion nanophosphors for small-animal imaging. *Chem. Soc. Rev.* **2012**, *41*, 1323–1349.

8. Wang, F.; Han, C.M.; Li, F.Y. Upconversion-nanophosphor-based functional nanocomposites. *Adv. Mater.* **2013**, *25*, 5287–5303.

9. Chen, J.; Guo, C.R.; Wang, M.; Wang, L.P.; Mao, C.B.; Xu, S.K. Controllable synthesis of NaYF₄:Yb,Er upconversion nanophosphors and their application to *in vivo* imaging of *Caenorhabditis elegans. J. Mater. Chem.* **2011**, *21*, 2632–2638.

10. Wang, F.; Han, Y.; Liu, X.G. Simultaneous phase and size control of upconversion nanocrystals through lanthanide doping. *Nature* **2010**, *463*, 1061–1065.

11. Hampl, J.; Hall, M.; Mufti, N.A.; Yao, Y.M.; Mac Queen, D.B.; Wright, W.H.; Cooper, D.E. Upconverting phosphor reporters in immunochromatographic assays. *Anal. Biochem.* **2001**, *288*, 176–187.

12. VandeRijke, F.; Zijlmans, H.; Li, S.; Vail, T.; Raap, A.K.; Niedala, R.S.; Tanke, H.J. Up-converting phosphor reporters for nucleic acid microarrays. *Nat. Biotechnol.* **2001**, *19*, 273–276.

13. Lim, S.F.; Riehn, R.; Ryu, W.S.; Khanarian, N.; Tung, C.-K.; Tank, D.; Austin, R.H. *In vivo* and scanning electron microscopy imaging of upconverting nanophosphors in *Caenorhabditis elegans. Nano Lett.* **2006**, *6*, 169–174.

14. Chatterjee, D.K.; Rufaihah, A.J.; Zhang, Y. Upconversion fluorescence imaging of cells and small animals using lanthanide doped nanocrystals. *Biomaterials* **2008**, *29*, 937–943.

15. Zhang, P.; Rogelj, S.; Nguyen, K.; Wheeler, D. Design of a highly sensitive and specific nucleotide sensor based on photon upconverting particles. *J. Am. Chem. Soc.* **2006**, *128*, 12410–12411.

16. Gai, S.L.; Yang, P.P.; Li, C.X.; Wang, W.X.; Dai, Y.L.; Niu, N.; Lin, J. Nanomorphology and charge generation in bulk heterojunctions based on low-bandgap dithiophene polymers with different bridging atoms. *Adv. Funct. Mater.* **2010**, *20*, 1166–1172.

17. Hinklin, T.R.; Rand, S.C.; Laine, R.M. Transparent, polycrystalline upconverting nanoceramics: Towards 3-D displays. *Adv. Mater.* **2008**, *20*, 1270–1273.

18. Zhang, Y.W.; Sun, X.; Si, R.; You, L.P.; Yan, C.H. Single-crystalline and monodisperse LaF₃ triangular nanoplates from a single-source precursor. *J. Am. Chem. Soc.* **2005**, *127*, 3260–3261.

19. Sun, X.; Zhang, Y.W.; Du, Y.P.; Yan, Z.G.; Si, R.; You, L.P.; Yan, C.H. From trifluoroacetate complex precursors to monodisperse rare-earth fluoride and oxyfluoride nanocrystals with diverse shapes through controlled fluorination in solution phase. *Chem. Eur. J.* **2007**, *13*, 2320–2332.

20. Boyer, J.C.; Cuccia, L.A.; Capobianco, J.A. Synthesis of colloidal upconverting NaYF₄:Er³⁺/Yb³⁺ and Tm³⁺/Yb³⁺ monodisperse nanocrystals. *Nano Lett.* **2007**, *7*, 847–852.

21. Wang, H.Q.; Thomas, N. Monodisperse upconverting nanocrystals by microwave-assisted synthesis. *ACS Nano* **2009**, *3*, 3804–3808.

22. Yi, G.S.; Lee, W.B.; Chow, G.M. Synthesis of LiYF$_4$, BaYF$_5$, and NaLaF$_4$ optical nanocrystals. *J. Nanosci. Nanotechnol.* **2007**, *7*, 2790–2794.

23. Naccache, R.; Vetrone, F.; Mahalingam, V.; Capobianco, J.A. Controlled synthesis and water dispersibility of hexagonal phase NaGdF$_4$:Ho/Yb nanoparticles. *Chem. Mater.* **2009**, *21*, 717–723.

24. Mahalingam, V.; Vetrone, F.; Naccache, R.; Speghini, A.; Capobianco, J.A. Colloidal Tm^{3+}/Yb^{3+}-doped LiYF$_4$ nanocrystals: Multiple luminescence spanning the UV to NIR regions via low-energy excitation. *Adv. Mater.* **2009**, *21*, 4025–4028.

25. Mahalingam, V.; Vetrone, F.; Naccache, R.; Speghini, A.; Capobianco, J.A. Structural and optical investigation of colloidal Ln^{3+}/Yb^{3+} co-doped KY$_3$F$_{10}$ nanocrystals. *J. Mater. Chem.* **2009**, *19*, 3149–3152.

26. Vetrone, F.; Mahalingam, V.; Capobianco, J.A. Near-infrared-to-blue upconversion in colloidal BaYF$_5$:Tm^{3+},Yb^{3+} nanocrystals. *Chem. Mater.* **2009**, *21*, 1847–1851.

27. Zhang, P.D.; Zhang, Y.W.; Zheng, Z.G.; Sun, L.D.; Yan, C.H. Highly luminescent self-organized sub-2-nm EuOF nanowires. *J. Am. Chem. Soc.* **2009**, *131*, 16364–16365.

28. Du, Y.P.; Zhang, Y.W.; Sun, L.D.; Yan, C.H. Luminescent monodisperse nanocrystals of lanthanide oxyfluorides synthesized from trifluoroacetate precursors in high-boiling solvents. *J. Phys. Chem. C* **2008**, *112*, 405–415.

29. Du, Y.P.; Zhang, Y.W.; Sun, L.D.; Yan, C.H. Atomically efficient synthesis of self-assembled monodisperse and ultrathin lanthanide oxychloride nanoplates. *J. Am. Chem. Soc.* **2009**, *131*, 3162–3163.

30. Stouwdam, J.W.; van Veggel, F.C.J.M. Near-infrared emission of redispersible Er^{3+}, Nd^{3+}, and Ho^{3+} doped LaF$_3$ nanoparticles. *Nano Lett.* **2002**, *2*, 733–737.

31. Yi, G.S.; Chow, G.M. Colloidal LaF$_3$:Yb,Er, LaF$_3$:Yb,Ho and LaF$_3$:Yb,Tm nanocrystals with multicolor upconversion fluorescence. *J. Mater. Chem.* **2005**, *15*, 4460–4464.

32. Heer, S.; Kompe, K.; Gudel, H.U.; Haase, M. Highly efficient multicolour upconversion emission in transparent colloids of lanthanide-doped NaYF$_4$ nanocrystals. *Adv. Mater.* **2004**, *16*, 2102–2105.

33. Heer, S.; Lehmann, O.; Haase, M.; Gudel, H.U. Blue, green, and red upconversion emission from lanthanide-doped LuPO$_4$ and YbPO$_4$ nanocrystals in a transparent colloidal solution. *Angew. Chem. Int. Ed.* **2003**, *42*, 3179–3182.

34. Zeng, J.H.; Su, J.; Li, Z.H.; Yan, R.X.; Li, Y.D. Synthesis and upconversion luminescence of hexagonal-phase NaYF$_4$:Yb^{3+},Er^{3+} phosphors of controlled size and morphology. *Adv. Mater.* **2005**, *17*, 2119–2123.

35. Teng, X.; Zhu, Y.H.; Wei, W.; Wang, S.C.; Huang, J.F.; Naccache, R.; Hu, W.B.; Han, Y.; Zhang, Q.C.; Fan, Q.L.; *et al.* Lanthanide-doped Na$_x$ScF$_{3+x}$ nanocrystals: Crystal structure evolution and multicolor tuning. *J. Am. Chem. Soc.* **2012**, *134*, 8340–8343.

36. Liang, L.F.; Xu, H.F.; Su, Q.; Konishi, H.; Jiang, Y.B.; Wu, M.M.; Wang, Y.F.; Xia, D.Y. Hydrothermal synthesis of prismatic NaHoF$_4$ microtubes and NaSmF$_4$ nanotubes. *Inorg. Chem.* **2004**, *43*, 1594–1596.

37. Zhang, M.F.; Fan, H.; Xi, B.J.; Wang, X.Y.; Dong, C.; Qian, Y.T. Synthesis, characterization, and luminescence properties of uniform Ln^{3+}-doped YF_3 nanospindles. *J. Phys. Chem. C* **2007**, *111*, 6652–6657.

38. Liang, X.; Wang, X.; Zhuang, J.; Peng, Q.; Li, Y.D. Branched $NaYF_4$ nanocrystals with luminescent properties. *Inorg. Chem.* **2007**, *46*, 6050–6055.

39. Zhao, J.W.; Suan, Y.J.; Kong, X.G.; Tian, L.J.; Wang, Y.; Tu, L.P.; Zhao, J.L.; Zhang, H. Controlled synthesis, formation mechanism, and great enhancement of red upconversion luminescence of $NaYF_4$:Yb^{3+},Er^{3+} nanocrystals/submicroplates at low doping level. *J. Phys. Chem. B* **2008**, *112*, 15666–15672.

40. Wang, X.; Zhuang, J.; Peng, Q.; Li, Y.D. Hydrothermal synthesis of rare-earth fluoride nanocrystals. *Inorg. Chem.* **2006**, *45*, 6661–6665.

41. Wang, X.; Li, Y.D. A general strategy for nanocrystal synthesis. *Nature* **2005**, *437*, 121–124.

42. Huo, Z.Y.; Chen, C.; Li, Y.D. Systematic synthesis of lanthanide phosphate nanocrystals. *Chem. Eur. J.* **2007**, *13*, 7708–7714.

43. Wang, L.Y.; Li, Y.D. Systematic synthesis of lanthanide phosphate nanocrystals. $Na(Y_{1.5}Na_{0.5})F_6$ single-crystal nanorods as multicolor luminescent materials. *Nano Lett.* **2006**, *6*, 1645–1649.

44. Wang, L.Y.; Li, Y.D. Controlled synthesis and luminescence of lanthanide doped $NaYF_4$ nanocrystals. *Chem. Mater.* **2007**, *19*, 727–734.

45. Wang, G.F.; Peng, Q.; Li, Y.D. Luminescence tuning of upconversion nanocrystals. *Chem. Eur. J.* **2010**, *16*, 4923–4931.

46. Wang, L.Y.; Li, P.; Li, Y.D. Down and upconversion luminescent nanorods. *Adv. Mater.* **2007**, *19*, 3304–3307.

47. Wang, X.; Li, Y.D. Synthesis and characterization of lanthanide hydroxide single-crystal nanowires. *Angew. Chem. Int. Ed.* **2002**, *41*, 4790–4793.

48. Wang, X.; Sun, X.M.; Yu, D.P.; Zou, B.S.; Li, Y.D. Rare earth compound nanotubes. *Adv. Mater.* **2003**, *15*, 1442–1445.

49. Wang, X.; Li, Y.D. Fullerene-like rare-earth nanoparticles. *Angew. Chem. Int. Ed.* **2003**, *42*, 3497–3500.

50. Hu, C.G.; Liu, H.; Wang, Z.L. $La(OH)_3$ and La_2O_3 nanobelts-synthesis and physical properties. *Adv. Mater.* **2007**, *19*, 470–474.

51. Zhang, F.; Wan, Y.; Yu, T.; Zhang, F.; Shi, Y.; Xie, S.; Li, Y.; Xu, L.; Tu, B.; Zhao, D. Uniform nanostructured arrays of sodium rare-earth fluorides for highly efficient multicolor upconversion luminescence. *Angew. Chem. Int. Ed.* **2007**, *46*, 7976–7979.

52. Liu, C.; Chen, D. Controlled synthesis of hexagon shaped lanthanide-doped LaF_3 nanoplates with multicolor upconversion fluorescence. *J. Mater. Chem.* **2007**, *17*, 3875–3880.

53. Pan, S.H.; Deng, R.P.; Feng, J.; Wang, S.Y.; Wang, S.; Zhu, M.; Zhang, H.J. Microwave-assisted synthesis and down and upconversion luminescent properties of $BaYF_5$:Ln (Ln = Yb/Er, Ce/Tb) nanocrystals. *CrystEngComm* **2013**, *15*, 7640–7643.

54. Yang, M.; You, H.P.; Zheng, Y.H.; Liu, K.; Jia, G.; Song, Y.H.; Huang, Y.J.; Zhang, L.H.; Zhang, H. Hydrothermal synthesis and luminescent properties of novel ordered sphere $CePO_4$ hierarchical architectures. *J. Inorg. Chem.* **2009**, *48*, 11559–11565.

55. Yang, M.; You, H.P.; Huang, Y.H.; Zhang, H.J. Synthesis and luminescent properties of orderly YPO$_4$:Eu^{3+} olivary architectures self-assembled by nanoflakes. *CrystEngComm* **2010**, *12*, 4141–4145.

56. Zheng, Y.H.; You, H.P.; Jia, K.; Liu, K.; Song, Y.H.; Yang, M.; Zhang, H.J. Highly uniform Gd$_2$O$_3$ hollow microspheres: Template-directed synthesis and luminescence properties. *Langmuir* **2010**, *26*, 5122–5128.

57. Jia, G.; You, H.P.; Liu, K.; Zheng, Y.H.; Guo, N.; Zhang, H.J. Facile hydrothermal synthesis and luminescent properties of large-scale GdVO$_4$:Eu^{3+} nanowires. *Cryst. Growth Des.* **2009**, *9*, 5101–5107.

58. Li, C.X.; Hou, Z.Y.; Zhang, C.M.; Yang, P.P.; Li, G.G.; Xu, Z.H.; Fan, Y.; Lin, J. Controlled synthesis of Ln^{3+} (Ln = Tb, Eu, Dy) and V^{5+} ion-doped YPO$_4$ nano-/microstructures with tunable luminescent colors. *Chem. Mater.* **2009**, *21*, 4598–4607.

59. Li, C.X.; Yang, J.; Yang, P.P.; Lian, H.Z.; Lin, J. Hydrothermal synthesis of lanthanide fluorides LnF$_3$ (Ln = La to Lu) nano/microcrystals with multiform structures and morphologies. *Chem. Mater.* **2008**, *20*, 4317–4326.

60. Li, C.X.; Yang, J.; Quan, Z.W.; Yang, P.P.; Kong, D.Y.; Lin, J. Different microstructures of ss-NaYF$_4$ fabricated by hydrothermal process: Effects of pH values and fluoride sources. *Chem. Mater.* **2007**, *19*, 4933–4942.

61. Wang, J.; Wang, F.; Wang, C.; Liu, Z.; Liu, X.G. Single-band upconversion emission in lanthanide-doped KMnF$_3$ nanocrystals. *Angew. Chem. Int. Ed.* **2011**, *50*, 10369–10372.

62. Tian, G.; Gu, Z.J.; Zhou, L.J.; Yin, W.Y.; Liu, X.X.; Yan, L.; Jin, S.; Ren, W.L.; G, M.; Li, S.G.; *et al.* Mn^{2+} dopant-controlled synthesis of NaYF$_4$:Yb/Er upconversion nanoparticles for *in vivo* imaging and drug delivery. *Adv. Mater.* **2012**, *24*, 1226–1231.

63. Zeng, S.J.; Yi, Z.G.; Lu, W.; Qian, C.; Wang, H.B.; Rao, L.; Zeng, T.M.; Liu, H.R.; Liu, H.G.; Fei, B.; *et al.* Simultaneous realization of phase/size manipulation, upconversion luminescence enhancement, and blood vessel imaging in multifunctional nanoprobes through transition metal Mn^{2+} doping. *Adv. Funct. Mater.* **2014**, *24*, 4051–4059.

64. Lin, J.; Yu, M.; Lin, C.K.; Liu, X.M. Multiform oxide optical materials via the versatile Pechini-type sol-gel process: Synthesis and characteristics. *J. Phys. Chem. C* **2007**, *111*, 5835–5845.

65. Patra, A.; Friend, C.S.; Kapoor, R.; Prasad, P.N. Upconversion in Er^{3+}:ZrO$_2$ nanocrystals. *J. Phys. Chem. B* **2002**, *106*, 1909–1912.

66. Cheng, Z.Y.; Xing, R.B.; Hou, Z.Y.; Huang, S.S.; Lin, J. Patterning of light-emitting YVO$_4$:Eu^{3+} thin films via inkjet printing. *J. Phys. Chem. C* **2010**, *114*, 9883–9888.

67. Zhu, Y.S.; Xu, W.; Zhang, H.Z.; Wang, W.; Xu, S.; Song, H.W. Inhibited long-scale energy transfer in dysprosium doped yttrium vanadate inverse opal. *J. Phys. Chem. C* **2012**, *116*, 2297–2302.

68. Wang, W.; Song, H.W.; Liu, Q.; Bai, X.; Wang, Y.; Dong, B. Modified optical properties in a samarium doped titania inverse opal. *Opt. Lett.* **2010**, *35*, 1449–1451.

69. Liu, Q.; Song, H.W.; Wang, W.; Bai, X.; Wang, Y.; Dong, B.; Xu, L.; Han, W. Observation of lamb shift and modified spontaneous emission dynamics in the YBO$_3$:Eu^{3+} inverse opal. *Opt. Lett.* **2010**, *35*, 2898–2900.

70. Zhu, Y.S.; Sun, Z.P.; Yin, Z.; Song, H.W.; Xu, W.; Wang, Y.F.; Zhang, L.G.; Zhang, H.Z. Self-assembly, highly modified spontaneous emission and energy transfer properties of $LaPO_4:Ce^{3+},Tb^{3+}$ inverse opals. *Dalton Trans.* **2013**, *42*, 8049–8057.

71. Patra, A.; Friend, C.S.; Kapoor, R.; Prasad, P.N. Fluorescence upconversion properties of Er^{3+}-doped TiO_2 and $BaTiO_3$ nanocrystallites. *Chem. Mater.* **2003**, *15*, 3650–3655.

72. Venkatramu, V.; Falcomer, D.; Speghini, A.; Bettinelli, M.; Jayasankar, C.K. Synthesis and luminescence properties of Er^{3+}-doped $Lu_3Ga_5O_{12}$ nanocrystals. *J. Lumin.* **2008**, *128*, 811–813.

73. Yang, K.S.; Zheng, F.; Wu, R.; Li, H.; Zhang, X. Upconversion luminescent properties of $YVO_4:Yb^{3+},Er^{3+}$ nano-powder by sol-gel method. *J. Rare Earth* **2006**, *24*, 162–166.

74. Lemyre, J.L.; Ritcey, A.M. Synthesis of lanthanide fluoride nanoparticles of varying shape and size. *Chem. Mater.* **2005**, *17*, 3040–3043.

75. Wang, G.F.; Qin, W.P.; Zhang, J.S. Synthesis, growth mechanism, and tunable upconversion luminescence of Yb^{3+}/Tm^{3+}-codoped YF_3 nanobundles. *J. Phys. Chem C* **2008**, *112*, 12161–12167.

76. Wang, G.F.; Qin, W.P.; Wei, G.D. Synthesis and upconversion luminescence properties of Yb^{3+}/Tm^{3+}-codoped $BaSiF_6$ nanorods. *J. Fluorine Chem.* **2009**, *130*, 158–161.

77. Vetrone, F.; Boyer, J.C.; Capobianco, J.A.; Speghini, A.; Bettinelli, M. Significance of Yb^{3+} concentration on the upconversion mechanisms in codoped $Y_2O_3:Er^{3+},Yb^{3+}$ nanocrystals. *J. Appl. Phys.* **2004**, *96*, 661–667.

78. Pandozzi, F.; Vetrone, F.; Boyer, J.C.; Naccache, R.; Capobianco, J.A.; Speghini, A.; Bettinellin, M. A spectroscopic analysis of blue and ultraviolet upconverted emissions from $Gd_3Ga_5O_{12}:Tm^{3+},Yb^{3+}$ nanocrystals. *J. Phys. Chem. B* **2005**, *109*, 17400–17405.

79. Luo, X.X.; Cao, W.H. Ethanol-assistant solution combustion method to prepare $La_2O_3:Yb,Pr$ nanometer phosphor. *J. Alloys. Compds.* **2008**, *46*, 529–534.

80. Xu, L.L.; Yu, Y.N.; Li, X.G. Synthesis and upconversion properties of monoclinic $Gd_2O_3:Er^{3+}$ nanocrystals. *Opt. Mater.* **2008**, *30*, 1284–1288.

81. Qin, X.; Yokomori, T.; Ju, Y.G. Flame synthesis and characterization of rare-earth (Er^{3+}, Ho^{3+}, and Tm^{3+}) doped upconversion nanocryphosphors. *Appl. Phys. Lett.* **2007**, *90*, 073104.

82. Dong, B.; Song, H.W.; Yu, H.Q.; Zhang, H.; Qin, R.F.; Bai, X.; Pan, G.H.; Lu, S.Z.; Wang, F.; Fan, L.B.; *et al.* Upconversion properties of Ln^{3+} doped $NaYF_4$/polymer composite fibers prepared by electrospinning. *J. Phys. Chem. C* **2008**, *112*, 1435–1440.

83. Ma, L.; Chen, W.X.; Zheng, Y.Z.; Zhao, J.; Xu, Z.D. Microwave-assisted hydrothermal synthesis and characterizations of PrF_3 hollow nanoparticles. *Mater. Lett.* **2007**, *61*, 2765–2768.

84. Chan, E.M. Combinatorial approaches for developing upconverting nanomaterials: High-throughput screening, modeling, and applications. *Chem. Soc. Rev.* **2014**, doi:10.1039/C4CS00205A.

85. Liu, H.; Jakobsson, O.; Xu, C.T.; Xie, H.; Laurell, T.; Andersson-Engels, S. Synthesis of $NaYF_4:Yb^{3+}/Er^{3+}$ Upconverting Nanoparticles in a Capillary-Based Continuous Microfluidic Reaction System. In *Colloidal Quantum Dots/Nanocrystals for Biomedical Applications*; SPIE: Bellingham, WA, USA, 2011; Volume 7909, p. 790917.

86. Xu, C.T.; Zhan, Q.; Liu, H.; Somesfalean, G.; Qian, J.; He, S.; Andersson-Engels, S. Upconverting nanoparticles for pre-clinical diffuse optical imaging, microscopy and sensing: Current trends and future challenges. *Laser Photon. Rev.* **2013**, *7*, 663–697.

87. Zhu, X.; Zhang, Q.; Li, Y.; Wang, H. Redispersible and water-soluble LaF₃:Ce,Tb nanocrystalsvia a microfluidic reactor with temperature steps. *J. Mater. Chem.* **2008**, *18*, 5060–5062.

88. Zhu, X.; Zhang, Q.; Li, Y.; Wang, H.J. Facile crystallization control of LaF₃/LaPO₄:Ce,Tb nanocrystals in a microfluidic reactor using microwave irradiation. *J. Mater. Chem.* **2010**, *20*, 1766–1771.

89. Yi, G.S.; Chow, G.M. Water-soluble NaYF₄:Yb,Er(Tm)/NaYF₄/polymer core/shell/shell nanoparticles with significant enhancement of upconversion fluorescence. *Chem. Mater.* **2007**, *19*, 341–343.

90. Liu, Y.S.; Tu, D.T.; Zhu, H.M.; Li, R.F.; Luo, W.Q.; Chen, X.Y. A strategy to achieve efficient dual-mode luminescence of Eu^{3+} in lanthanides doped multifunctional NaGdF₄ nanocrystals. *Adv. Mater.* **2010**, *22*, 3266–3271.

91. Vetrone, F.; Naccache, R.; Mahalingam, V.; Morgan, C.G.; Capobianco, J.A. The active-core/active-shell approach: A strategy to enhance the upconversion luminescence in lanthanide-doped nanoparticles. A*dv. Funct. Mater.* **2009**, *19*, 2924–2929.

92. Liu, X.M.; Kong, X.G.; Zhang, Y.L.; Tu, L.P.; Wang, Y.; Zeng, Q.H.; Li, C.G.; Shi, Z.; Zhang, H. Breakthrough in concentration quenching threshold of upconversion luminescence via spatial separation of the emitter doping area for bio-applications. *Chem. Commun.* **2011**, *47*, 11957–11959.

93. Wang, Z.L.; Quan, Z.W.; Jia, P.Y.; Lin, C.K.; Luo, Y.; Chen, Y.; Fang, J.; Zhou, W.; O'Connor, C.J.; Lin, J. A facile synthesis and photoluminescent properties of redispersible CeF₃,CeF₃:Tb^{3+}, and CeF₃:Tb^{3+}/LaF₃ (core/shell) nanoparticles. *Chem. Mater.* **2006**, *18*, 2030–2037.

94. Mai, H.X.; Zhang, Y.W.; Yan, C.H. Highly efficient multicolor up-conversion emissions and their mechanisms of monodisperse NaYF₄:Yb,Er core and core/shell-structured nanocrystals. *J. Phys. Chem. C* **2007**, *111*, 13721–13729.

95. Boyer, J.C.; Gagnon, J.; Capobianco, J.A. Synthesis, characterization, and spectroscopy of NaGdF₄:Ce^{3+},Tb^{3+}/NaYF₄ core/shell nanoparticles. *Chem. Mater.* **2007**, *19*, 3358–3360.

96. Helmut, S.; Pavel, P.; Otmanef, Z.; Markus, H. Synthesis and optical properties of KYF₄/Yb,Er nanocrystals, and their surface modification with undoped KYF₄. *Adv. Funct. Mater.* **2008**, *18*, 2913–2918.

97. Xie, X.J.; Gao, N.Y.; Deng, R.R.; Sun, Q.; Xu, Q.H.; Liu, X.G. Mechanistic investigation of photon upconversion in Nd^{3+}-sensitized core-shell nanoparticles. *J. Am. Chem. Soc.* **2013**, *135*, 12608–12611.

98. Huang, P.; Zheng, W.; Zhou, S.Y.; Tu, D.T.; Chen, Z.; Zhu, H.M.; Li, R.F.; Ma, E.; Huang, M.D.; Chen, X.Y. Lanthanide-doped LiLuF₄ upconversion nanoprobes for the detection of disease biomarkers. *Angew. Chem. Int. Ed.* **2014**, *53*, 1252–1257.

99. Wang, F.; Deng, R.R.; Liu, X.G. Preparation of core-shell NaGdF₄ nanoparticles doped with luminescent lanthanide ions to be used as upconversion-based probes. *Nat. Protoc.* **2014**, *9*, 1634–1644.

100. Wang, Y.F.; Sun, L.D.; Xia, J.W.; Feng, W.; Zhou, J.C.; Shen, J.; Yan, C.H. Rare-earth nanoparticles with enhanced upconversion emission and suppressed rare-earth-ion leakage. *Chem. Eur. J.* **2012**, *18*, 5558–5564.

101. Darbandi, M.; Nann, T. One-pot synthesis of YF$_3$@silica core/shell nanoparticles. *Chem. Commun.* **2006**, *7*, 776–778.

102. Li, Z.Q.; Zhang, Y. Monodisperse silica-coated polyvinylpyrrolidone/NaYF$_4$ nanocrystals with multicolor upconversion fluorescence emission. *Angew. Chem. Int. Ed.* **2006**, *45*, 7732–7735.

103. Li, Z.Q.; Zhang, Y.; Jiang, S. Multicolor core/shell-structured upconvers fluorescent nanoparticles. *Adv. Mater.* **2008**, *20*, 4765–4769.

104. Chen, B.T.; Dong, B.; Wang, J.; Zhang, S.; Xu, L.; Yu, W.; Song, H.W. Amphiphilic silane modified NaYF$_4$:Yb,Er loaded with Eu(TTA)$_3$(TPPO)$_2$ nanoparticles and their multi-functions: Dual mode temperature sensing and cell imaging. *Nanoscale* **2013**, *5*, 8541–8549.

105. Sivakumar, S.; Diamenteand, P.R.; van Veggel, F.C.J.M. Silica-coated Ln^{3+}-doped LaF$_3$ nanoparticles as robust down and upconverting biolabels. *Chem. Eur. J.* **2006**, *12*, 5878–5884.

106. Liu, Z.Y.; Yi, G.S.; Zhang, H.T.; Ding, J.; Zhang, Y.W.; Xue, J.M. Monodisperse silica nanoparticles encapsulating upconversion fluorescent and superparamagnetic nanocrystals. *Chem. Commun.* **2008**, *6*, 694–696.

107. Wang, M.; Mi, C.C.; Zhang, Y.Z.; Liu, J.L.; Li, F.; Mao, C.B.; Xu, S.K. NIR-responsive silica-coated NaYbF$_4$:Er/Tm/Ho upconversion fluorescent nanoparticles with tunable emission colors and their applications in immunolabeling and fluorescent imaging of cancer cells. *J. Phys. Chem. C* **2009**, *113*, 19021–19027.

108. Yu, X.F.; Chen, L.D.; Li, M.; Xie, M.Y.; Zhou, L.; Li, Y.; Wang, Q.Q. Highly efficient fluorescence of NdF$_3$/SiO$_2$ core/shell nanoparticles and the applications for *in vivo* NIR detection. *Adv. Mater.* **2008**, *20*, 4118–4123.

109. Li, X.H.; Wu, Y.Q.; Liu, Y.; Zou, X.M.; Yao, L.M.; Lia, F.Y.; Feng, W. Cyclometallated ruthenium complex-modified upconversion nanophosphors for selective detection of Hg^{2+} ions in water. *Nanoscale* **2014**, *6*, 1020–1028.

110. Wang, C.; Cheng, L.; Liu, Z. Drug delivery with upconversion nanoparticles for multi-functional targeted cancer cell imaging and therapy. *Biomaterials* **2011**, *32*, 1110–1120.

111. Budijono, S.J.; Shan, J.N.; Yao, N.; Miura, Y.; Hoye, T.; Austin, R.H.; Ju, Y.G.; Prud'homme, R.K. Synthesis of stable block-copolymer-protected NaYF$_4$:Yb^{3+},Er^{3+} up-converting phosphor nanoparticles. *Chem. Mater.* **2010**, *22*, 311–318.

112. Chen, Z.G.; Chen, H.L.; Hu, H.; Yu, M.X.; Li, F.Y.; Zhang, Q.; Zhou, Z.G.; Yi, T.; Huang, C.H. Versatile synthesis strategy for carboxylic acid-functionalized upconverting nanophosphors as biological labels. *J. Am. Chem. Soc.* **2008**, *130*, 3023–3029.

113. Bogdan, N.; Vetrone, F.; Ozin, G.A.; Capobianco, J.A. Synthesis of ligand-free colloidally stable water dispersible brightly luminescent lanthanide-doped upconverting nanoparticles. *Nano Lett.* **2011**, *11*, 835–840.

114. Yi, G.S.; Chow, G.M. Synthesis of hexagonal-phase NaYF$_4$:Yb,Er and NaYF$_4$:Yb,Tm nanocrystals with efficient up-conversion fluorescence. *Adv. Funct. Mater.* **2006**, *16*, 2324–2329.

115. Chen, G.Y.; Ohulchanskyy, T.Y.; Law, W.C.; Agren, H.; Prasad, P.N. Monodisperse NaYbF$_4$: Tm^{3+}/NaGdF$_4$ core/shell nanocrystals with near-infrared to near-infrared upconversion photoluminescence and magnetic resonance properties. *Nanoscale* **2011**, *3*, 2003–2008.

116. Boyer, J.C.; Manseau, M.P.; Murray, J.I.; van Veggel, F.C.J.M. Surface modification of upconverting NaYF$_4$ nanoparticles with PEG-phosphate ligands for NIR (800 nm) biolabeling within the biological window. *Langmuir* **2010**, *26*, 1157–1164.

117. Nyk, M.; Kumar, R.; Ohulchanskyy, T.Y.; Bergey, E.J.; Prasad, P.N. High contrast *in vitro* and *in vivo* photoluminescence bioimaging using near Infrared to near infrared up-conversion in Tm^{3+} and Yb^{3+} doped fluoride nanophosphors. *Nano Lett.* **2008**, *8*, 3834–3838.

118. Cao, T.Y.; Yang, T.S.; Gao, Y.; Yang, Y.; Hu, H.; Li, F.Y. Water-soluble NaYF$_4$:Yb/Er upconversion nanophosphors: Synthesis, characteristics and application in bioimaging. *Inorg. Chem. Commun.* **2010**, *13*, 392–394.

119. Zhou, J.; Yu, M.X.; Sun, Y.; Zhang, X.Z.; Zhu, X.J.; Wu, Z.H.; Wu, D.M.; Li, F.Y. Fluorine-18-labeled Gd^{3+}/Yb^{3+}/Er^{3+} co-doped NaYF$_4$ nanophosphors formultimodality PET/MR/UCL imaging. *Biomaterials* **2011**, *32*, 1148–1156.

120. Naczynski, D.J.; Andelman, T.; Pal, D.; Chen, S.; Richard, E.; Riman, R.E.; Roth, C.M.; Moghe, P.V. Albumin nanoshell encapsulation of near-infrared-excitable rare-earth nanoparticles enhances biocompatibility and enables targeted cell imaging. *Small* **2010**, *6*, 1631–1640.

Permissions

All chapters in this book were first published in Nanomaterials, by MDPI; hereby published with permission under the Creative Commons Attribution License or equivalent. Every chapter published in this book has been scrutinized by our experts. Their significance has been extensively debated. The topics covered herein carry significant findings which will fuel the growth of the discipline. They may even be implemented as practical applications or may be referred to as a beginning point for another development.

The contributors of this book come from diverse backgrounds, making this book a truly international effort. This book will bring forth new frontiers with its revolutionizing research information and detailed analysis of the nascent developments around the world.

We would like to thank all the contributing authors for lending their expertise to make the book truly unique. They have played a crucial role in the development of this book. Without their invaluable contributions this book wouldn't have been possible. They have made vital efforts to compile up to date information on the varied aspects of this subject to make this book a valuable addition to the collection of many professionals and students.

This book was conceptualized with the vision of imparting up-to-date information and advanced data in this field. To ensure the same, a matchless editorial board was set up. Every individual on the board went through rigorous rounds of assessment to prove their worth. After which they invested a large part of their time researching and compiling the most relevant data for our readers.

The editorial board has been involved in producing this book since its inception. They have spent rigorous hours researching and exploring the diverse topics which have resulted in the successful publishing of this book. They have passed on their knowledge of decades through this book. To expedite this challenging task, the publisher supported the team at every step. A small team of assistant editors was also appointed to further simplify the editing procedure and attain best results for the readers.

Apart from the editorial board, the designing team has also invested a significant amount of their time in understanding the subject and creating the most relevant covers. They scrutinized every image to scout for the most suitable representation of the subject and create an appropriate cover for the book.

The publishing team has been an ardent support to the editorial, designing and production team. Their endless efforts to recruit the best for this project, has resulted in the accomplishment of this book. They are a veteran in the field of academics and their pool of knowledge is as vast as their experience in printing. Their expertise and guidance has proved useful at every step. Their uncompromising quality standards have made this book an exceptional effort. Their encouragement from time to time has been an inspiration for everyone.

The publisher and the editorial board hope that this book will prove to be a valuable piece of knowledge for researchers, students, practitioners and scholars across the globe.

List of Contributors

Samet Kocabey
Faculty of Physics and Center for Nanoscience, Ludwig-Maximilians University, Munich 80799,Germany

Hanna Meinl
Division of Clinical Pharmacology, Department of Internal Medicine IV, Klinikum der Universität München, Munich 80336, Germany

Iain S. MacPherson
Faculty of Physics and Center for Nanoscience, Ludwig-Maximilians University, Munich 80799,Germany

Valentina Cassinelli
Baseclick GmbH, Tutzing 82327, Germany

Antonio Manetto
Baseclick GmbH, Tutzing 82327, Germany

Simon Rothenfusser
Division of Clinical Pharmacology, Department of Internal Medicine IV, Klinikum der Universität München, Munich 80336, Germany

Tim Liedl
Faculty of Physics and Center for Nanoscience, Ludwig-Maximilians University, Munich 80799,Germany

Felix S. Lichtenegger
Division of Clinical Pharmacology, Department of Internal Medicine IV, Klinikum der Universität München, Munich 80336, Germany
Department of Internal Medicine III, Klinikum der Universität München, Munich 81377, Germany

Keitel Cervantes-Salguero
Department of Bioengineering and Robotics, Tohoku University, Sendai 980-8579, Japan

Shogo Hamada
Kavli Institute at Cornell for Nanoscale Science, Cornell University, Ithaca, NY 14853, USA

Shin-ichiro M. Nomura
Department of Bioengineering and Robotics, Tohoku University, Sendai 980-8579, Japan

Satoshi Murata
Department of Bioengineering and Robotics, Tohoku University, Sendai 980-8579, Japan

Yunfei Shang
School of Chemical Engineering and Technology, Harbin Institute of Technology, Harbin 150001, China

Shuwei Hao
School of Chemical Engineering and Technology, Harbin Institute of Technology, Harbin 150001, China

Jing Liu
School of Chemical Engineering and Technology, Harbin Institute of Technology, Harbin 150001, China

Meiling Tan
School of Chemical Engineering and Technology, Harbin Institute of Technology, Harbin 150001, China

Ning Wang
School of Chemical Engineering and Technology, Harbin Institute of Technology, Harbin 150001, China

Chunhui Yang
School of Chemical Engineering and Technology, Harbin Institute of Technology, Harbin 150001, China
Harbin Huigong Technology Co. Ltd, Harbin 150001, China

Guanying Chen
School of Chemical Engineering and Technology, Harbin Institute of Technology, Harbin 150001, China
Institute for Lasers, Photonics and Biophotonics, University at Buffalo, The State University of New York, Buffalo, NY 14260, USA

Gabriela Mera
Institute für Materialwissenschaft, Technische Universität Darmstadt, Jovanka-Bontschits-Strasse 2, D-64287 Darmstadt, Germany

Markus Gallei
Ernst-Berl-Institut für Technische und Makromolekulare Chemie, Technische Universität Darmstadt, Alarich-Weiss-Strasse 4, D-64287 Darmstadt, Germany

Samuel Bernard
Institut Européen des Membranes (UMR 5635-CNRS/ENSCM-UM2) CC 047-Place E. Bataillon, 34095 Montpellier Cedex 05, France

Emanuel Ionescu
Institute für Materialwissenschaft, Technische Universität Darmstadt, Jovanka-Bontschits-Strasse 2, D-64287 Darmstadt, Germany
Department Chemie, Institut für Anorganische Chemie, Universität zu Köln, Greinstrasse 6, D- 50939 Köln, Germany

Dušan Galusek
Joint Glass Centre of the Institute of Inorganic Chemistry, Slovak Academy of Sciences, Alexander Dubček University of Trenčín, Študentská 2, 91150 Trenčín, Slovak Republic
Faculty of Chemical and Food Technology, Slovak University of Technology in Bratislava, Vazovova 5, 81243 Bratislava, Slovak Republic

Dagmar Galusková
Joint Glass Centre of the Institute of Inorganic Chemistry, Slovak Academy of Sciences, Alexander Dubček University of Trenčín, Študentská 2, 91150 Trenčín, Slovak Republic
Faculty of Chemical and Food Technology, Slovak University of Technology in Bratislava, Vazovova 5, 81243 Bratislava, Slovak Republic

Andrey V. Nomoev
Institute of Physical Materials Science, Siberian Branch of the Russian Academy of Sciences, Sakhyanovoy str., 6, Ulan-Ude 670047, Russia
Department of Physics and Engineering, Buryat State University, Smolina str., 24a, Ulan-Ude 670000, Russia

Sergey P. Bardakhanov
Department of Physics and Engineering, Buryat State University, Smolina str., 24a, Ulan-Ude 670000, Russia
Institute of Applied and Theoretical Mechanics, Siberian Branch of the Russian Academy of Sciences, Institutskaya str., 4/1, Novosibirsk 630090, Russia
Department of Physics, Novosibirsk State University, Pirogova str., 2, Novosibirsk 630090, Russia

Makoto Schreiber
Department of Physics and Engineering, Buryat State University, Smolina str., 24a, Ulan-Ude 670000, Russia

Dashima Zh. Bazarova
Department of Physics and Engineering, Buryat State University, Smolina str., 24a, Ulan-Ude 670000, Russia

Boris B. Baldanov
Department of Physics and Engineering, Buryat State University, Smolina str., 24a, Ulan-Ude 670000, Russia

Nikolai A. Romanov
Department of Physics and Engineering, Buryat State University, Smolina str., 24a, Ulan-Ude 670000, Russia

Iftikhar Ahmad
Center of Excellence for Research in Engineering Materials, Advanced Manufacturing Institute, King Saud University, Riyadh 11421, Saudi Arabia

Bahareh Yazdani
College of Engineering, Mathematics and Physical Sciences, University of Exeter, Exeter EX4 4QF, UK

Yanqiu Zhu
College of Engineering, Mathematics and Physical Sciences, University of Exeter, Exeter EX4 4QF, UK

Hongjin Chang
Key Laboratory of Flexible Electronics (KLOFE) and Institute of Advanced Materials (IAM), Jiangsu National Synergistic Innovation Center for Advanced Materials (SICAM), Nanjing Tech University (NanjingTech), Nanjing 211816, China

Juan Xie
Key Laboratory for Organic Electronics and Information Displays and Institute of Advanced Materials (IAM), National Synergistic Innovation Center for Advanced Materials (SICAM), Nanjing University of Posts and Telecommunications, Nanjing 210023, China

Baozhou Zhao
Key Laboratory of Flexible Electronics (KLOFE) and Institute of Advanced Materials (IAM), Jiangsu National Synergistic Innovation Center for Advanced Materials (SICAM), Nanjing Tech University (NanjingTech), Nanjing 211816, China

Botong Liu
Key Laboratory of Flexible Electronics (KLOFE) and Institute of Advanced Materials (IAM), Jiangsu National Synergistic Innovation Center for Advanced Materials (SICAM), Nanjing Tech University (NanjingTech), Nanjing 211816, China

Shuilin Xu
Key Laboratory of Flexible Electronics (KLOFE) and Institute of Advanced Materials (IAM), Jiangsu National Synergistic Innovation Center for Advanced Materials (SICAM), Nanjing Tech University (NanjingTech), Nanjing 211816, China

Na Ren
Key Laboratory of Flexible Electronics (KLOFE) and Institute of Advanced Materials (IAM), Jiangsu National Synergistic Innovation Center for Advanced Materials (SICAM), Nanjing Tech University (NanjingTech), Nanjing 211816, China

Xiaoji Xie
Key Laboratory of Flexible Electronics (KLOFE) and Institute of Advanced Materials (IAM), Jiangsu National Synergistic Innovation Center for Advanced Materials (SICAM), Nanjing Tech University (NanjingTech), Nanjing 211816, China

Ling Huang
Key Laboratory of Flexible Electronics (KLOFE) and
Institute of Advanced Materials (IAM), Jiangsu National
Synergistic Innovation Center for Advanced Materials
(SICAM), Nanjing Tech University (NanjingTech),
Nanjing 211816, China

Wei Huang
Key Laboratory of Flexible Electronics (KLOFE) and
Institute of Advanced Materials (IAM), Jiangsu National
Synergistic Innovation Center for Advanced Materials
(SICAM), Nanjing Tech University (NanjingTech),
Nanjing 211816, China